INFRARED AND MILLIMETER WAVES

VOLUME 6 SYSTEMS AND COMPONENTS

Edited by *KENNETH J. BUTTON*

NATIONAL MAGNET LABORATORY
MASSACHUSETTS INSTITUTE OF TECHNOLOGY
CAMBRIDGE, MASSACHUSETTS

1982

ACADEMIC PRESS

A Subsidiary of Harcourt Brace Jovanovich, Publishers

New York   London
Paris   San Diego   San Francisco   São Paulo   Sydney   Tokyo   Toronto

ACADEMIC PRESS, INC.
111 Fifth Avenue, New York, New York 10003

United Kingdom Edition published by
ACADEMIC PRESS, INC. (LONDON) LTD.
24/28 Oval Road, London NW1 7DX

Library of Congress Cataloging in Publication Data
Main entry under title:

Infrared and millimeter waves.

Includes bibliographies and indexes.
CONTENTS: v. 1. Sources of radiation.--v. 2. Instrumentation.--v. 3. Submillimeter techniques.
--v. 6. Systems and components.
1. Infra-red apparatus and appliances. 2. Millimeter
wave devices. I. Button, Kenneth J.
TA1570.I52   621.36'2   79-6949
ISBN 0-12-147706-1 (v. 6)   AACR1

INFRARED AND MILLIMETER WAVES

VOLUME 6 SYSTEMS AND COMPONENTS

CONTRIBUTORS

PAUL F. GOLDSMITH

L. T. GREENBERG

J. E. HARRIES

G. D. HOLAH

D. H. MARTIN

P. L. RICHARDS

M. V. SCHNEIDER

CONTENTS

LIST OF CONTRIBUTORS

Numbers in parentheses indicate the pages on which the authors' contributions begin.

PAUL F. GOLDSMITH (277), *Five College Radio Astronomy Observatory, Department of Physics and Astronomy, University of Massachusetts, Amherst, Massachusetts*

L. T. GREENBERG (149), *Space Sciences Laboratory, The Aerospace Corporation, El Segundo, California 90245*

J. E. HARRIES* (1), *National Physical Laboratory, Teddington, Middlesex, United Kingdom*

G. D. HOLAH (345), *Engineering Experiment Station, Georgia Institute of Technology, Atlanta, Georgia 30332*

D. H. MARTIN (65), *Physics Department, Queen Mary College, University of London, London E1 4NS, United Kingdom*

P. L. RICHARDS (149), *Department of Physics, University of California, Berkeley, California 94720*

M. V. SCHNEIDER (209), *Bell Laboratories, Crawford Hill Laboratory, Holmdel, New Jersey 07733*

* Present Address: Remote Sounding Group, Rutherford and Appleton Laboratories, Chilton, Didcot, Oxfordshire OX11 OQX, United Kingdom.

PREFACE

This book consists of chapters prepared by the most eminent authorities in their specialties that it has been possible for us to find. The plans for this book were made more than two years before its publication. We are rewarded, however, by chapters that are not only outstanding in the description of each topic—they also contain significant and valuable material to be used in other areas of millimeter and submillimeter technology.

Volume 6 opens with a timely review of the infrared and submillimeter spectroscopy of the atmosphere by John E. Harries, who is so well known for this work. There is a great deal of interest now in this topic for practical reasons as well. Chapter 2 is unique in the sense that it is the only complete treatment of the most modern interferometric spectrometers. The thorough description of the principles of the dispersive Fourier transform spectrometer was given by J. R. Birch and T. J. Parker in Volume 2, and now we have the polarizing version of the Martin–Puplett interferometers written by D. H. Martin himself. It will hardly be possible to find more complete treatments than these for years to come. The performance of these instruments and a wide variety of applications have been summarized by D. H. Martin in the later sections of his chapter.

It is quite apparent that Paul Richards should write a book on incoherent and coherent detection devices from one micrometer to one millimeter. In the meantime, however, Richards and Greenberg have provided us with Chapter 3 dealing with the general principles of infrared detectors, treatments of photon detectors, bolometers, coherent detectors, and much more on the subject of detection. The reader is urged to look through this chapter now to see how this exacting subject has been described so systematically. Please do not miss Appendix I, where heterodyne detection is compared briefly and simply with direct detection.

For the rapidly growing ranks of millimeter-wave enthusiasts, Martin Schneider's Chapter 4 on "Metal–Semiconductor Junctions as Frequency Converters" is a valuable bonus in this book. Dr. Schneider shared the IEEE Microwave Prize last year for the paper "Subharmonically Pumped Millimeter-Wave Mixers" (IEEE Transactions, Vol. MTT-26, pp. 706–715, October, 1978). If there were a prize for excellence in writing, he would have certainly won that as well; his Chapter 4 is actually entertaining. To be practical, however, I refer you to the first two pages of his Introduction, where he discusses briefly the wide variety of applications of his subject.

In these books, we should include as many chapters as possible on the problems of the transition between optical techniques and microwave techniques. If we have a thorough treatment of the application of some of these systems and components, we may solve some of the "near-millimeter" problems that seem so difficult to visualize now. Paul Goldsmith in "Quasi-Optical Techniques at Millimeter and Submillimeter Wavelengths" and Geoffrey Holah in "Far-Infrared and Submillimeter-Wavelength Filters" have dealt with these problems for several years and have summarized their subjects very well.

Several additional volumes in this treatise are in various stages of publication, preparation, and planning. We are grateful to our authors and to our readers for their general encouragement and also for their specific suggestions and contributions. Additional suggestions will be welcomed and considered seriously.

CONTENTS OF OTHER VOLUMES

CHAPTER 1

Infrared and Submillimeter Spectroscopy of the Atmosphere*

J. E. Harries†

National Physical Laboratory
Teddington, Middlesex
United Kingdom

* Written in part while the author was on leave of absence at the National Center for Atmospheric Research, Boulder, Colorado. The National Center for Atmospheric Research is sponsored by the National Science Foundation. This review was completed in 1979.

† *Present address:* Remote Sounding Group, Rutherford and Appleton Laboratories, Chilton, Didcot, Oxfordshire, United Kingdom.

1

ISBN 0-12-147706-1

I. Introduction

The study of the earth's atmosphere using infrared (IR) spectroscopy has a long and respectable history; investigations using submillimeter (SM) spectroscopy have a shorter, but arguably no less valuable, background. It will be our purpose in this chapter to present a survey of this combined subject, giving some discussion of the basic theory behind these studies, the instrumental techniques used, the nature of the atmospheric spectrum, and the information concerning the atmosphere that spectroscopic investigations have revealed.

Because we shall be dealing with a multidimensional, large-scale problem, involving several decades of wavelength, including an enormous quantity of spectral structure, the four dimensions of the atmosphere (x, y, z, and t), and considerable structure (vertical distributions of absorber amount and temperature and horizontal and temporal variations), it is, unfortunately, necessary to limit our discussion at the outset.

Thus, we shall discuss only the neutral atmosphere, below, say, 100 km altitude; we shall, in fact, concentrate on the stratosphere, where most of the work has been done in recent years, though we shall extend our considerations on occasion to the troposphere below and the mesosphere above. We shall consider only the "thermal" IR and SM regions, i.e., we shall not consider nonthermally excited spectra [e.g., airglow or auroral emissions; see Vallance-Jones (1971, 1973) for reviews]. We shall limit our range of wavelengths arbitrarily, but according to widely accepted custom, by defining the IR and SM as follows:

IR: from 1 μm to 30 μm   (10,000–333 cm^{-1}),

SM: from 30 μm to 1000 μm   (333–10 cm^{-1}).

The discussion of techniques used must also be limited, otherwise a book, not one chapter of a book, would be our aim. Thus, we shall concentrate primarily on methods that do not use laser techniques: mention will be made of these techniques, but only in so far as they have already contributed to our subject. Because, in the author's view, laser techniques are more a feature of the future than of the past, a thorough discussion of them would be an chapter of quite a different nature than the present one [for a review of laser methods, see Hinkley (1976)]. Finally, we shall not consider the extensive spectroscopy that has been carried out from spacecraft; although almost invariably of excellent quality, the long lead times in space activities usually mean that such experiments are not at the forefront of new spectroscopic research, which must remain the main thrust of this chapter. An excellent review of spaceborne spectroscopy may be found in Houghton and Taylor (1973).

These are the main constraints to our discussion. No doubt, others will creep in consciously or unconsciously. Therefore, apologies are made, in the usual way, at the outset for exclusions that may irritate the reader. A review, by its nature, must be something of a personal view.

Even with these exclusions, we are faced with a formidable range to cover. Before beginning, therefore, it may help the reader to have some indication of the general route to be taken. Section II will be a summary of some of the more important physics associated with atmospheric spectroscopy. In Section III, instrumental techniques will be considered. A discussion of some detailed aspects of spectroscopic atmospheric research in the IR and SM regions, respectively, will be presented in Sections IV and V. Our conclusions will be given in Section VI. (This review was completed in 1979.)

II.   Introductory Radiative Transfer and Spectroscopy

The physical processes involved in the transfer of electromagnetic radiation through a real atmosphere are complex. The basic radiative transfer process is usually treated for the case of a nonscattering atmosphere in order to reduce complexity and for the simplified geometrical model of a plane-parallel stratified atmosphere with no refraction of rays. We shall follow these conventions here.

A.   Equation of Radiative Transfer

For a plane-parallel nonscattering atmosphere, our basic equation is the equation of transfer (Goody, 1964):

$$dI_{\bar{\nu}}/d\mathbf{s} = -\kappa_{\bar{\nu}}\rho(I_{\bar{\nu}} - J_{\bar{\nu}}) \tag{1}$$

in which $I_{\bar{\nu}}$ is the intensity of radiation (energy per unit spectral interval transported across unit area in unit time and within unit solid angle defined at a wavenumber $\bar{\nu}$), $J_{\bar{\nu}}$ the atmospheric source function, which is equal to the locally defined Planck function $B(\bar{\nu}, T)$ if the atmosphere is in local thermodynamic equilibrium (LTE; see Goody, 1964), $\kappa_{\bar{\nu}}$ the spectral absorption coefficient (to be discussed later), ρ the density, and s the linear dimension along the ray path.

The complex nature of the radiative transfer process becomes clear when it is remembered that absorber density ρ and atmospheric temperature T are generalized nonanalytical functions of space and time and that $\kappa_{\bar{\nu}}$ is a highly complicated function of $\bar{\nu}$, which may or may not be regarded as well known, depending on our understanding of molecular absorption processes.

The solution to Eq. (1) may be written as

$$I_{\bar{\nu}}(s) = I_{\bar{\nu}}(0) \exp[-\sigma_{\bar{\nu}}(s, 0)] + \int_0^s J_{\bar{\nu}}(s') \exp[-\sigma_{\bar{\nu}}(s, s')] \, \kappa_{\bar{\nu}}\rho \, ds \qquad (2)$$

in which $\sigma_{\bar{\nu}}(s, s')$ represents the optical depth, or thickness, of the atmosphere between points s and s' at wave number $\bar{\nu}$ and may be written as

$$\sigma_{\bar{\nu}}(s, s') = \int_{s'}^s \kappa_{\bar{\nu}}\rho \, ds. \qquad (3)$$

It is instructive to form a physical picture of the transfer process, using Eq. (2). The first term, the constant formed from the integration of Eq. (1), is the intensity at s due to a radiative source $I_{\bar{\nu}}(0)$ external to the atmosphere but reduced by atmospheric absorption represented by the fractional transmittance from 0 to s:

$$\tau_{\bar{\nu}}(s, 0) = \exp[-\sigma_{\bar{\nu}}(s, 0)]. \qquad (4)$$

The latter expression arises from Lambert's law,

$$dI_{\bar{\nu}} = -I_{\bar{\nu}}\kappa_{\bar{\nu}}\rho \, ds. \qquad (5)$$

The second term in Eq. (2) gives the intensity of emission from each small element of optical path ds' (see Fig. 1):

$$dI_{\bar{\nu}} = J_{\bar{\nu}}(s')\kappa_{\bar{\nu}}(s')\rho(s') \, ds', \qquad (6)$$

modified by transmission from s' to s (where "the observer" is situated). Equation (2) may be rewritten in an equally useful form by using Eqs.

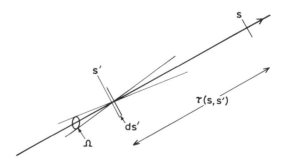

FIG. 1 Geometry for discussion of the radiative transfer equation. A bundle of rays propagating from position s' to position s is considered, passing through an element of area dA at s', with a solid angle Ω, ds' being an elemental increment of path and $\tau(s, s')$ the transmittance between s' and s as defined by Eq. (7).

(3) and (4), which, taken together for a generalized path from s to s', give

$$\tau_{\bar{\nu}}(s, s') = \exp\left[-\int_{s'}^{s} \kappa_{\bar{\nu}}(s)\rho(s)\,ds\right], \tag{7}$$

so that

$$\partial\tau_{\bar{\nu}}(s, s')/\partial s = -\kappa_{\bar{\nu}}(s)\rho(s)\tau_{\bar{\nu}}(s, s'). \tag{8}$$

Using Eq. (8), Eq. (2) may then be rewritten as

$$I_{\bar{\nu}}(s) = I_{\bar{\nu}}(0)\tau_{\bar{\nu}}(s, 0) + \int_{s}^{0} J_{\bar{\nu}}(s')\,d\tau_{\bar{\nu}}(s, s'), \tag{9}$$

or, equivalently,

$$I_{\bar{\nu}}(s) = I_{\bar{\nu}}(0)\tau_{\bar{\nu}}(s, 0) + \int_{s}^{0} J_{\bar{\nu}}(s')[\partial\tau_{\bar{\nu}}(s, s')/\partial s']\,ds'. \tag{10}$$

Equation (9) reduces the solution of the radiative transfer equation to a very simple expression, showing directly the dependence of I on τ; this form may very conveniently be used for numerical computations of radiative transfer. Equation (10) shows an equivalent form, useful in a different way for atmospheric physics: we see that the term $\partial\tau_{\bar{\nu}}(s, s')/\partial s'$ in a sense "weights" the $J_{\bar{\nu}}(s')$ source term along the ray path. It tells us what contributions to $I_{\bar{\nu}}(s)$ are obtained from each element ds' of path as we progress along that path. For this reason, $\partial\tau_{\bar{\nu}}(s, s')/\partial s'$ is often called the weighting function $K_{\bar{\nu}}(s, s')$, defined as

$$K_{\bar{\nu}}(s, s') = \partial\tau_{\bar{\nu}}(s, s')/\partial s'. \tag{11}$$

The function $K_{\bar{\nu}}(s, s')$ may be used to show from which part of a ray path the major contribution to a detected signal $I_{\bar{\nu}}(s)$ originates. The calculation of the shape of $K_{\bar{\nu}}(s, s')$ is, therefore, a rapid way of answering the popular question "but from where in the atmosphere does the emission come?"

B. THE FREQUENCY DEPENDENCE

The frequency dependences in Eq. (2) arise from the $J_{\bar{\nu}}(s)$ and $\kappa_{\bar{\nu}}(s)$ terms. In the case of LTE, which is the only case we shall consider (Goody, 1964), we have

$$J_{\bar{\nu}}(s) = B(\bar{\nu}, T(s)), \tag{12}$$

where

$$B(\bar{\nu}, T(s)) = 2\bar{\nu}^2 \frac{h\bar{\nu}c}{\exp(h\bar{\nu}c/kT) - 1}, \tag{13}$$

and the terms have their usual meanings.

The $\kappa_{\bar{\nu}}(s)$ term, the absorption coefficient, consists, at each value of $\bar{\nu}$, of the contributions from numerous single lines, which, in general, will be spaced randomly and will have a range of strengths and widths.

There are several line shapes that are appropriate for different atmospheric situations (Burch, 1968), although here, for the purpose of illustration, we shall consider only the simplest and most widely used, the pressure-broadened line shape due to Lorentz, given, for a single line, by

$$\kappa(\bar{\nu}) = S\gamma/\pi[(\bar{\nu} - \bar{\nu}_0)^2 + \gamma^2], \tag{14}$$

where the half-width γ is related to the value γ_0 at a standard pressure P_0 and temperature T_0 by

$$\gamma/\gamma_0 = (P/P_0)(T/T_0)^n, \tag{15}$$

$n = -0.5$ for a hard spheres collision model of molecular interactions, but may be different for a more realistic model (Townes and Schawlow, 1975, p. 369).

The term S is the integrated line strength

$$S = \int_0^\infty \kappa(\bar{\nu}) \, d\bar{\nu} \tag{16}$$

and is a measure of the total intensity of the line (Townes and Schawlow, 1975).

A calculation of atmospheric transmission or emission by a spectral band containing many lines, each having a profile like Eq. (14) plus a variation of $J(\bar{\nu})$ through Eqs. (12) and (13), is, therefore, complex and lengthy.

C. The Atmospheric Dependence

Because pressure P, temperature T, and mixing ratio μ vary with height in the atmosphere and because dependence on these parameters enters in several places in Eqs. (2), (13), (14), and (15), the calculation of an atmospheric spectrum is further complicated. It is a logically trivial matter to carry the calculation through, because all dependences are, in principle, known, but this can be extremely expensive in computing time and it is necessary to approximate the atmospheric structure in some way. One such method is to use the Curtis–Godson approximation, which allows the derivation of a mean pressure $\overline{P}$ for a layer of the atmosphere between altitudes s_1 and s_2 by

$$\overline{P} = \int_{s_1}^{s_2} P(s)\mu(s)\rho(s) \, ds \Big/ \int_{s_1}^{s_2} \mu(s)\rho(s) \, ds, \tag{17}$$

where $\overline{P}$ is mean pressure weighted in terms of the absorber density, which has a mixing ratio $\mu(s)$.

D.   BAND MODELS

Although we shall mainly be concerned in this chapter with high-resolution spectroscopy and, consequently, calculations that consider individual lines, there also exist important approximation methods for dealing with wide bands containing many individual lines. These are the so-called band models (Rodgers, 1976a, b), which seek to represent a real spectral band by a small number of parameters (usually a mean strength $\overline{S}$ and a mean width $\overline{\gamma}$). The best-known model for dealing with the randomly positioned bands of asymmetric top molecules is that due to Goody (1964). Over the spectral range $\Delta\tilde{\nu}$, this model yields a mean transmittance given by

$$\tau = \exp(-W/\Delta\tilde{\nu}), \tag{18}$$

where W is the equivalent width, or equivalent area, of the band. After expansion, $\overline{\tau}$ is usually given as

$$\overline{\tau} = \exp\left\{-\overline{S}a\Big/\left[\left(\frac{a}{\pi}\frac{\overline{S}^2}{\overline{R}^2}\right) + 1\right]^{1/2}\right\}, \tag{19}$$

where

$$\overline{S} = \sum_i S_i, \qquad \overline{R} = \sum_i (S_i\gamma_i)^{1/2}, \tag{20}$$

and S_i and γ_i are the integrated line strengths and half-widths, respectively, of the i individual lines comprising the band in question and a the absorber amount (density × path length units).

Other details that are important in a study of radiative transfer such as the Ladenburg–Reiche equation, curves of growth, Doppler effects, etc., have been omitted here because they will not be used directly in what follows. More complete discussions may, of course, be found in standard textbooks (Goody, 1964; Houghton, 1977).

E.   INFORMATION RETRIEVAL

An important subject is the question of how to retrieve the information concerning the atmosphere once we have recorded spectral data using an instrument viewing the atmosphere. The observed emission or absorption spectrum is a convolution of atmospheric temperature (density) structure, composition distribution, spectral structure, and instrumental effects. It is found that measurements at different wavelengths, for example, using a downward-looking spectroradiometer, perhaps on a satellite, are not mutually orthogonal. In general, we may find that in the presence of measurement noise a given spectral distribution of energy may be produced

by a wide (perhaps infinite) variety of atmospheric states (combinations of
$T(h)$ and $\mu(h)$, where h represents altitude).

Thus the study of retrieval techniques is not trivial, and a great deal of
work has been done to elucidate the optimum retrieval method for a given
type of measurement. The problem divides naturally into two parts: first,
actually how to mathematically derive a profile of the required parameter
(T or μ) as a function of height in the atmosphere from measurements at
different frequencies or at different zenith angles; second, how to con-
strain the many possible solutions in a consistent way to obtain the "correct" result.

The details of such methods are of considerable interest but would take
an unfair proportion of this chapter to treat properly, so the reader is referred to recent works that can provide far more information than we can
here (Twomey, 1977; Rodgers, 1976a, b).

F. General Features of the IR and SM Spectrum of the Atmosphere

As a further guide to the reader, we shall briefly describe the main features of the atmospheric spectrum. Figure 2 is taken from the work of
Kyle and Goldman (1975) and shows the atmospheric transmission spectrum for a vertical path through the atmosphere in the $0-2500$-cm^{-1} range
from two altitudes, 4 and 14 km. The principle absorption bands are
marked on the 4-km curves. More detail of the vibration–rotation bands
that occur in this range are given in Table I. In addition, it should be remembered that pure rotation bands due to H_2O, O_3, O_2, HNO_3, NO_2,
N_2O, CO, HCl, and OH are known to exist in the $0-150$-cm^{-1} range (see
Harries, 1977; Bangham, 1978a; Kendall, 1979), though these become important only in the stratosphere where the H_2O absorption or emission is
considerably reduced, and at high spectral resolution (see Section V for
further discussion).

III.   Instruments and Techniques

In this section, we shall briefly consider the several techniques that
have been utilized in atmospheric spectroscopic studies to which we shall
refer to later. Thus the purpose is not to undertake a detailed description
of a list of experimental methods but rather to give the reader who may
not have wide experience in this area some idea of the merits and disadvantages of the different techniques and of the practicalities of their implementation. As a good general reference, we refer to Carleton (1974),
where details of many of these techniques may be found.

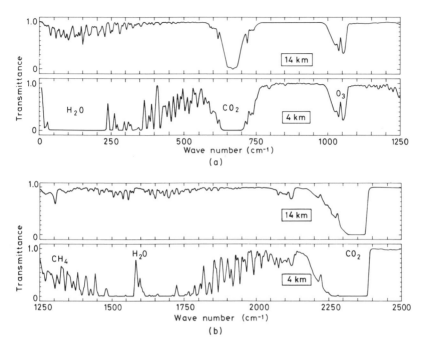

FIG. 2 Calculations of atmospheric transmittance at a spectral resolution of 5 cm^{-1} for a vertical path from two altitudes, 4 and 14 km, for the spectral ranges (a) 0–1250 cm^{-1} and (b) 1250–2500 cm^{-1}. (From Kyle and Goldman, 1975.)

A. FILTER PHOTOMETER/RADIOMETER

At a fairly simple level, one may consider an instrument that employs some fixed, wideband, spectral selection component to isolate a chosen spectral band. Measurements of atmospheric emission or absorption are then integrated over this complete spectral band. In the case of absorption measurements, a natural source such as the sun or an artificial source such as a lamp or a laser is used. As an example of this type of device, it is possible to use multilayer interference filters, built by depositing successive layers of dielectric materials onto a transparent substrate, to provide spectral selection (Smith *et al.*, 1972).

Figure 3 shows a schematic section of such a wideband radiometer, used for measurements of the water vapor distribution in the stratosphere. This device employs dielectric multilayer filters to isolate the 0–300-cm^{-1} band, which includes the pure rotation band of H_2O (Swann *et al.*, 1982). The basic advantages of this type of instrument are its simplicity, ruggedness, and good light-gathering efficiency; the main disadvantage, of course, is its poor spectral selectivity.

TABLE I

BAND CENTERS[a]

Gas	Band	Wave number (cm^{-1})	Wavelength (μm)
H_2O	ν_2	1595	6.27
	ν_1	3657	2.73
	ν_3	3756	2.66
CO_2	ν_2	677	15.0
	ν_3	2349	4.26
O_3	ν_2	701	14.3
	ν_3	1042	9.60
	ν_1	1103	9.07
N_2O	ν_2	589	17.0
	ν_1	1285	7.78
	ν_3	2224	4.50
CO	$1-0$	2145	4.66
	$2-0$	4260	2.35
	$3-0$	6350	1.57
CH_4	ν_4	1306	7.66
	ν_2	1534	6.52
	ν_3	3019	3.31
O_2	$0-1$ $(^1\Delta - {}^3\Sigma)$	6325	1.58
SO_2	ν_2	518	19.30
	ν_1	1151	8.69
	ν_3	1362	7.34
NO	$1-0$	1876	5.33
NO_2	ν_3	1621	6.17
	$\nu_1 + \nu_3$	2910	3.44
NH_3	ν_2	933	10.70
HCl	$1-0$	2886	3.46
	$2-0$	5668	1.76
HF	$1-0$	3962	2.52
HNO_3	$2\nu_9$	896	11.16
CH_3Cl	ν_4	3039	3.29
	$3\nu_6$	3042	3.29

[a] From Park (1977).

B. GRATING SPECTROMETER

The best-known device for obtaining some degree of spectral scanning is the dispersion spectrometer based on a prism or grating as the dispersive element. We shall consider only the latter here in view of its wider range of applicability and greater intrinsic accuracy.

The general layout of a grating spectrometer is shown in Fig. 4a. It con-

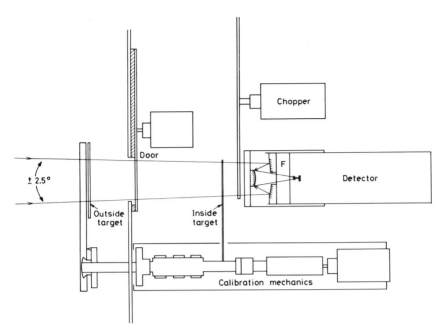

FIG. 3 An example of a fixed bandwidth filter radiometer, in this case designed for operation in the far-infrared between 30- and 1000-μm wavelengths (Swann *et al.*, 1982). A protective door is opened once the device is above the cloud level and a 5° beam of radiation, thermally emitted by the atmosphere, is accepted by a small Cassegrain telescope after chopping and focused onto a triglycine–sulphate pyroelectric detector. Periodically, a (warm) inside target and a (colder) outside target are moved into the beam for calibration purposes.

sists of an entrance slit, a collimating mirror or lens, the grating, a focusing mirror or lens, and a detector. The basis of this method lies in the properties of the grating, which is essentially a reflecting plate scored with closely spaced parallel lines. All undergraduate physics courses will have contained a description of the theory of the diffraction grating (e.g., see Schroeder, 1974), so we shall only mention here the governing equation, which states that a wavelength λ will be transmitted by the spectrometer if the following relation is satisfied:

$$n\lambda = 2d(\sin \theta + \sin \phi), \tag{21}$$

where d is the grating spacing, θ and ϕ the incidence and diffraction angles at the grating, respectively, and n the order number. These terms are defined in Fig. 4b, in which the grating is shown with a faceted surface rather than simply scored grooves. This is done to take advantage of specular re-

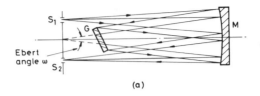

(a)

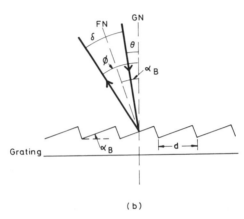

(b)

FIG. 4   (a) The layout of an Ebert–Fastie dispersive grating spectrometer, where S_1 is the entry slit, mirror M provides both collimation onto grating G and focusing onto exit slit S_2, and ω is the so-called Ebert angle. Grating G is usually rotated about an axis perpendicular to the plane of the paper to achieve spectral scanning. (b) The parameters involved in the grating equation [Eqs. (21) and (22)], where FN and GN represent the normals to the facet and grating base, respectively, and δ is the angle between the incident and reflected rays. This case illustrates the use of the blaze to throw more energy into a particular order and angle by taking advantage of specular reflection from the facets. (From J. F. James and R. S. Sternberg, 1969, "The Design of Optical Spectrometers." Chapman & Hall Ltd, publisher.)

flection in order to throw most of the intensity of the incident beam into a desired wavelength and order. The blaze angle α_B is chosen to achieve this, using the expression (valid for $n = 1$, first order)

$$\lambda_B = 2d \sin \alpha_B \cos(\alpha_B - \theta). \tag{22}$$

Spectral scanning is usually achieved by rotation of the grating.

    The chief virtue of the grating spectrometer, or, more generally, the dispersion spectrometer, is that it is a fairly simple device that, in a straightforward manner, allows a spectrum to be scanned. The spectral resolution can be simply controlled by adjusting the width of the entrance and exit slits, subject to the limitation imposed by the size of the grating

(Schroeder, 1974). This can be seen by consideration of Fig. 4 and Eq. (21). Using blaze techniques, good transmission efficiency can be achieved.

For these reasons, the grating spectrometer remains the basic spectroscopic tool. However, atmospheric spectroscopy is distinguishable by two main characteristics: the signal-to-noise ratio is usually not good enough and the time available for an observation is usually not long enough (these two parameters are, of course, interconnected). Thus efficient use of time and spectral energy are important requirements, and it is possible to devise instruments that are better than the grating in these respects, though usually the experimenter pays for such an advantage in increased difficulties of manufacture or use or in increased data analysis time.

A further difficulty with the grating spectrometer is that it is, in practice, somewhat limited in maximum resolving power. This is determined by the size of the grating (Schroeder, 1974) and by the angular size of the slits (i.e., the focal length of the spectrometer). Neither of these parameters can be easily increased beyond values equivalent to a resolving power of $\sim 5.10^4$ at 10 μm, or a spectral resolution of 0.02 cm^{-1}, and even this is difficult to achieve. For higher resolving power, quite different methods must be used.

C.   The Grille Spectrometer

A novel variant of the grating spectrometer is the grille spectrometer invented and developed by Girard (1963). In this device, the entrance slit is replaced by a grille, or an array with a hyperbolic structure having alternate opaque and transparent bands (Fig. 5a). The radiation passes through this grille, then through a conventional Littrow mounting grating system, and finally through a copy of the first grille. The presence of the two grilles causes the instrumental function due to a slit spectrometer (Fig. 5b) to be modified (Fig. 5c) to include a roughly triangular function whose width is set by d, a typical width of an element of the grille, and a second function whose width is set by D, the width of the whole grille.

For a simplistic view of how the device works, imagine that the opaque and transparent regions of the grille are interchanged. Then the D function will not change, but the d function will be inverted (Fig. 5d). If the difference is taken between these two results, then we obtain the curve in Fig. 5e, i.e., a narrow function characteristic of the small dimension d of the grille. However, the throughput of the spectrometer is dictated by the large dimension D, and we have a high-resolution spectrometer whose energy throughput is independent of the spectral resolution (unlike the classical grating spectrometer).

This device has been extensively used by Girard and co-workers for

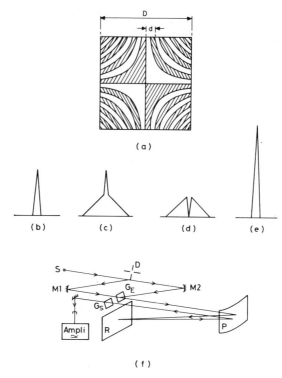

Fig. 5 The grille spectrometer. (a) A schematic representation of a grille: the shaded portions are opague and the clear portions are transparent, with d a typical dimension of one of the elements of the grille and D the overall size. Parts (b)–(e) show how the instrumental profile (e) of the device is obtained from those [(c) and (d)] due to complementary grilles; details are given in the text. (f) A typical layout of a grille spectrometer: S is the source, D the chopper, M_1, M_2 the mirrors, G_E the entrance grille, G_S the exit grille, R the grating, and P the off-axis parabolic mirror. (From Girard, 1963.)

stratospheric spectroscopy. Of course, our description here has been entirely pictorial and a more mathematical treatment is possible (Girard, 1963).

Figure 5f shows a configuration that illustrates the method of use (although the system actually employed differs slightly from this). The incident beam is chopped between the two arms M_1 and M_2, passes through grille G_E, the grating spectrometer, and finally through the exit grille G_S. The effect of G_E is complementary between arms 1 and 2 because in the first G_E is used in transmission and in the second it is used in reflection. Demodulation at the chopping frequency gives the difference signal.

D. Michelson Interferometer

The Michelson interferometer developed as a widely used spectrometric instrument in the middle of this century, long after its first use by Michelson as a means of measuring the separation of a pair of spectral lines (Michelson, 1881). Its development became possible through one major practical achievement: the advent of modern digital computers with which the lengthy Fourier transform operation could be speedily accomplished. Coupled with this came some impressive new thinking about the generalized nature of spectrometers as information processing and transmitting devices, led by people such as Fellgett, Jacquinot, and Mertz (see Chamberlain, 1979).

The beautiful work of Janine and Pierre Connes (1966) illustrated beyond any question that these fundamental advantages were, in practice, realizable. In her doctoral thesis, J. Connes pioneered the theoretical and computational aspects of Fourier spectroscopy (J. Connes, 1961, 1971), and in his experimental work on planetary atmospheres, P. Connes and his co-workers produced some of the most beautiful and elegant atmospheric spectra ever reported, which revealed a wealth of spectral detail (P. Connes, 1971; Connes et al., 1967, 1968; Connes and Michel, 1974).

The Michelson interferometer operates (see Fig. 6) by dividing the am-

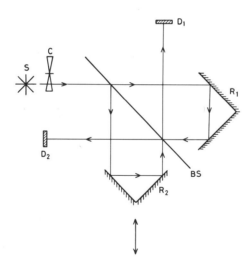

Fig. 6 The Michelson interferometer (shown, for simplicity, without collimation and with "rooftop," or 90°, reflectors R_1 and R_2), where S is the source, C the chopper, BS the beam splitter, and D_1, D_2 are the detectors. One of the reflectors (R_2 in this case) is moved along the local optical axis in order to produce a difference of optical path between the two beams.

plitude of a beam of radiation and varying the path difference between the two beams so produced by introducing optical delay in one arm. This delay is quite simply achieved by moving the reflector in one arm in and out along the optical axis (x axis).

The intensity of the combined beam measured by detector D_1 is then given (in units of intensity, not amplitude) by

$$I_1(x) = \tfrac{1}{2}I_0 + \int_{-\infty}^{\infty} B(\bar{\nu}) \exp(2\pi i \bar{\nu} x)\, d\bar{\nu}, \tag{23}$$

where

$$I_0 = \int_{-\infty}^{\infty} B(\bar{\nu})\, d\bar{\nu}. \tag{23a}$$

Similarily, the intensity at D_2 is

$$I_2(x) = \tfrac{1}{2}I_0 - \int_{\infty}^{\infty} B(\bar{\nu}) \exp(2\pi i \bar{\nu} x)\, d\bar{\nu}. \tag{24}$$

Quite clearly, from the reciprocity of Fourier relations,

$$B(\bar{\nu}) = \int_{-\infty}^{\infty} (I_1(x) - \tfrac{1}{2}I_0) \exp(-2\pi i \bar{\nu} x)\, dx, \tag{25}$$

with a similar equation for $I_2(x)$:

$$B(\bar{\nu}) = \int_{-\infty}^{\infty} -(I_2(x) - \tfrac{1}{2}I_0) \exp(-2\pi i \bar{\nu} x)\, dx. \tag{26}$$

A Fourier transform of the interferogram $I(x)$, therefore, yields the spectrum $B(\bar{\nu})$. But why go to such lengths? The answer to this question is that, for the study of weak sources,[1] the Michelson interferometer is highly efficient in its use of available energy, because of the following two fundamental advantages:

1. *Fellgett's Multiplex Advantage* (Fellgett, 1958). For a detector-noise-limited system, a Michelson interferometer exhibits a $\sqrt{M}/2\sqrt{2}$ signal-to-noise ratio advantage over a conventional (single-channel) dispersive spectrometer (e.g., Section III.B) for equal observation times, where M is the number of spectral elements in the spectrum being studied (Treffers, 1977).

Physically, this advantage arises because in a dispersive spectrometer each spectral element is observed consecutively (in "series"), whereas in

---

[1] Here the definition of a weak source is that the noise intrinsic in the detector (the most common noise source in the IR) should dominate over noise due to statistical photon fluctuations in the beam of energy from the source.

an interferometer all spectral elements fall onto the detector all the time (observation in "parallel"). The reason the $1/2\sqrt{2}$ factor multiplies the widely accepted $\sqrt{M}$ value is that the modulation efficiency of a two-beam interferometer is not perfect (see Treffers, 1977; Tai and Harwit, 1976).

2. *Jacquinot's Throughput Advantage.* A dispersive spectrometer is limited in energy throughput by the presence of slits; furthermore, this limitation increases as the slits are narrowed to increase resolution. In an interferometer, circular symmetry prevails because no slits are needed; moreover, as optical path difference is increased to increase resolution, no stopping down occurs. This latter statement does, in fact, need some qualification, because the solid angle of the rays passing through an interferometer can limit the achievable spectral resolution (physically, because rays following different paths in the ray bundle traverse paths of slightly different length, leading to a "blurring" effect in the spectrum). However, a definite throughput advantage does arise from the avoidance of the use of slits, which was first recognized by Jacquinot (1958).

A full discussion of the details of the theory of the Michelson interferometer is too long for the present text but can be found in several excellent references (e.g., Chamberlain, 1979). One usually finds that, in practice, some degradation of the basic advantages of Fourier spectroscopy occur, so care must be taken in the detailed design of an interferometer to attend to all such problems.

Because of the requirement that the dominant source of noise be independent of the incident photon flux, Michelson interferometry has found its most important (though not exclusive) application in the infrared, particularily the submillimeter, regions. To illustrate that it is dangerous to assume that the well-known advantages automatically apply, consider the following example.

Suppose the intention is to build a high-resolution device for atmospheric spectroscopy. New, more sensitive detectors are available, which naturally we wish to use and which have such very low noise properties that, in our application, "scintillation," or source noise, dominates over detector noise. It may be shown in this case (Chamberlain, 1979) that a $(\sqrt{M}/2\sqrt{2})^{-1}$ advantage occurs, i.e., a *disadvantage* in using an interferometer. It would be better to build a dispersive spectrometer to use with our new super detectors. (This ignores the Jacquinot advantage.)

Although it is possible for this example to be realized today,[2] even in

---

[2] The new generation of doped silicon photoconductors, developed in the United States, is said to have noise equivalent power ratings of $\sim 10^{-16}$ to 10^{-17} W/Hz$^{1/2}$ when surrounded by a thermal environment at a temperature below about 10°K.

the IR, it is unlikely to arise often in most practical atmospheric situations. It is given, not so much as a specific case, but more as a warning that a Michelson interferometer is not a panacea for all difficult spectroscopic measurement problems. Properly used, however, it is an extremely powerful device.

Several variants of the Michelson interferometer have been used in atmospheric research. The one to make the most significant contribution recently is the polarization interferometer developed by Martin and Puplett (1969) for use in the SM region. In this device, the beam splitter is replaced by a wire grid, which acts as a linear polarizer. A prepolarizing grid and an analyzer grid precede and follow the beam splitter, respectively. It can be shown that two complementary interferograms may be produced, depending on the orientation of the analyzer. A combination of those two interferograms eliminates any "dc" term in the interferogram [see Chamberlain (1979) and Eqs. (23) and (24)] and thus a potential noise "carrier." Furthermore, the spectral transmission of this device is markedly better than that achieved with a conventional thin-film beam splitter. The polarization interferometer has recently been used to good advantage for very high resolution studies of the SM stratospheric emission spectrum, as we shall see in Section V.

E. Gas Correlation Radiometer

The basic idea behind the gas correlation radiometer class of instruments was first discussed, to this author's knowledge, in a somewhat unknown paper by Smith and Pidgeon (1964).

If we are trying to achieve high spectral resolution, say, for the detection of a specific atmospheric constituent in a complex spectrum containing lines of many other gases, Smith and Pigeon's suggestion was to utilize a sample of that constituent itself within our spectrometer and to modulate the absorption lines of that gas so as to produce an instrument sensitive only to those wavelengths.

Several arrangements have been proposed for doing this, and for the present purpose, we shall in fact use as an example not the one proposed by Smith and Pigeon but the more recent version developed and used for atmospheric remote sensing by Houghton and colleagues at Oxford University. This device is known as the pressure modulator radiometer [PMR; see Taylor *et al.* (1972)], and its operation is illustrated in simple terms in Fig. 7.

A cell containing a sample of gas is introduced into a beam of radiation. The pressure of the gas is modulated by means of a piston arrangement, at a frequency f. The maximum and minimum pressures are P_1 and P_2,

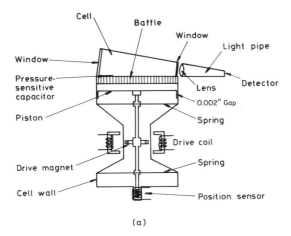

(a)

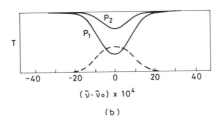

(b)

Fig. 7 The pressure modulator radiometer, an example of the class of gas correlation selective radiometers. (a) Radiation traverses the cell before being focused onto the detector. Pressure modulation is effected by a resonant piston driven electromagnetically, and the resultant motion of the piston is sensed, also electromagnetically. The pressure in the cell is monitored capacitatively. (b) Cell transmittance (dashed curve) for the case of extreme pressures P_1 and P_2 (curves P_1 and P_2 show spectral line transmission at these two pressures; the dashed curve is essentially the difference between these two). Note the highly expanded abscissa (i.e., high spectral resolution), where $\bar{\nu}$ and $\bar{\nu}_0$ are in units of reciprocal centimeters. (From Houghton *et al.*, 1971.)

respectively. The transmission near a single absorption line of the gas at pressure P_1 looks like the curve labeled P_1 in Fig. 7b; at pressure P_2, the transmission is shown as curve P_2. If the signal falling on the detector is demodulated at f, the result is proportional to the difference between these two transmission curves, or the dashed curve in Fig. 7b.

If we now consider a generalized situation, observing emission in a complicated spectral region from the atmosphere, then using a pressure modulator radiometer will allow the automatic selection (transmission) by the gas cell of only those lines we wish to detect and rejection of all others. Furthermore, this process occurs at the linewidth of the lines, unbroadened by instrumental effects, and gives rise to an essentially high-resolution selective filter. A problem associated with such devices is the containment of the gases (not all atmospheric constituents can be contained easily or for long periods without deterioration): a further problem in the case of the PMR, of course, is the actual implementation of the pressure modulation, although systems for this have been utilized with great success on several occasions in the demanding environment of a spacecraft experiment (e.g., Houghton, 1978).

Other variants, as previously mentioned, have been proposed or developed, including types employing beam switching between full and empty cells (Houghton and Smith, 1970), and a type based on a Michelson interferometer, using the two arms to compare a full or empty cell, but not requiring any mechanical motion for its operation (Harries and Chamberlain, 1976).

F. Fabry–Perot Interferometer

The Fabry–Perot interferometer is a classic device, the theory of which is taught in all college physics courses. It operates, of course, by utilizing the interference properties of two highly polished plane-parallel reflecting plates.

The Fabry–Perot transmits an infinite series of peaks at wavelengths corresponding to constructive interference between the plates (Fig. 8). The width of the peaks is determined by the reflectivity of the plates, as well as the spacing.

It is not necessary to reproduce here all the equations governing the operation of this interferometer (see, for example, Roesler, 1974): rather we should here discuss its applicability to atmospheric spectroscopy.

The interferometer has good energy efficiency because of its circular symmetry and lack of slits. The user must be aware, however, that, like the Michelson, the Fabry–Perot requires quite strict limits on the solid angle (and therefore optical throughput) of radiation passing through the device if spectral resolution is not to be degraded.

Spectral resolution can, at least in the IR where suitable materials exist, be made very high indeed, e.g., 10^{-3} cm^{-1} at 10^3 cm^{-1} (Hernandez, 1978). In the SM region, where metal mesh reflectors have been used, somewhat poorer resolutions (e.g., 10^{-2} cm^{-1}) have been achieved (Chanin and Lecullier, 1978). A major problem lies in rejecting all orders but the desired one (see Fig. 8). At high resolutions, of course, these orders come very close indeed. As an example, a realizable value of the finesse is ~50, which implies an interorder spacing (free spectral range, FSR), for a spec-

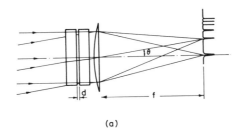

(a)

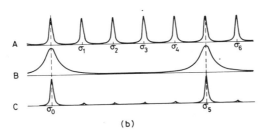

(b)

FIG. 8   The Fabry–Perot interferometer. (a) The schematic layout, where d is the plate spacing, f the focal length, and θ the field of view. The image plane pattern is shown. (From J. F. James and R. S. Sternberg, 1969. "The Design of Optical Spectrometers. Chapman & Hall Ltd, publisher.) (b) The combination of a high-resolution (curve A) etalon with a low-resolution etalon (curve B) to provide selection of only some of the high-resolution peaks (curve C). Further selection may clearly be obtained by using more etalons and bandpass interference filters (which act as multielement etalons). (From F. L. Roesler, 1974.)

tral resolution of 10^{-3} cm^{-1}, of only 5×10^{-2} cm^{-1}. Often these close-by orders are removed by using a series of Fabry–Perot interferometers at successively lower resolutions (see Fig. 8b). At high spectral resolution, also, the problem of wavelength stability becomes more serious; thermal drifts of the plate spacing, for example, can be a very difficult problem to overcome. Note that scanning of the spectrum can be achieved using commercial piezoelectric devices (Roesler, 1974).

Thus, for IR and SM atmospheric spectroscopy, the principle area of application of the Fabry–Perot interferometer arises when a measurement of only one or a few spectral lines is required at high spectral resolution. In other regimes, dispersive spectrometers or Michelson interferometers are more appropriate, e.g., for lower resolution or for wide spectral bandwidth at high resolution.

G.   Cryogenic Instruments

As a further, distinct class of instrumentation, we should point out that in recent years several investigators have developed spectrometric instruments for atmospheric observations that are cooled using cryogenic liquids. The reasons for cooling a complete spectrometer are twofold: first, the thermal emission from the optical elements and other components can act as a fluctuating intensity noise source, which can mask very weak atmospheric emission bands, and therefore such emission represents a limit to sensitivity; second (a quite different point), the thermal flux from the instrument carries its own photon noise even if the intensity of this flux remains perfectly steady, and this "photon noise" can degrade the performance of the cold detectors that are available today.

Clearly, although the advantage may be very significant, the technical problems associated with operating a complete spectrometer at a very low temperature are not trivial. However, a number of authors have now reported the successful operation of such devices. Stair et al. (1974) have successfully built and flown liquid-^{4}He-cooled Michelson interferometers on rockets and aircraft; Grossman and Offerman (1978) have flown a liquid ^{4}He grating spectrometer on a rocket; Murcray and co-workers (Williams et al., 1976a) have flown liquid-nitrogen- and liquid-helium-cooled grating spectrometers for aircraft and balloon measurements; and at the NPL, Pollitt et al. (1982; see also Bangham et al., 1980) have recently flown a liquid-nitrogen-cooled grating spectrometer on a high-altitude balloon platform.

H.   Measurement Techniques

Atmospheric spectroscopy in the IR has been carried out using most of the instruments previously mentioned in a variety of configurations, which we shall briefly note here.

First, we must ask what observational platforms are available to the spectroscopist who wishes to study the atmosphere? Many early measurements were made from mountain observatories. It is a general property of the IR spectrum that at sea level large portions of the IR spectrum are very heavily, if not totally, attenuated, both because of the increasing atmospheric density and thus absorber amount and the increasing pressure-broadened half-widths (see Section II). Thus, more of the spectrum becomes accessible at higher altitude where these effects are diminished.

Realizing this fact, spectroscopists have been able to place their instruments higher in the atmosphere by using aircraft and balloons. Aircraft have the advantages of (usually) allowing an observer on board and of (usually) being able to provide relatively large amounts of space, weight capability, and electrical power. They also have very considerable range, usually several thousand kilometers, which permits, for instance, studies of the horizontal variability of atmospheric parameters. Aircraft are, though, somewhat limited in altitude capability compared with balloons. Except for a few aircraft available in the United States (and, presumably, also in the USSR), such as the WB57 weather reconnaissance or the U2 reconnaissance airplanes, which have ceilings of about 20 km, most research aircraft are limited to an ~14-km ceiling or below, which means that truly stratospheric heights may not be reached where the tropopause is high, e.g., at low latitudes. Balloons, on the other hand, are capable of raising large payloads (e.g., 1–2 tons) to heights of 40 km or more. Balloon experiments, therefore, have been used for investigations of vertical profiles but are of rather limited value for latitudinal or longitudinal studies unless they are small enough to be hand-launched. Balloon experiments must also be either completely automatic or controllable via a radio link, because no observer can be carried on board. The electrical and mechanical noise conditions, however, are much quieter on a balloon than on an aircraft. So the comparison can go on, of course; the selection of flight platform can only be done by a complete analysis of the requirements of each specific case.

Continuing to increase our altitude, the next vehicle at our disposal is the rocket. Widely used by upper atmosphere physicists to study the ionosphere, rockets can reach apogees of several hundred kilometers. Experiment durations are, however, very short, on the order of minutes, and instruments designed for rocket experiments have to be able to acquire data very rapidly as well as be of very rugged construction so as to withstand the shock of a rocket launch. On the face of it, given these facts plus the difficulty of accurately steering and stabilizing a rocket, this would seem not to be a very attractive proposition. For studies of atmospheric levels above about 50 km, however, rockets provide the only device for research groups whose budgets will not extend to satellite experiments. And, of course, for in situ, rather than remote-sensing, measurements, rockets provide the only vehicle at altitudes above balloons yet below orbital altitudes.

Satellites are quite clearly in a class of their own. Aircraft, balloons, and rockets do, of course, require substantial organizational support for successful operation but nothing like the massed ranks of managers, engineers, and scientists required to implement a satellite experiment. For this reason and others, a satellite mission is very expensive and can only be attempted on a national scale by the more wealthy nations. Furthermore, a satellite experiment design must be "frozen" several years before flight, to allow time for all the quality assurance testing and flight integration to be performed. In a sense, therefore, all satellite experiments (in a time of rapid scientific and technological advance) are doomed to be somewhat out of date. However, having negatively listed some disadvantages, we can also note that, on the positive side, a satellite experiment provides near-global coverage and a long time-base experiment (typically 1–2 years). Both these aspects are often essential in atmospheric science and meteorology and are important enough to convince politicians in many countries to support scientific satellite programs. The influence of militaristic interests here is evident but is not, of course, the subject of this chapter.

The forthcoming Space Shuttle/Spacelab program offers rather shorter duration missions than a free-flying satellite, with (in principle) a far more rapid turnaround and the ability to recover apparatus. It should, also, be cheaper in terms of unit costs of carrying instruments in orbit, though some scientists are beginning to question whether this may be true in reality. The actual value or impact of this new facility, in short, is not easily assessable at the present time and will need to be evaluated largely on the basis of experience.

Turning our attention from the vehicles available to us, we shall now consider what are the modes of operation that have been used for atmospheric spectroscopy.

In recent years, most investigators have chosen to operate in one of two modes, though, or course, others exist. The first of these is to direct the instrument toward the sun and to measure the absorption due to the intervening atmosphere. The second is to measure the thermal (or in some cases nonthermal) emission from the atmospheric constituents. The former method has advantages in terms of good signal-to-noise ratios owing to the high brightness temperature of the sun, the relative independence of atmospheric temperature in the retrieval process, and also the relative simplicity of building an optical device that can track the sun and ensure that it fills the input aperture of the instrument at all times. The latter method suffers from poorer signal levels but can be used if measurements are required by night as well as day; difficulties can be experienced in this case, however, in controlling or even knowing the exact location of the line of sight because of the nature of the atmosphere as an extended source. Solutions to this problem have been devised, however, for satellites (Gille and House, 1971) and for balloons and aircraft (Moss, 1974, 1982). The emission method, furthermore, requires an accurate knowledge of atmospheric temperature for composition retrieval.

The observational geometry must also be decided. Generally, de-

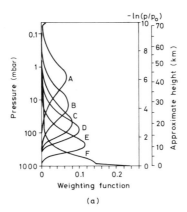

(a)

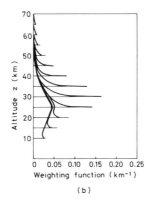

(b)

FIG. 9 The weighting functions $K_{\tilde{\nu}}$ defined in Eq. (11) for two different observational geometries: (a) vertical sounding (from Houghton, 1977); (b) limb sounding (from Gille and House, 1971). In part (a), the different curves A–F correspond to different frequencies across the 15-μm CO_2 band (see Barnett, 1974); in part (b), a single spectral CO_2 band has been used (585–705 cm^{-1}), and the different curves correspond to different observational zenith angles (θ in Fig. 10).

pending on the observer's position within or outside the atmosphere, we may direct our instrument along any line of sight. Recently, however, it was realized that directing the line of sight toward the limb of the atmosphere (i.e., that region immediately above the horizon where line-of-sight zenith angles will lie in the 90° plus range) can be advantageous (Gille and House, 1971). In this case, for detection of weak emission or absorption lines, the number of molecules integrated along the line of sight is considerably larger than for other directions of view; also, it may be shown that most of the absorption or emission is located in a rather narrow layer roughly one scale height thick above the point along the line of sight where the latter is tangential to the local earth's surface. Other advantages also occur, such as the fact that in emission spectral features are viewed against the cold background of space rather than, for example, the hot surface of the earth as in a downward-looking configuration. The principal disadvantages of this "limb-sounding" technique are that a rather large horizontal scale applies to the measurement, several hundred kilometers, owing to the near-horizontal line of sight. Also, tropospheric sounding is usually impossible because of cloud attenuation: for a near-horizontal line of sight, the probability of intersecting a cloud is considerably higher than for a vertical line of sight.

For these reasons, this method has been applied primarily from rather high vehicles, such as balloons, rockets, and satellites, and for studies of the stratosphere and above.

As a final comment in this section concerning observational geometries, Fig. 9 shows the weighting functions (see Section II) for IR frequencies around the 15-μm CO_2 band (used for atmospheric temperature sounding) for the case of a downward-looking (Fig. 9a) and a limb-sounding (Fig. 9b) satellite instrument. Clearly, the weighting functions in Fig. 9b allow a much sharper definition in the vertical. Figure 10 shows, however, that the horizontal definition of the vertical sounder is much finer than the limb sounder. Such differences must be borne in mind when designing a specific experiment.

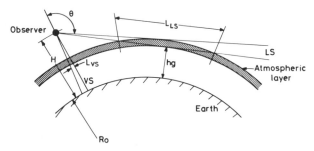

FIG. 10 Comparison of vertical sounding (VS) and limb sounding (LS) geometries, where θ is the zenith angle, H the observer altitude, R_0 the radius of the earth, hg the tangent height, and L_{LS} and L_{VS} the horizonatal scale for limb sounding and vertical sounding, respectively.

IV. Investigations of the Infrared Spectrum

A. INTRODUCTION

Because our intention is to provide an up-to-date survey, we shall treat only briefly the more historical aspects of the subject.

As reported in Kuiper's (1949) book, the first observations of the IR spectrum of the atmosphere were made by Langley in 1884. Until the middle of the twentieth century, progress in infrared atmospheric spectroscopy was slow, but in the postwar years, progress in infrared technology, based mainly on military-supported research,[3] permitted very rapid improvements in the sensitivity and accuracy with which the IR spectrum could be studied. In the 1950s, several excellent pieces of work appeared, which attempted to provide atlases and reviews of the IR atmospheric spectrum. Among these we might mention the works by Shaw et al. (1951) and Mohler et al. (1950) in the United States, Migeotte et al. (1956) in Belgium, and Houghton et al. (1961) in Britain. As an example of our understanding of the general properties of the spectrum at about this time, Fig. 11 shows a composite spectrum due to J. H. Shaw (from Shaw and Garing, 1962, and Houghton and Smith, 1966). The diagram shows, in the bottom frame, the transmission spectrum of the atmosphere between 1 and 15 μm from sea level for a vertical path. In the upper frames, the spectra of those atmospheric constituents contributing significantly to the overall spectrum are shown separately, based on laboratory measurements. The diagram, in this way, clearly demonstrates the main spectral structural features of the region (cf., Fig. 2).

After about 1960, a number of groups continued to work to improve the spectral resolution and sensitivity of spectrometers, so as to probe more deeply into the detailed structure. This process would very likely have progressed at a rather steady pace had it not been realized, in the late 1960s, that IR spectroscopy held enormous potential for providing quantitative constituent concentration data, particularily at high resolution. This, coupled with the rather sudden emergence of concern about the fate of the layer of ozone that exists in the stratosphere because of potential photochemical decomposition by anthropogenic trace constituents, has led to extremely rapid progress in the development of more sensitive instrumental techniques and in our detailed knowledge of atmospheric spectral structure. Of course, much remains to be learned, and certainly as soon as this book is published, new results will supersede the best data

---

[3] This is a sad example of the human proclivity for seeking military power rather than knowledge for its own sake.

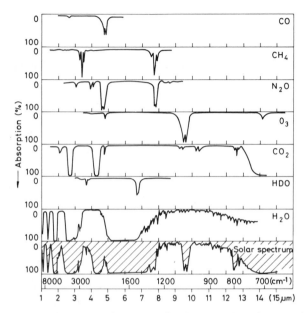

FIG. 11   Atmospheric transmission spectra for the atmosphere from sea level, together with spectra of separate atmospheric constituents that make up the former (see text for further discussion) (From Shaw and Garing, 1962, published in Houghton and Smith, 1966, p. 284.)

currently available. Nevertheless, because of this explosive increase in activity, our review will naturally be heavily weighted toward recent years.

B. MODERN INVESTIGATIONS

The subject of atmospheric spectroscopy has grown in such a remarkable way, for the reasons just given, that we cannot realistically attempt a detailed review of all measurements. Our approach, therefore, will be to list those groups that have made contributions and to select examples from the work of only some of those groups, in order to illustrate specific points concerning the IR atmospheric spectrum. Table II provides such a list of groups active in the subject. Both IR and SM groups are included: the latter will be considered in Section V.

1.   *Modern High-Resolution IR Atmospheric Spectroscopy Begins: Discovery of HNO₃*

We shall take as our starting point the publication in 1969 by D. G. Murcray and his associates at Denver University of an atlas of high-resolution

TABLE II

ATMOSPHERIC IR AND SM SPECTROSCOPY: A SELECTION
OF GROUPS/TECHNIQUES

Group	Country	Technique[a]	Platform
Ackerman/Girard	Belgium, France	Grille spectrometer/IR solar occ.	Balloon, aircraft
Baluteau and Bussoletti	France/Italy	FIR MI/emission	Aircraft
Buijs et al.	Canada	MI/solar occ.	Balloon
Clark and Kendall	Canada	FIR MI/atmospheric emission	Balloon
Evans et al.	Canada	GS/MI/UV and IR solar occ. and atmospheric emission	Balloon, rocket
Farmer et al.	U.S.	MI/solar occ.	Balloon, aircraft
Gille et al.	U.S.	Filter radiometer/atmospheric emission	Balloon, satellite
Harries et al.	U.K.	FIR MI/Cryogenic GS/atmospheric emission	Balloon, aircraft
Houghton et al.	U.K.	SCR/PMR/atmospheric emission	Balloon, satellite, aircraft
Mankin and Coffey	U.S.	FIR MI/emission; IR MI/solar trans.	Aircraft
Marten	France	FIR MI/emission	Aircraft
Menzies et al.	U.S.	IR laser heterodyne/solar occ.	Balloon, aircraft
Murcray et al.	U.S.	GS (cooled and uncooled)/solar occ/atmospheric emission	Balloon, aircraft
Offerman et al.	West Germany	Cryogenic GS/atmospheric emission	Rocket
Patel	U.S.	Laser absorption using spin-flip Raman laser/local measurement	Balloon
Stair et al.	U.S.	GS/Cryogenic MI/atmospheric emission	Rocket
Waters	U.S.	Microwave spectrometer/atmospheric emission	Balloon, aircraft, ground
Zander	Belgium	GS/solar occ.	Balloon

[a] Key to terms: IR, infrared; UV, ultraviolet; solar occ., solar occultation; FIR, far-infrared; solar trans., solar transmission; SCR, selective chopper radiometer; GS, grating spectrometer; PMR, pressure modulator radiometer; MI, Michelson interferometer.

IR solar absorption spectra measured from a balloon (Murcray et al., 1969). This work represents the results of some of the earliest activities employing limb-sounding techniques (see Section III), which also means that these were spectra that were capable of yielding data on numerous weak absorption bands and constituents previously inaccessible.

Figure 12 illustrates some of the data from this work, showing a series

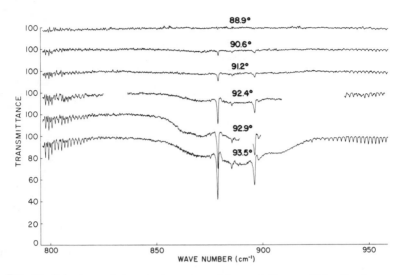

FIG. 12 Solar transmission spectra measured from a balloon at 30 km for the range of solar zenith angles marked on the curves. The appearance of absorption features resulting from the HNO_3 molecule ($2\nu_9$ band) at angles below the local horizontal is clearly seen. The curves are displaced from one another for clarity. (From Murcray *et al.*, 1969.)

of absorption spectra measured from a balloon at about 30-km altitude, using the setting sun as a source. The spectral resolution is 0.3 cm^{-1}, and this section of the IR spectrum is from 800 to 950 cm^{-1}. These data were recorded using a grating spectrometer with a focal length of 50 cm and a liquid-helium-cooled copper-doped Ge detector.

At solar zenith angles $< 90°$, the spectrum shows very little absorption, and this was essentially the state of knowledge in the mid-1960s. Indeed, this spectral region was commonly referred to as the "8–14-μm atmospheric window." However, as the setting sun caused the line of sight to sweep through deeper layers of the atmosphere, numerous absorption features could be seen building up because of a number of different spectral bands, the most exciting, perhaps, being the $2\nu_9$ band of nitric acid in the 890-cm^{-1} region. Actually, the existence of HNO_3 in the stratosphere had earlier been confirmed by the Denver group, as a result of balloon-borne limb-sounding measurements in the 1300-cm^{-1} range. Figure 13 shows this detection: an atmospheric spectrum is compared with a laboratory absorption spectrum of HNO_3 vapor, confirming the observation, even in the presence of quite intense absorption lines from CH_4, H_2O, and N_2O.

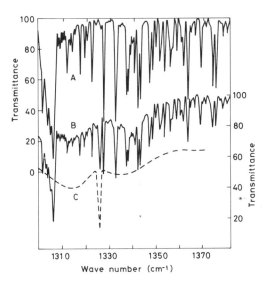

FIG. 13 Measurements of solar transmission spectra in the 1325-cm^{-1} region (resolution 0.2 cm^{-1}). Curve A, altitude 11.0 km, solar zenith angle 58.7°; curve B, altitude 30 km, solar zenith angle 92.4°; curve C shows the outline of the P, Q, and R banches of an absorption band due to HNO_3. The curves are displaced from one another for clarity. (From Murcray *et al.*, 1968.)

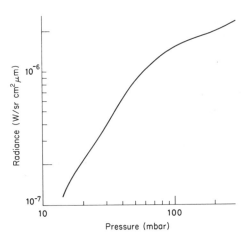

FIG. 14 A profile of measured radiance in the 890-cm^{-1} region as a function of pressure (altitude) measured during the ascent of a balloon-borne experiment. (From Murcray *et al.*, 1972.)

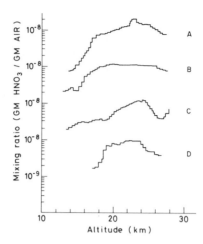

FIG. 15 Profiles of HNO₃ mass mixing ratio as functions of altitude derived from radiance profiles like the one shown in Fig. 14: A, $V = 2.7 \times 10^{-4}$ atm·cm; B, $V = 3.3 \times 10^{-4}$ atm·cm; C, $V = 2.1 \times 10^{-4}$ atm·cm; D, $V = 1.4 \times 10^{-4}$ atm·cm. Comments on the method of analysis are given in the text. (From Murcray *et al.*, 1972.)

Further observations of HNO₃ in the stratosphere were made by Murcray and co-workers using, on separate occasions, a liquid-helium-cooled emission radiometer and a similarly cooled grating spectrometer for solar absorption. Using the former device, the change of radiance in the 890-cm⁻¹ region was measured as the balloon ascended through the atmosphere with the instrument looking up at a variety of zenith angles (Williams *et al.*, 1972; Murcray *et al.*, 1972). Figure 14 shows a typical profile of radiance measured as a function of altitude, and Figure 15 shows the derived distribution of nitric acid. In this case, the concentration retrieval process is carried out starting at the top of the ascent and working down one layer at a time, determining at each step the total column density above each layer. Thus the derivation of concentration depends, not on the radiance–height profile $R(h)$, but on the first derivative $R'(h) = \partial R(h)/\partial h$. This leads to a requirement of high signal-to-noise ratios if small changes in $R'(h)$ are to be distinguished. Further studies (Murcray *et al.*, 1975) showed a strong latitudinal and seasonal dependence of HNO₃ amounts.

The work by the Denver group showed clearly by about 1970 that the high-resolution IR spectrum of the atmosphere (particularly at stratospheric heights) contained a great deal of composition information, particularly when observations were made in limb-sounding geometry.

2. NO_y Spectroscopy[4]

Once the problem of NO_x—O_3 catalysis had been recognized, (e.g., Crutzen, 1970) and linked to the exhaust emissions from high-flying aircraft such as the proposed supersonic transport (SST) fleets, infrared spectroscopy was immediately turned to the search for NO and NO_2 in the stratosphere. Chronologically, the first published report of the detection of NO came from the JPL group under C. B. Farmer (Toth *et al.*, 1973). Figure 16 shows a spectrum measured by that group of solar absorption in the 1900-cm^{-1} region measured during sunset from an aircraft. The instrument used was a fast-scanning Michelson interferometer, with a nitrogen-cooled detector. Clearly, the detection was near the limit of sensitivity, but an estimate of the NO column density was possible from the data, yielding an equivalent mean volume mixing ratio (assumed to be constant with height) of $1.0 \pm 0.2 \times 10^{-9}$.

Further measurements by other groups soon followed (Ackerman *et al.*, 1973, 1974; Girard *et al.*, 1973; Chaloner *et al.*, 1978; Jarnot, 1976; Drummond and Jarnot, 1978), many from balloon platforms, which allowed the estimation of the vertical distribution of NO. The importance of allowing for absorption lines due to solar CO in the 1900-cm^{-1} range was

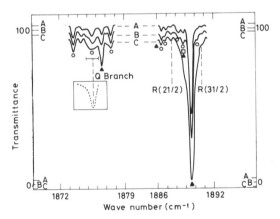

FIG. 16 Aircraft measurements of solar transmission for a limb path with solar zenith angles of 82° (A), 86° (B), and 91° (C). Aircraft altitude was 12 km, and the tropopause altitude was 10 km. The NO absorption features occur at 1875.8, 1887.56, and 1890.76 cm^{-1}, where ▲ is H_2O, □ is CO_2, and ○ is solar CO. Inset shows calculated NO Q branch for temperature of 220 °K. (From Toth *et al.*, 1973).

---

[4] Note: NO_y refers to the family NO, NO_2, HNO_3, and N_2O_5; NO_x refers to NO + N⌐

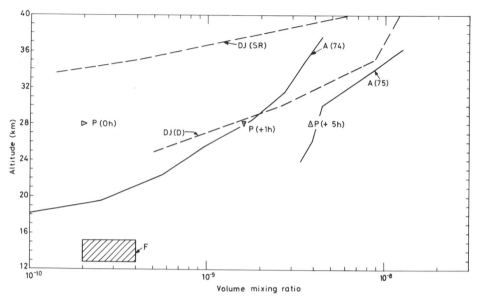

FIG. 17   Results of some IR spectroscopic measurements of stratospheric NO, where DJ(SR) refers to Drummond and Jarnot (1978) at sunrise, DJ(D) to Drummond and Jarnot (1978) at daytime, A(74) to Ackerman *et al.* (1973) at sunset, A(75) to Ackerman *et al.* (1975) at sunset, P(Oh) to Patel *et al.* (1974) at sunrise, P(+1h) to Patel *et al.* (1974) at 1 hr after sunrise, P(+5h) to Patel *et al.* (1974) at 5 hr after sunrise, and F to Farmer (1974).

emphasized by Murcray *et al.* (1974). Figure 17 shows a sample of the measurements of NO using IR spectroscopic methods. All these measurements were made using the 1 − 0 fundamental band of NO at 1876 cm^{-1}, and with two exceptions, they utilized solar occultation techniques. The first of these two exceptions was the measurements by Patel *et al.* (1974) who flew a spin-flip Raman laser plus a multipass absorption cell on a balloon payload. Ambient air at 28-km altitude was passed through the absorption cell, and absorption lines due to NO and H$_2$O were observed. The experiment was of considerable complexity and was consequently heavy, weighing about 1 ton. Also, because only local ambient air was investigated, the measurement was in situ rather than remotely sensed.

The results were of very good quality, as the example in Fig. 18 shows. This diagram illustrates the absorption in two single lines due to NO at a float level of 28 km. Spectral scanning of the spin-flip Raman laser was achieved by varying a magnetic field, which passed through a nonlinear InSb crystal in which a CO pump laser gave rise to Stokes and anti-Stokes lines obeying the relation

$$\nu = \nu_0 \pm g\mu B, \tag{27}$$

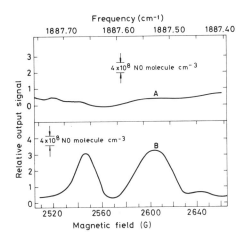

FIG. 18 Balloon-borne measurements of NO using a spin-flip Raman laser and an in situ absorption cell opened to the ambient atmosphere. Curve A shows a spectrum recorded during the night (19 October 1973 at 11:55:00–12:27:00 G.M.T.), and Curve B a similar result during daylight (19 October 1973 at 17:21:50–17:52:05 G.M.T.), at altitudes of 28 km and temperatures of 225 °K. The two strong features in curve B are assigned to NO. (From C. K. N. Patel, E. G. Burkhardt, and C. A. Lambert (1974). *Science* **184** (June 14), pp. 1173–1176. Copyright 1974 by the American Association for the Advancement of Science.)

where ν_0 is the pump frequency, B the magnetic field strength, g the electron spin g-factor, and μ the Bohr magneton (Born, 1962).

The second method used for nonsolar occultation detection of NO was employed by the Oxford group (Chaloner *et al.*, 1978), who used the pressure modulation gas correlation method (discussed in Section III.E) to make measurements of NO thermal emission at 5.3 μm by the limb-sounding method. Because of the use of emission, this group was able to make measurements during both day and night and confirmed the disappearance of stratospheric NO as the sun sets because of the removal of the NO photolytic production process (Crutzen, 1970):

$$NO_2 + h\nu \rightarrow NO + O. \tag{28}$$

In this work by the Oxford group, they were also able to measure the emission from NO_2 at 6.2 μm, again using gas correlation methods, and to measure the diurnal variation of this species (Chaloner *et al.*, 1978; Drummond and Jarnot, 1978; Drummond, 1976). In fact, NO_2 had first been detected in spectra from the Denver group (Goldman *et al.*, 1970), and an analysis of this data was reported by Ackerman and Muller (1972). Subsequently, a reanalysis of the same data was presented by Murcray *et al.* (1974), which produced a slightly different profile for NO_2 in the stratosphere.

A number of independent measurements followed (Ackerman *et al.*, 1974, 1975; Chaloner *et al.*, 1978). Figure 19 shows the 1610-cm⁻¹ region used for measurements of NO_2 by several groups, in this case taken from observations by Ackerman *et al.* (1975) [a similar result being reported by Mankin (1978)]. Even though the ν_3 band of NO_2 is mixed with the ν_2 band of H_2O at 6.3 μm, a number of features due to NO_2 can quite clearly be seen at high resolution (say, better than 0.3 cm⁻¹) between the H_2O lines. Figure 20 illustrates the results of a number of measurements of NO_2 made using a variety of IR spectroscopic techniques.

In Canada, Evans and co-workers employed a variety of instruments on board a balloon gondola to make measurements of NO, NO_2, and HNO_3 in an attempt to discuss the total NO_x amount in the stratosphere, although only HNO_3 was measured in the IR. This was successfully done (Evans *et al.*, 1976), although it was noted that the HNO_3 measurements were made during the ascent phase of the flight, at about midday, whereas the NO and NO_2 data were obtained using solar occultation methods during the subsequent sunset. Thus, although corrections can possibly be applied to the data to allow for the time difference, the separate data were obtained at different times and, perhaps, not on the same air mass. Never-

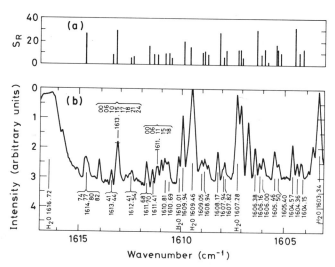

Fig. 19   Solar transmission spectrum (b) measured in the 1610-cm⁻¹ region from an aircraft at 16-km altitude with the sun at a zenith angle of 91°. Lines due to H_2O (labeled) and NO_2 (not labeled) are observed. Part (a) shows the calculated relative intensities for lines of NO_2. (From Ackerman *et al.*, 1974.) Reproduced by permission of the National Research Council of Canada from the Canadian Journal of Chemistry, Volume 52, pp. 1532–1535, 1974.)

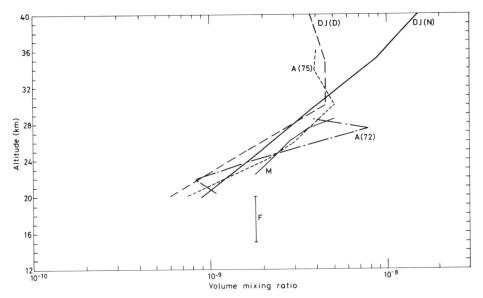

FIG. 20 Results of some IR spectroscopic measurements of stratospheric NO_2, where DJ(D) refers to Drummond and Jarnot (1978) at daytime, DJ(N) to Drummond and Jarnot (1978) at nighttime, A(72) to Ackerman and Muller (1972) at sunset, A(75) to Ackerman *et al.* (1975) at sunset, M to Murcray *et al.* (1974) at sunset, and F to Farmer (1974).

theless, these spectroscopic measurements provided a valuable test of photochemical theory.

Measurements of NO, NO_2, and HNO_3, made simultaneously using a grille spectrometer in solar occultation mode from Concorde 001 were reported by Fontanella *et al.* (1974). By measuring the variation of absorption intensity with solar zenith angles, these authors were able to deduce mixing ratio profiles above the flight level (which was about 16 km) for all three constituents as well as total overburdens. These measurements were made over the Bay of Biscay area of the northeastern Atlantic.

To date, no confirmed infrared detection of N_2O_5 in the stratosphere, to the author's knowledge, has been made (though a possible detection has been reported by Evans *et al.* 1978).

3. ClO_x-Related Spectroscopy[5]

The arrival of a second major scare about the stratospheric environment (Molina and Rowland, 1974), involving the potential decomposition

---

[5] ClO_x refers to the sum of all odd chlorine-containing molecules, i.e., Cl + ClO + HOCl + etc.

of ozone through chlorine catalysis, signaled a further major effort in IR atmospheric spectroscopy to detect and measure the concentration of Cl-containing molecules in the stratosphere.

The primary anthropogenic source of chlorine was thought to be due to the chlorofluoromethane (CFM) propellants used in a wide variety of applications (National Academy of Sciences, 1976). The CFMs most widely used have been CF_2Cl_2 and $CFCl_3$, and these and their stratospheric decay products (following ozone reduction reactions), for example, ClO, HCl, HF, and $ClONO_2$, have been and are being sought using IR spectroscopy.

The presence of CFMs in the atmosphere, previously known through in situ sampling measurements (for example, see Schmeltekopf et al., 1975), was confirmed using IR spectroscopy, again by D. G. Murcray and co-workers, using a grating spectrometer that made solar occultation measurements from a balloon (Williams et al. 1976b). When measurements were made to very low sun angles, corresponding to a line of sight that actually intersected the troposphere, absorptions at 10.8 and 11.8 μm were observed. These absorptions were assigned to CF_2Cl_2 and $CFCl_3$, respectively. The very weak absorptions at stratospheric altitudes are the result of the rapid decrease of CFM density above the tropopause due to photolysis. These bands represented further new absorptions in the so-called atmospheric window. Figure 21 shows an example of a series of solar absorption spectra measured during sunset that illustrates the growth of the CFM bands. Mixing ratios for CF_2Cl_2 and $CFCl_3$ determined from these data (Williams et al., 1976b) agreed well with independent measurements.

Perhaps of more interest in a spectroscopic sense were efforts to measure HCl and HF in the stratosphere. These molecules play a role equivalent to that of HNO_3 in the NO_x cycle in representing rather stable end products of a series of chemical reactions, including the important catalytic cycle. The first reported measurement of HCl was by Farmer et al. (1976), based on an analysis of aircraft solar absorption spectra in the 2925-cm^{-1} region. This was followed by solar absorption results from several workers who had made measurements from balloon altitudes (Williams et al., 1976c; Ackerman et al., 1976; Eyre and Roscoe, 1977; Buijs et al., 1977; Raper and Farmer, 1977), all using lines in the HCl fundamental band at 3 μm. Figure 22 illustrates some of the spectral features utilized in these determinations. As can be seen from Figure 23, which shows the HCl profiles derived from the various experimenters, good agreement is found among the various results.

Measurements of HF began with the detection by Zander of HF absorption lines in balloon-borne solar spectra at a wave number of 4039 cm^{-1}

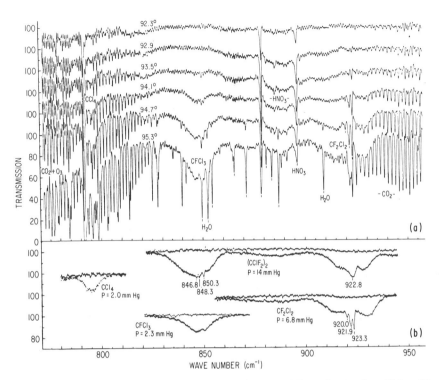

FIG. 21　Solar transmission spectra recorded in the 800–950-cm^{-1} range from 30-km altitude at the solar zenith angles marked on each curve in part (a) at a resolution of 0.2 cm^{-1}. The appearance of absorption bands due to CCl_4, CF_2Cl_2, and $CFCl_3$ in the lower stratosphere and upper troposphere is made clear by comparison with the laboratory spectra for these gases shown in part (b) at $T = 26\ °C$, $L = 0.5\ cm$, and the resolution is 0.4 cm^{-1}. (from W. J. Williams, J. J. Kosters, A. Goldman, and D. G. Murcray, 1976b. *Geophysical Research Letters* **3**, 379–382, 1976, copyrighted by American Geophysical Union.)

(Zander, 1975). Further measurements were reported later (Zander *et al.*, 1977). The simultaneous measurement of HF and HCl was made by Raper and Farmer (1977) and Buijs *et al.* (1977). In the former case, an almost constant value of [HF]/[HCl] of 0.1 was obtained between 10 and 40 km. In the latter case, the measurements yielded about 0.2 at 15 km, 0.13 at 20 km, and 0.28 at 28 km (see Fig. 24).

Recently, further progress has been reported by Murcray *et al.* in the identification of chlorofluoro compounds in the stratosphere using high-resolution infrared spectroscopy. Murcray *et al.* (1977) reported an upper limit for the mixing ratio of $ClONO_2$ of 2×10^{-11} at 10 km up to 2×10^{-9} at 30 km, based on studies of the 780-cm^{-1} region of the spectrum. Pre-

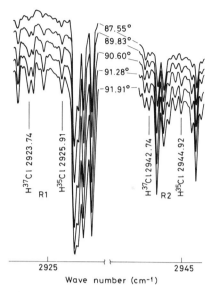

FIG. 22   Solar transmission spectra at ~3.3 μm measured from a high-altitude aircraft (20–21 km) at a variety of zenith angles (marked) showing the appearance of R1 and R2 lines of $H^{35}Cl$ and $H^{37}Cl$. (From C. B. Farmer, O. F. Raper, and R. H. Norton. *Geophysical Research Letters* **3**, pp. 13–16, 1976, copyrighted by American Geophysical Union.)

viously, the detection of CCl_4 had been reported near 800 cm^{-1} (see Fig. 21), and an analysis of the spectra obtained had yielded a constant mixing ratio for this molecule of ~10^{-10} in the 10–20-km range (Williams *et al.*, 1976b).

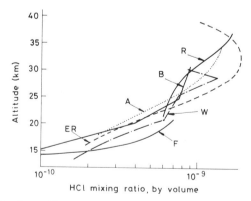

FIG. 23   Results of some IR spectroscopic measurements of stratospheric HCl, where R refers to Raper and Farmer (1977), A to Ackerman *et al.* (1976), W to Williams *et al.* (1976c), ER to Eyre and Roscoe (1977), F to Farmer *et al.* (1976), and B to Buijs *et al.* (1977).

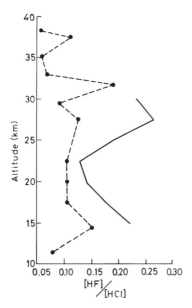

FIG. 24 Simultaneous measurements of HCl and HF yielding the ratio [HF]/[HCl]. In both cases, measurements were made using IR absorption spectroscopy from balloons: dashed line (Raper and Farmer, 1977) was measured in March 1977 over Australia (30.5° S); solid line (Buijs *et al.*, 1977) was measured in May 1976 over Alaska (65° N).

4. N_2O, CH_4, CO, H_2O, NH_3

The measurement of gases N_2O, CH_4, CO, H_2O, and NH_3 has excited less interest in recent years, partly because their spectral characteristics have been known in a general way for a number of years and partly because they do not figure prominently or directly in ozone photochemical reactions. Often, however, lines due to these and other species (e.g., CO_2, O_3) appear in spectral regions where other trace species are being sought, so a considerable quantity of data exists for these gases that is probably largely unanalyzed at the present time. Because of space limitations, we shall here merely refer to some recent work on these molecules and avoid any detailed discussion. However, in passing, we should note that it has frequently been found that even for these "well-known" molecules, our knowledge of spectral line data is far from perfect and that serious discrepancies are sometimes found, for example, between measured and calculated transmission spectra in regions away from band centers where high rotational energy state transitions occur. These high-*J* lines are more likely to be associated with line position or strength errors than are low-*J* lines because of the greater importance of distortion effects; also, "hot" bands and overtone bands can influence the weaker parts of a fundamental band.

The gas CH_4 has been studied by several authors, particularly Ackerman and co-workers (Kyle et al., 1969; Lowe and McKinnon, 1972; Ackerman et al., 1972, 1977). Both the ν_3 band at 3018 cm^{-1} and the ν_4 band at 1304 cm^{-1} have been utilized, in all cases using solar absorption spectroscopy. Recently, Ackerman et al. (1977) have considered the implications of a variation in the [CH_4] density profile in the stratosphere on the [HCl]/[ClO] ratio (i.e., the ratio of stable-to-active chlorine).

Measurements of N_2O have been carried out by Goldman et al. (1973a), using the ν_3 fundamental and the $\nu_3 + \nu_2 - \nu_2$ "hot" band, at about 4.5 μm. Detailed line-by-line calculations were carried out, yielding a volume mixing ratio of $\sim 0.33 \times 10^{-6}$ between 3 and 12 km, falling to $\sim 0.1 \times 10^{-6}$ at about 19 km. Farmer (1974) has reported measurements using the $2\nu_1$ (2560-cm^{-1}) and $\nu_1 + \nu_2$ (1880-cm^{-1}) bands from aircraft with no discernible variation in the amount of N_2O with latitude.

In the case of CO, again, only a few studies have been carried out, using the $1 - 0$ fundamental at 2143 cm^{-1} (Murcray et al., 1972; Farmer, 1974). A pronounced latitudinal variation of the total amount of CO above 16-km altitude (the flight level) was reported by Farmer (1974).

In the case of H_2O, only a few publications give analyzed data, although H_2O lines must have been measured in spectroscopic determinations of other constituents on many occasions. Measurements of solar absorption on the short-wavelength side of the pure rotation band of water vapor, between 24 and 29 μm, from a balloon platform were reported by Goldman et al. (1973b). Kuhn et al. (1976) have reported H_2O overburden measurements from aircraft on several occasions. Ackerman (1974) reported measurements using the grille spectrometer in the solar absorption mode, using the ν_2 band of H_2O at about 1600 cm^{-1}, also from a balloon. Farmer (1974) discussed measurements made at 1594 cm^{-1} in the height range 14–20 km. The data from these investigations and other measurements of stratospheric water vapor have been discussed by Harries (1976).

The ν_2 band of ammonia NH_3 has been identified in high-resolution observations of the sun from ground level (1.6 km above sea level in Denver, Colorado, where the measurements were made) by Murcray et al. (1978). By careful comparison with laboratory spectra of NH_3 in the 750–950-cm^{-1} range, a sunset path (~ 15 air masses) column amount of 7×10^{-3} atm·cm of NH_3 was deduced.

5. IR Spectroscopy at Higher Altitudes

The development of our understanding of the detailed IR spectrum of the atmosphere above about 40 km has been slower than at lower elevations mainly because the center of interest has been the ozone layer. However, in recent years, work by A. T. Stair and associates using rocket-borne cryogenically cooled spectrometers and interferometers has greatly improved our knowledge in this area, though it would seem that there is much more to be learned, particularly about excited-state radiative processes.

As examples of this work, Stair et $al.$ (1974) reported measurements of O_3 (at 9.6 μm) and CO_2 (at 15 μm) as a function of altitude; and Rogers et $al.$ (1977) discussed rocket-borne measurements in the 6.7–12.5-μm range using a liquid-helium-cooled circular variable filter spectrometer from which data the authors deduced a constant mixing ratio for H_2O of 3.5 $\pm$ 2.2 $\times$ 10^{-6} by volume between altitudes of 49 and 70 km.

6. *Latitude Surveys*

A particular application of IR spectroscopy is worth mentioning at this stage because of the significant contribution it can make to our understanding of the photochemistry of the atmosphere: This is the use of aircraft to carry IR spectrometers over large distances to study the atmosphere on a nearly global scale.

In particular, studies of latitudinal variations have been carried out under the Latitude Survey Mission, in which an instrumented aircraft was flown from the North to the South poles, making measurements of a number of atmospheric parameters at latitudes in between. In particular, we might note here that Girard and Besson (1977) reported measurements of several stratospheric species above the aircraft flight level, using solar absorption with a grille spectrometer. In a subsequent paper, Girard et $al.$ (1978–1979) further discussed the analysis of these data. Figure 25 shows some typical spectra recorded during this mission, with NO_2 absorption features at 1604 cm^{-1} and HNO_3 absorption features at 1326 cm^{-1}. Figure 26 shows some of the composition data obtained from the spectra for the NO, NO_2, and HNO_3 family. These data refer to total column density above the flight level and illustrate an important point in such measurements: for those molecules that exist in the stratosphere as a layer with decreasing density toward the troposphere, the interpretation of "overburden" measurements is rather straightforward. However, for other constituents with, say, constant mixing ratio distributions, the measured overburden depends strongly on the absolute height or perhaps on the relative height above the local tropopause, and the interpretation and value of such measurements must be investigated critically.

Other workers who were concerned with the Latitude Survey Mission and carried out IR observations were Kuhn (1977), Peyton et $al.$ (1977), and Barker et $al.$ (1977). Results of IR atmospheric spectroscopy studies on aircraft were also discussed by Mankin (1978).

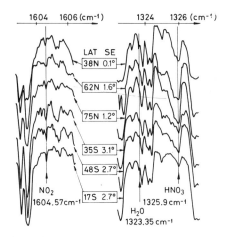

FIG. 25   A series of solar absorption spectra measured from an aircraft at levels between 9.3 and 11.7 km over the range of latitudes marked on each curve, where SE represents the solar elevation angle. The spectral traces are displaced from one another for clarity, and the two regions show evidence for absorption by NO_2 and HNO_3. (From Girard and Besson, 1977.)

7. *Recent Studies*

In this section, we wish to discuss some experiments that are at the moment producing new results and that may, therefore, be an indication of the direction in which IR atmospheric spectroscopy will be going in the immediate future.

The first of these refers to the latest improvement in spectral resolution

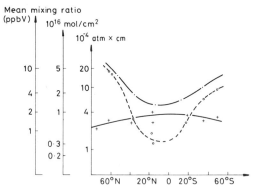

FIG. 26   The vertical column densities of $NO + NO_2$ (——), $NO + NO_2 + HNO_3$ (—·—·), and HNO_3 (– – –) deduced as a function of latitude from spectra like those shown in Fig. 25. (From Girard *et al.*, 1978–1979.)

in broad-band IR spectroscopy. Recent balloon flights of a Michelson interferometer system have been made by the Denver University group, with an interferometer capable of a spectral resolution of 0.01 cm⁻¹ in the 5–15-μm region, operated in solar occultation limb-sounding mode. The early results from this experiment are of very high quality and so far have been processed to a resolution of 0.02 cm⁻¹. At such high resolutions, a large amount of spectral detail becomes available, which should lead to a number of new detections of stratospheric constituents. One of the first results to emerge from this work was the identification of the ν_3 band of CF_4 in solar absorption spectra obtained at balloon altitudes, at 1283 cm⁻¹. A preliminary analysis yielded a mixing ratio for CF_4 of about 75×10^{-12} near 25-km altitude (Goldman et al., 1979). This increase in resolution of the atmospheric data, of course, makes even more critical the need to have equally high resolution measurements of atmospheric gases available in the laboratory so that the necessary spectral data (line positions, widths, strengths) can be obtained with which to analyze the atmospheric results.

As a second example of current progress, consider the work of Menzies (1979), who recently flew a laser heterodyne spectrometer (Menzies, 1978) on a balloon in solar occultation limb-sounding mode and reported

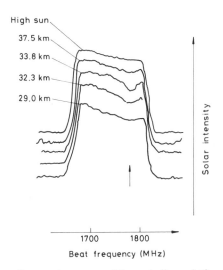

FIG. 27 Solar absorption spectra measured from a balloon platform using a laser heterodyne radiometer for the tangent (grazing) heights marked. The position of a possible ClO line at 853.122 cm⁻¹ is marked by the arrow. The beat frequency on the abscissa is measured between the local oscillator and the source. (From Menzies, *Geophysical Research Letters* **6**, pp. 151–154, 1979, copyrighted by American Geophysical Union).

detecting the $R(13/2)$, $\Omega = 3/2$ transition in the $^{35}Cl^{16}O$ isotope of ClO at 853.122 cm^{-1} [see Margolis et al. (1978) for details of the laboratory measurement of the line position]. The recorded spectral scans, looking at the sun from a balloon at ~ 40 km, are shown in Fig. 27. The spectral resolution achieved is roughly 1×10^{-3} cm^{-1}, and the measured equivalent width is $\sim 10^{-4}$ cm^{-1}. This result powerfully demonstrates the exciting potential of laser techniques for atmospheric spectroscopy.

The third example concerns the work of Farmer et al. (1980), in which the results of balloon flights in northern and southern midlatitudes were reported. These experiments were carried out on a high-resolution rapid-scan Michelson interferometer (0.13 cm^{-1}) using solar occultation methods in the 1800–3600-cm^{-1} region. Results for the vertical distributions of H_2O, O_3, N_2O, HCl, HF, CH_4, CO, and CO_2 illustrate the ability of powerful wide-bandwidth Fourier spectrometers to obtain data on many species simultaneously.

The fourth example concerns a recent submillimeter infrared balloon experiment (SIBEX) carried out as a cooperative exercise between the National Physical Laboratory (NPL) in Britain and the Institute for Research on Electromagnetic Waves (IROE) in Italy. An early report of this experiment can be found in Bangham et al. (1980). In this experiment, a cryogenically cooled 8–20-μm grating spectrometer (Pollitt et al. 1982) and an uncooled, but very high resolution (0.004-cm^{-1}) Michelson interferometer (Carli et al., 1980) for the 100–1000-μm region were used together to make measurements on the same part of the atmosphere, at the same time, throughout the flight. The two instruments were equipped with optical systems for scanning the atmospheric limb synchronously and common stabilization against motions of the balloon or gondola. Each instrument was capable of making measurements of a different set of atmospheric constituents, and the combined results (Table III) provide a powerful basis for testing photochemical and dynamical models of the atmosphere. The basic point raised by this experiment, therefore, is the need, in future spectroscopic investigations, to make measurements of families of related constituents at the same time and place, possibly using a number of different instruments. Evans et al. (1976) have already made progress in this direction, though in this case measurements of different species were sometimes made at different times of day.

Figure 28 illustrates a typical emission spectrum measured using the cryogenic spectrometer on SIBEX (Pollitt et al., 1982). These data were obtained from an altitude of 35.7 km at an observation zenith angle of 92°. Three orders of the grating spectrometer are shown in this uncalibrated spectrum, and the detection of several atmospheric bands is illustrated.

TABLE III

STRATOSPHERIC TRACE CONSTITUENTS TO BE
MEASURED BY SIBEX

Spectral region	Constituent	Estimated minimum detectable amounts for SIBEX at 25 km tangent height (by volume)
Submillimeter	H_2O	10^{-9}
(10–60 cm^{-1};	O_3	10^{-8}
1000–170 μm)	HNO_3	5×10^{-10}
	NO_2	10^{-10}
	N_2O	10^{-8}
	CO	10^{-8}
	HF	2×10^{-11}
	HCl	1×10^{-10}
	ClO	2×10^{-10}
	NH_3	2×10^{-11}
Infrared	O_3	10^{-9}
(500–1250 cm^{-1})	CF_2Cl_2	3×10^{-11}
	$CFCl_3$	1×10^{-11}
	HNO_3	5×10^{-10}
	CCl_4	5×19^{-11}
	Aerosols	$<0.1^a$

[a] Measured in units of particles per cubic centimeter.

From similar spectra recorded at other zenith angles, measurements of radiance as a function of tangent height (see Fig. 10) were derived for each constituent, and an "onion peeling" retrieval process (Twomey, 1977; Rodgers, 1976a, b) was used to obtain the concentration profiles. Figure 29 (S. Pollitt, private communication, 1979) illustrates this process for the case of HNO_3 measurements at 890 cm^{-1}. Figure 29a shows a measured radiance profile during a flight, and Fig. 29b shows the retrieved HNO_3 mixing ratio (the atmospheric temperature profile measured during the flight was used as input data in this last part of the calculation).

This discussion of the SIBEX experiment provides a convenient link to the next section, which will deal with submillimeter atmospheric spectroscopy and will include a discussion of the initial submillimeter results from SIBEX.

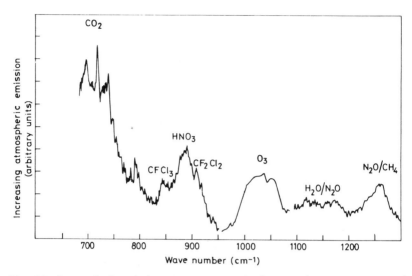

FIG. 28 Stratospheric emission spectra measured using a grating spectrometer cooled to 77 °K as part of the SIBEX experiment (see text). the spectra are normalized and show three grating orders with breaks at 950 and at 1100 cm⁻¹. Spectral bands are identified. (From Bangham *et al.*, 1980.)

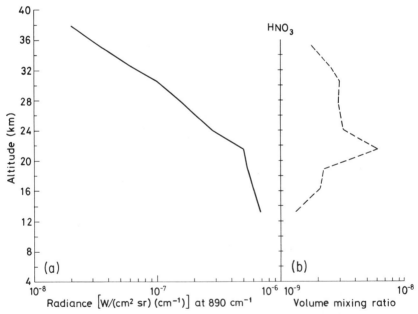

FIG. 29 An example of reduced data from the cryogenic spectrometer on the SIBEX experiment (see text): (a) The profile of radiance at 890 cm⁻¹ (HNO₃) as a function of tangent height; (b) the derived mixing ratio profile of HNO₃ deduced using a retrieval calculation (Section II) from the radiance data. (From S. Pollitt, private communication, 1979.)

V. Investigations of the Submillimeter Spectrum

A. INTRODUCTION

The history of submillimeter atmospheric research is much shorter than in the case of the IR, beginning as it did in 1957 with the publication by H. A. Gebbie (1957) of the first measurements of atmospheric transmission at wave numbers between 10 and 35 cm^{-1} at a spectral resolution of 0.2 cm^{-1}. These data were recorded using a large-aperture (300-mm) lamellar grating interferometer and a Golay cell detector at the Jungfraujoch mountain observatory in Switzerland. The spectrum observed in this work is reproduced in Fig. 30 and quite clearly shows, for the first time, the existence of several transmission windows at between 12 and 28 cm^{-1}. Many of the observed spectral features were identified as due to water vapor H_2O and others were provisionally assigned to ozone O_3. Subsequent laboratory spectroscopy (Stone, 1964; Gebbie et al., 1966) confirmed that numerous O_3 lines were present in the observations along

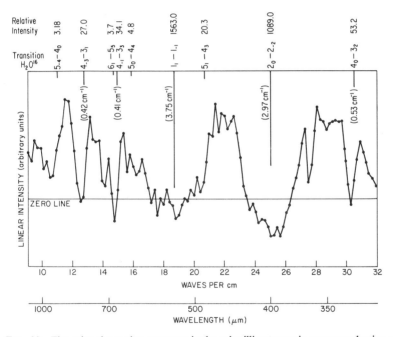

FIG. 30   The solar absorption spectrum in the submillimeter region measured using a lamellar grating interferometer at an altitude of 3.58 km, a temperature of 2 °C, a pressure of 503 torr, precipitable water of 2 ± 0.5 mm, and a mean zenith angle of 49.5°. Recording time 35 min. This represents the first observation of the transmission windows at 12, 14, 16, 22, and 28 cm^{-1}. (From Gebbie, 1957.)

with magnetic dipole transitions in molecular oxygen (Gebbie *et al.*, 1969).

The development of submillimeter atmospheric research from 1957 until about the end of 1976 has been reviewed by Harries (1977). It is not necessary, therefore, to repeat such a review here; rather, we shall extract a few of the most significant results from that paper in order to provide continuity within the present work.

Following the early activities just described, a number of groups carried out studies of the submillimeter emission spectrum of the lower atmosphere from aircraft (Harries *et al.*, 1972; Bussoletti and Baluteau, 1974; Marten and Chauvel, 1975; Mankin, 1975; see Table II). During these studies, new methods such as phase modulation (Chamberlain, 1971, 1979) and new helium-cooled detectors were introduced, and spectral resolutions of ~ 0.06 cm^{-1} were achieved for the first time. Figure 31 illustrates two stratospheric emission spectra in the 12–27-cm^{-1} range recorded at an altitude of 12 km, a zenith angle of 75°, and with a spectral resolution of 0.06 cm^{-1} (Harries *et al.*, 1972). These observations showed the existence of highly reproducible spectral structure due to O_3 and other molecules (Harries, 1977).

Experiments followed that were mounted on balloon platforms (Harries *et al.*, 1973, 1976; Clark and Kendall, 1976). At the higher elevations ac-

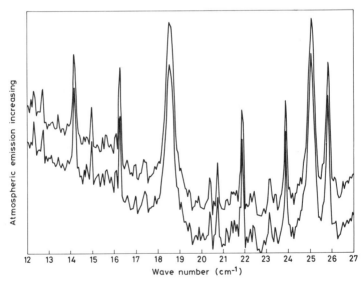

FIG. 31   Stratospheric emission spectra in the 12–27-cm^{-1} region measured from an aircraft at an altitude of 12 km, a zenith angle of 75°, and a spectral resolution of 0.06 cm^{-1}. (From Harries *et al.*, 1972.)

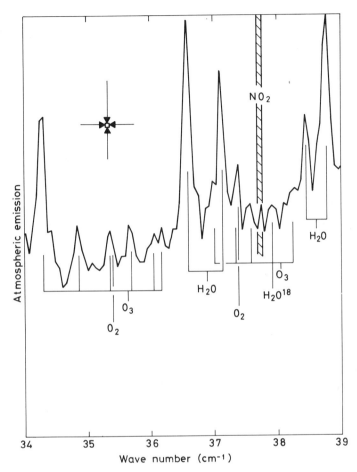

FIG. 32   Stratospheric emission spectrum in the 34–39-cm^{-1} region measured from a balloon at an altitude of 35 km, a zenith angle of 92°, and a spectral resolution of 0.07 cm^{-1}. (From Harries *et al.*, 1976.)

cessible by such methods, lines were narrow (particularily H_2O, which causes considerable opacity still at aircraft altitudes), and it became possible to detect weak lines of other species such as NO_2 and to use limbsounding methods to derive concentration–height profiles throughout the stratosphere for a number of species. Figure 32 shows an emission spectrum measured from a balloon floating at 35 km with a zenith angle of 92° (Harries *et al.*, 1976). This result was the first observation of an NO_2 Q branch at 37.8 cm^{-1}, possibly confirmed by other workers (e.g., Kendall, 1979). Figure 33 illustrates simultaneous measurements of H_2O, NO_2, and

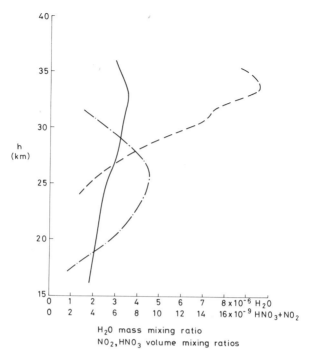

FIG. 33   Vertical profiles of H₂O (——), HNO₃ (—·—·), and NO₂ (– – –) deduced from balloon-borne measurements of the stratospheric submillimeter spectrum (cf., Fig. 32) at an altitude of 35 km, latitude 44° N. (From Harries *et al.*, 1976.)

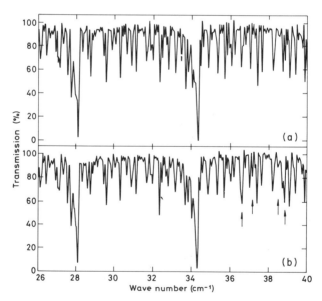

FIG. 34   Comparison of synthetic (a) and experimental (b) transmittance spectra of ozone for laboratory conditions (room temperature ~296 °K; path length 1.0 m; pressure 18 torr O₃ + 12 torr O₂; spectral resolution 0.05 cm⁻¹). The arrows indicate the positions of strong H₂O lines, which influence the experimental spectrum. (From Fleming and Wayne, 1975.)

HNO_3 derived from submillimeter spectral data like that shown in the previous illustration.

Considerable progress was made, also, in the early 1970s in the availability of laboratory spectra of stratospheric trace gases, mainly through the work of Fleming and co-workers (e.g., Fleming and Wayne, 1975; Fleming, 1976a, b). Figure 34 shows a beautiful result from Fleming and Wayne (1975): a laboratory measurement of the transmission spectrum of 18 torr of ozone in a 1-m cell (Fig. 34b) in comparison with a calculated spectrum (Fig. 34a), showing excellent agreement between the two.

Since the time of the Harries (1977) review, progress has occurred mainly on two fronts. The first is in the area of theoretical calculations of the submillimeter spectrum, and the second is in improved experimental results obtained from balloons. Both aspects will now be considered in turn.

B. THEORETICAL CALCULATIONS

In France, Rabache and Rebours continued their studies of the submillimeter spectrum of the lower stratosphere. Rabache (1977) studied the details of the pure rotation spectrum of nitric oxide NO and provided a listing of line positions and line strengths. He concluded, however, that the NO lines made such a weak contribution to the spectrum by contrast with lines of O_3, H_2O, and O_2 that NO is essentially undetectable at these wavelengths, given current instrumental performances.

In a separate study, Rebours and Rabache (1979) discussed the method proposed by Burroughs and Harries (1970) for making quantitative estimates of stratospheric H_2O concentrations using an "internal calibration" procedure against O_2 lines. Through this study, Rebours and Rabache were able to show that the method gave reasonably accurate results (within $\pm 20\%$) in the case of H_2O, which generally decreases in density with height in the stratosphere in a monotonic way. However, much larger errors ($\pm 100\%$ and more) were found for the case of O_3, which is distributed in the atmosphere in the form of a layer: in this case, clearly, the method breaks down and cannot be used.

At the NPL, Bangham (1978a, b) carried out a number of line-by-line spectral emission computations, generally for the case of limb-sounding geometry with an observer on a balloon or spacecraft. These calculations were carried out in order to deduce the amount of retrievable information contained in the submillimeter spectrum at a resolution of about 0.01 cm^{-1} (the resolution achievable at the time) as well as to analyze existing data. Two examples of this work are shown in Figs. 35 and 36, two calculations for a high-resolution limb-scanning submillimeter spectrometer mounted on a spacecraft at 250 km and looking at the atmospheric limb at a tangent

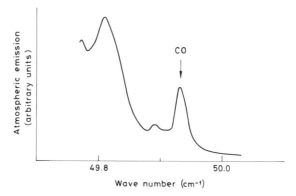

FIG. 35　A calculated spectrum of the submillimeter stratospheric emission spectrum in the 50-cm^{-1} region for a standard atmosphere containing H_2O, O_2, and O_3 plus CO completely mixed at a mole fraction of 10^{-7}. The calculation was performed at a spectral resolution of 0.01 cm^{-1}, assuming a tangent height of 20 km, and an observer altitude of 250 km. (From Bangham, 1978a.)

height of 20 km. Figure 35 represents a calculation including H_2O, O_2, O_3, and CO near 50 cm^{-1} and shows the CO line (arrowed) clearly defined against the background formed by the other three molecules: a mixing ratio of 10^{-7} was used for the CO in this case. Figure 36 shows the spectrum near 41 cm^{-1}: in this case, the solid line is for H_2O, O_2, and O_3 only: the broken line shows the effect of adding HCl and HF with a relative concentration of 1 part in 10^9 by volume (constant with height). Even bearing

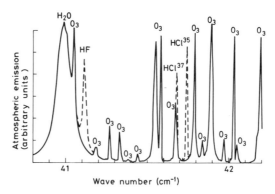

FIG. 36　A calculated spectrum of the submillimeter atmospheric emission spectrum in the 41–42-cm^{-1} region for a standard atmosphere containing H_2O, O_2, and O_3 plus HCl and HF completely mixed at a mole fraction of 10^{-9}. The calculation was undertaken for the same observational geometry and spectral resolution as that in Fig. 35. (From Bangham, 1978a.)

in mind that the measured HF mixing ratio is nearer 10^{-10} (Zander *et al.*, 1977), we can see that both HCl and HF should be easily detectable in this region under the conditions of the calculation.

More calculations of this type will be required in the future, both for data analysis and for prediction purposes. At present, one of the most severe limitations on the accuracy of the computations is the low accuracy of the spectral line data available, particularly for the molecules with more complex spectra (e.g., HNO_3, SO_2, and O_3). A major effort to provide better spectral data from laboratory measurements and calculations is needed.

C. Recent Experimental Results

Kendall and Clark at the University of Calgary have reported further measurements of the stratospheric emission spectrum from balloon altitudes, using a rapid-scan Michelson interferometer and a cooled detector (Kendall and Clark, 1978, 1979a; Kendall, 1979). These data were measured at resolutions up to 0.07 cm^{-1} over the 30–110-cm^{-1} range. A number of flights were made, details of which may be found in Kendall (1979).

In this work, the most notable results have been the possible detection of HCl and OH and the tentative confirmation of the detection of an NO_2 Q branch at 37.8 cm^{-1} [first suggested by Harries *et al.* (1976); e.g., see Kendall and Clark (1979)].

The question of OH emission has been investigated in some detail by these authors (Kendall, 1979; Kendall and Clark, 1979b), including a calculation of line strengths and positions, and it appears that the spectral resolution of their measurements was just on the limit for the detection of OH. Figure 37 shows an example of an emission spectrum from Kendall and Clark (1979b): the top curve is the experimental results (see caption for details), and the lower curve is a line-by-line calculation including H_2O and O_3 only. The arrows mark the predicted OH line positions, and it can be seen that features have been detected that might be due to OH.

Clearly, a more certain assignment for OH can only be achieved through measurements at higher spectral resolution in the 50–200-cm^{-1} region where the most intense lines of OH occur. Table IV gives values of the line position and strength for OH rotational lines in the ground vibrational state as calculated by Kendall and Clark (1979b); four series of transitions occur because of spin-doublet and lambda-doubling effects.

Recently, Grossman and Offerman (1978) reported rocket-borne measurements of the 63-μm emission from thermospheric OI ($^3P_1 - {}^3P_2$ fine structure transition in the ground state). An interesting interpretation of the data is given by these authors in terms of the breakdown of local ther-

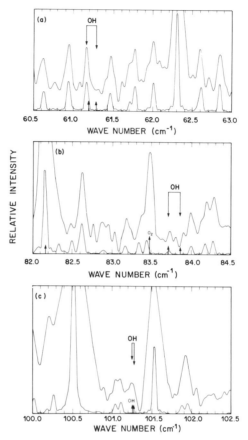

FIG. 37   Measured submillimeter stratospheric emission spectrum obtained using a Michelson interferometer at a spectral resolution of 0.08 cm^{-1} from a balloon at approximately 35-km altitude (top curve) compared with a calculated spectrum for a standard atmosphere of H_2O and O_3 on which the positions of OH lines have been marked by arrows; an O_2 line is similarly indicated. (From Kendall and Clark, 1979b.)

modynamic equilibrium at the height of the measurements (above 80 km), which results in much lower emission intensity than previously assumed. The significance of this result, of course, is that this emission line represents a much less effective radiative cooling mechanism than previous calculations (assuming LTE) had indicated.

More recently, Carli *et al.* (1980) flew a very high resolution polarizing Michelson interferometer (Martin and Puplett, 1969) on a high-altitude balloon to measure the emission spectrum of the atmosphere in the 20–60-cm^{-1} range. The limiting unapodized spectral resolution of this de-

TABLE IV

CALCULATED VALUES OF LINE POSITIONS $\bar{\nu}_0$ AND INTEGRATED LINE STRENGTHS S_0 FOR OH[a]

	$\bar{\nu}_0$ (cm^{-1})				S_0 at 220° K (cm^{-2} atm^{-1})	
Energy state	$F(1c)$	$F(1d)$	$F(2c)$	$F(2d)$	$F(1)$	$F(2)$
$v = 0$, $J'' = 0.5$			61.29	61.19		109.4
1.5	83.71	83.85	101.27	101.26	710.3	372.4
2.5	118.18	118.43	140.39	140.47	1406.2	489.5
3.5	153.16	153.50	178.70	178.87	1374.0	361.4
4.5	188.44	188.87	216.30	216.56	840.8	172.4
5.5	223.85	224.35	253.29	253.61	352.5	57.0
6.5	259.24	259.80	289.73	290.11	106.1	13.6
7.5	294.49	295.09	325.64	326.06	23.6	2.40
8.5	329.51	330.15	361.04	361.49	3.94	0.321
9.5	364.23	364.88	395.92	396.37	0.503	0.0328

[a] From Kendall and Clark (1979b).

vice is ~ 0.004 cm^{-1}. Liquid-helium-cooled germanium bolometer detectors were used, and the polarizing elements (beam splitter, polarizer, analyzer) were constructed from stretched metallic wire grids (Costley *et al.*, 1977).

This device was flown as part of the SIBEX payload (see Section IV). The early results indicate that the experiment worked very successfully, and to date, spectra in the 20–40-cm^{-1} range have been processed to a spectral resolution of about 0.008 cm^{-1}. Although detailed assignments have not yet been completed, we show in Fig. 38 an example of the data, which illustrates some of the exciting spectroscopic results being obtained. This spectrum was recorded at an altitude of ~ 40 km with an observational zenith angle of 91.6°. In the figure, we can see numerous emission lines due to H_2O and O_3, and the positions of lines due to N_2O, HCl, CO, and HF are also marked: particularly in the 41.60–42.20-cm^{-1} region, lines due to each of these species may have been observed.

These examples of recent progress should serve to illustrate the important contribution to atmospheric spectroscopy that can still be made at submillimeter wavelengths. Without doubt, it is essential to achieve high spectral resolution to extract all the information contained in the spectrum (Harries, 1977). Obtaining very high spectral resolution using Michelson interferometers or dispersive spectrometers does, of course, require large instruments, and work is needed in the optical design of these

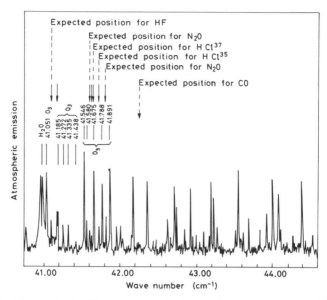

FIG. 38   Measured submillimeter stratospheric emission spectrum at very high resolution (0.008 cm⁻¹) obtained during the SIBEX II experiment from an altitude of 40.0 km with an atmospheric zenith angle of 91.6°. The considerable amount of spectral structure is discussed in the text. (From Carli *et al.*, 1980.)

new instruments to minimize the overall size as much as possible so that flight experiments are not so unwieldy as to become unusable.

VI. Conclusion

In this chapter, we have sought to provide a survey of the subject of IR and SM atmospheric spectroscopy that will at least provide the reader with a basic understanding of the principles and methods utilized, along with some feeling for the current status of research. Through the many references, it is hoped that the interested reader can delve deeper into areas of particular interest.

The infrared spectroscopic method is a very powerful one for remote determination of the state of the atmosphere, and the current intensive activities briefly described indicate that these techniques will continue to be applied to atmospheric studies with success for some time to come.

In concluding, we might perhaps consider in what particular areas progress is likely to be the most interesting. In the IR, the current "contest" is between, on the one hand, the high-resolution broad-band methods such as grating spectrometers and, particularly, Michelson interferometers, which are capable of recording over wide spectral ranges, with resolutions of 0.1 cm^{-1} being quite easily achievable and those of 0.01 cm^{-1} and better now becoming possible and, on the other hand, the monochromatic heterodyne techniques using tunable lasers as local oscillators, which are capable of even higher spectal resolutions, e.g., 10^{-3} cm^{-1}, but which typically are not able to scan such a large spectral range. Experimental difficulties exist with both types of approaches, so neither has a clear advantage in terms of ease of use in the atmosphere. It seems fair to predict that each type of method will make its contribution and that neither will obviate the need for observations with the other. A third type of spectrometer, which is intrinsically capable of very high spectral resolution and which has contributed significant results in recent years, is the gas correlation device. One can foresee a continuation of the use of such devices, with particularly valuable results coming from studies of high altitudes in the atmosphere (say, from spacecraft), where the very narrow atmospheric lines do not suffer any overlap with "contaminant" lines and can be well matched with absorption lines in the gas correlator when low gas pressures are used.

In the SM range, less choice exists. Until a tunable monochrometic source becomes available, the heterodyne techniques used with such success in the IR (and the microwaves) will not be possible. Future progress seems to rely on the development of very high resolution, but physically not too large, Michelson interferometers designed to obtain maximum throughput efficiency with ultrasensitive detectors. (Cooling with liquid ^{3}He is one possibility currently being considered.) Shortage of energy will always represent a major disadvantage at SM wavelengths, but the richness of information in the spectrum would seem to argue for continuing efforts to exploit the spectroscopy of this region for atmospheric studies.

We have exclusively considered the use of spectroscopy to study composition. Studies of atmospheric temperature do not generally require high resolution, so they have not been considered [although note the PMR techniques used by Houghton *et al;* see Houghton (1978)]. However, as achievable spectral resolutions improve, it might become possible to measure Doppler shifts of atmospheric emission or absorption lines due to mean atmospheric motions along a line of sight and, therefore, to develop a technique for remotely sensing winds. In the IR and SM regions, however, current techniques are such that this remains a possibility for the future.

ACKNOWLEDGMENTS

The author would like to thank Drs. A. Girard, D. J. W. Kendall, and N. W. B. Stone and Mr. G. Toon for their comments on the manuscript.

REFERENCES

Ackerman, M. (1974). *Planet. Space Sci.* **22**, 1265–1267.

Ackerman, M., and Muller, C. (1972). *Nature (London)* **240**, 300–301.

Ackerman, M., Frimout, D., Lippens, C., and Muller, C. (1972). *Aeron. Acta, A* **97**.

Ackerman, M., Frimout, D., Muller, C., Nevejens, D., Fontanella, J. C., Girard, A., and Louisnard, N. (1973). *Nature (London)* **245**, 205–206.

Ackerman, M., Frimout, D., Muller, C., Nevejens, D., Fontanella, J. C., Girard, A., Gramont, L., and Louisnard, N. (1974). *Can. J. Chem.* **52**, 1532–1535.

Ackerman, M., Fontanella, J. C., Frimout, D., Girard, A., Louisnard, N., and Muller, C. (1975). *Planet. Space Sci.* **23**, 651–660.

Ackerman, M., Frimout, D., Girard, A., Gottiginies, M., and Muller, C. (1976). *Geophys. Res. Lett.* **3**, 81–83.

Ackerman, M., Frimout, D., and Muller, C. (1977). *Nature (London)* **269**, 226–227.

Bangham, M. J. (1978a). *Infrared Phys.* **18**, 357–369.

Bangham, M. J. (1978b). *Infrared Phys.* **18**, 799–801.

Bangham, M. J., Bonetti, A., Bradsell, R. H., Carli, B., Harries, J. E., Mencaraglia, F., Moss, D. G., Pollitt, S., Rossi, E., and Swann, N. R. (1980). *Proc. Quadrennial Intern. Ozone Symp., Boulder, Colorado,* pp. 759–764.

Born, M. (1962). "Atomic Physics." Blackie and Sons, London.

Barker, D. B., Williams, W. J., and Murcray, D. G. (1977). *NASA Tech. Memo.* **NASA TM-X-73630**, 132–136.

Barnett, J. J. (1974). *Q. J. R. Meteorol. Soc.* **100**, 505–530.

Buijs, H. J., Vail, G., and Tremblay, G. (1977). "Simultaneous Measurement of the Volume Mixing Ratios for HCl and HF in the Stratosphere" (unpublished). Available from BOMEM Inc, 910 Place Dufour, Vanier, Quebec, Canada.

Burch, D. E. (1968). *J. Opt. Soc. Am.* **58**, 1383–1394.

Burroughs, W. J., and Harries, J. E. (1970). *Nature (London)* **227**, 824–825.

Bussoletti, E., and Baluteau, J. P. (1974). *Infrared Phys.* **14**, 293–302.

Carli, B., Mencaraglia, F., and Bonetti, A. (1980). *Int. J. Infrared and Millimeter Waves* **1**, 263–272.

Carleton, N., ed. (1974). "Methods of Experimental Physics," Vol. 12A. Academic Press, New York.

Chaloner, C. P., Drummond, J. R., Houghton, J. T., Jarnot, R. F., and Roscoe, H. K. (1978). *Proc. R. Soc. London, Ser. A* **364**, 145–159.

Chamberlain, J. (1971). *Infrared Phys.* **11**, 25–55.

Chamberlain, J. (1979). "The Principles of Interferometric Spectroscopy." Wiley, New York.

Chanin, G., and Lecullier, J. C. (1978). *Infrared Phys.* **18**, 589–594.

Clark, T. A., and Kendall, D. J. W. (1976). *Nature (London)* **260**, 31–32.

Connes, J. (1961). *Rev. Opt.* **40**, 45–79, 116–140, 171–190, 231–265.

Connes, J. (1971). *Aspen Int. Conf. Fourier Spectrosc.* [*Proc.*], AFCRL 71-0019, pp. 83–115.

Connes, J., and Connes, P. (1966). *J. Opt. Soc. Am.* **56**, 896–910.

Connes, P. (1971). *Aspen Int. Conf. Fourier Spectrosc.* [*Proc.*], AFCRL 71-0019, pp. 121–125.

Connes, P., and Michel, G. (1974). *Astrophys. J.* **190**, L29–L32.

Connes, P., Connes, J., Benedict, W. S., and Kaplan, L. D. (1967). *Astrophys. J.* **147**, 1230–1237.

Connes, P., Connes, J., Kaplan, L. D., and Benedict, W. S. (1968). *Astrophys. J.* **152**, 731–743.

Costley, A. E., Hursey, K. H., Neill, G. F., and Ward, J. W. M. (1977). *J. Opt. Soc. Am.* **67**, 979–981.

Crutzen, P. J. (1970). *Q. J. R. Meteorol. Soc.* **96**, 320–325.

Drummond, J. R. (1976). D. Phil. Thesis, University of Oxford.

Drummond, J. R., and Jarnot, R. F. (1978). *Proc. R. Soc. London, Ser. A* **364**, 237–254.

Evans, W. F. J., Kerr, J. B., Wardle, D. I., McConnell, J. C., Ridley, B. A., and Schiff, H. I. (1976). *Atmosphere* **14**, 189–198.

Evans, W. F. J., Fast, H., Kerr, J. B., McElroy, C. T., O'Brien, R. S., Wardle, D. I., Mc-Connell, J. C., and Ridley, B. A. (1978). *WMO* [*Publ.*] **511**.

Eyre, J. R., and Roscoe, H. K. (1977). *Nature (London)* **266**, 243–244.

Farmer, C. B. (1974). *Can. J. Chem.* **52**, 1544–1559.

Farmer, C. B., Raper, O. F., and Norton, R. H. (1976). *Geophys. Res. Lett.* **3**, 13–16.

Farmer, C. B., Raper, O. F., Robbins, B. D., Toth, R. A., and Muller, C. (1980). *J. Geophys. Res.* **85(C3)**, 1621–1632.

Fellgett, P. B. (1958). *J. Phys. Radium* **19**, 187–237.

Fleming, J. W. (1976a). *J. Quant. Spectrosc. Radiat. Transfer* **16**, 63–68.

Fleming, J. W. (1976b). *Spectrochim. Acta, Part A* **32A**, 787–795.

Fleming, J. W., and Wayne, R. P. (1975). *Chem. Phys. Lett.* **32**, 135–140.

Fontanella, J. C., Girard, A., Gramont, L., and Louisnard, N. (1974). Office National d'Etudes et de Recherches Aerospatiole Rep. TP 1441. Paris, France.

Gebbie, H. A. (1957). *Phys. Rev.* **107**, 1194.

Gebbie, H. A., Stone, N. W. B., Topping, G., Gora, E. K., Clough, S. A., and Kneizys, F. X. (1966). *J. Mol. Spectrosc.* **19**, 7–24.

Gebbie, H. A., Burroughs, W. J., and Bird, G. R. (1969). *Proc. R. Soc. London, Ser. A* **310**, 579–590.

Gille, J. C., and House, F. B. (1971). *J. Atmos. Sci.* (American Meteorological Society, Publisher) **28**, 1427–1442.

Girard, A. (1963). *Appl. Opt.* **2**, 79–87.

Girard, A., and Besson, J. (1977). *NASA Tech. Memo.* **NASA TMX-73630**, 69–78.

Girard, A., Fontanella, J. C., and Gramont, L. (1973). *C. R. Hebd. Seances Acad. Sci., Ser. B* **276**, 845–846.

Girard, A., Besson, J., Giraudet, R., and Gramont, L. (1978–1979). *Pure Appl. Geophys.* **117**, 381–394.

Goldman, A., Murcray, D. G., Murcray, F. H., Williams, W. J., and Bonomo, F. S. (1970). *Nature (London)* **225**, 443–444.

Goldman, A., Murcray, D. G., Murcray, F. H., and Williams, W. J. (1973a). *J. Opt. Soc. Am.* **63**, 843–846.

Goldman, A., Murcray, D. G., Murcray, F. H., Williams, W. J., and Brooks, J. N. (1973b). *Appl. Opt.* **12**, 1045–1053.

Goldman, A., Murcray, D. G., Murcray, F. J., Cook, G. R., Van Allen, J. W., Bonomo, F. S., and Blaitherwick, R. D. (1979). *Geophys. Res. Lett.* **6**, 609–612.

Goody, R. M. (1964). "Atmospheric Radiation," Vol. I. Oxford Univ. Press, London and New York.

Grossman, K. U., and Offerman, D. (1978). *Nature (London)* **276**, 594–595.

Harries, J. E. (1976). *Rev. Geophys. Space Phys.* **14**, 565–575.

Harries, J. E. (1977). *J. Opt. Soc. Am.* **67**, 880–894.

Harries, J. E., and Chamberlain, J. (1976). *Appl. Opt.* **15**, 2667–2672.

Harries, J. E., Swann, N. R., Beckman, J. E., and Ade, P. A. R. (1972). *Nature (London)* **236**, 159–161.

Harries, J. E., Swann, N. R., Carruthers, G. P., and Robinson, G. A. (1973). *Infrared Phys.* **13**, 149–155.

Harries, J. E., Moss, D. G., Swann, N. R., Neill, G. F., and Gildwarg, P. (1976). *Nature (London)* **259**, 300–302.

Hernandez, G. (1978). *Appl. Opt.* **17**, 2967–2972.

Hinkley, E. D., ed. (1976). "Laser Monitoring of the Atmosphere," Top. Appl. Phys., Vol. 14. Springer-Verlag, Berlin and New York.

Houghton, J. T. (1977). "The Physics of Atmospheres." Cambridge Univ. Press, London and New York.

Houghton, J. T. (1978). *Q. J. R. Meteorol. Soc.* **104**, 1–29.

Houghton, J. T., and Smith, S. D. (1966). "Infrared Physics." Oxford Univ. Press, London and New York.

Houghton, J. T., and Smith, S. D. (1970). *Proc. R. Soc. London, Ser. A* **320**, 23–33.

Houghton, J. T., and Taylor, F. W. (1973). *Rep. Prog. Phys.* **36**, 827–919.

Houghton, J. T., Hughes, N. D. P., Moss, T. S., and Seeley, J. S. (1961). *Philos. Trans. R. Soc. London* **254**, 47–123.

Houghton, J. T., Peskett, G. D., and Rodgers, C. D. (1971). "A Scientific Proposal to Employ the Pressure Modulator Radiometer to Sound the Temperature and Composition of the Stratosphere and Mesosphere." Oxford Univ.

Jacquinot, P. (1958). *J. Phys.* **19**, 223–229.

James, J. F., and Sternberg, R. S. (1969). "The Design of Optical Spectrometers." Chapman & Hall, London.

Jarnot, R. F. (1976). D. Phil. Thesis, University of Oxford.

Kendall, D. J. W. (1979). Ph.D. Thesis, University of Calgary, Canada.

Kendall, D. J. W., and Clark, T. A. (1978). *Infrared Phys.* **18**, 803–813.

Kendall, D. J. W., and Clark, T. A. (1979a). *Appl. Opt.* **18**, 346–353.

Kendall, D. J. W., and Clark, T. A. (1979b). *J. Quant. Spectrosc. Radiat. Transfer* **21**, 511–526.

Kuhn, P. M. (1977). *NASA Tech. Memo.* **NASA TMX-73630**, 45–54.

Kuhn, P. M., Magaziner, E., and Stearns, L. P. (1976). *Geophys. Res. Lett.* **3**, 529–532.

Kuiper, G., ed. (1949). "The Atmospheres of the Earth and Planets." Univ. of Chicago Press, Chicago, Illinois.

Kyle, T. G., and Goldman, A. (1975). *NCAR Tech. Note* **NCAR-TN/STR-112**.

Kyle, T. G., Murcray, D. G., Murcray, F. H., and Williams, W. J. (1969). *J. Geophys. Res.* **74**, 3421–3425.

Lowe, R. P., and McKinnon, D. (1972). *Can. J. Phys.* **50**, 668–673.

Mankin, W. G. (1975). *Proc. Soc. Photo-Opt. Instrum. Eng.* **67**, 69–75.

Mankin, W. G. (1978). *Opt. Eng.* **17**, 39–43.

Margolis, J. S., Menzies, R. T., and Hinkley, E. D. (1978). *Appl. Opt.* **17**, 1680–1682.

Marten, A., and Chauvel, Y. (1975). *Infrared Phys.* **15**, 205–209.

Martin, D. H., and Puplett, E. F. (1969). *Infrared Phys.* **10**, 105–109.

Menzies, R. T. (1978). *Opt. Eng.* **17**, 44–49.

Menzies, R. T. (1979). *Geophys. Res. Lett.* **6**, 151–154.

Michelson, A. A. (1881). *Am. J. Sci.* **22**, 120.

Migeotte, M., Neven, L., and Swensson, J. (1956). "The Solar Spectrum for 2.8 to 23.7 microns," *Mem. Soc. R. Soc. Liege* Parts I and II, Spec. Vol.

Mohler, O. C., Pierce, A. K., McMath, R. R., and Goldberg, L. (1950). "Photometric Atlas of the Near Infrared Solar Spectrum, λ8465 to λ25242." Univ. of Michigan Press, Ann Arbor.

Molina, M. J., and Rowland, F. S. (1974). *Nature (London)* **249**, 810–812.

Moss, D. G. (1974). *NPL Rep.* **Mat. App. 35**.

Moss, D. G. (1982). To be published.

Murcray, D. G., Kyle, T. G., Murcray, F. H., and Williams, W. J. (1968). *J. Opt. Soc. Am.* **59,** 1131–1134.

Murcray, D. G., Murcray, F. H., Williams, W. J., Kyle, T. G., and Goldman, A. (1969). *Appl. Opt.* **8,** 2519–2536.

Murcray, D. G., Goldman, A., Murcray, F. H., Williams, W. J., Brooks, J. N., and Barker, D. B. (1972). *Proc. Conf. Clim. Impact Assess. Program, 2nd,* pp. 86–98. Washington, DC.

Murcray, D. G., Goldman, A., Williams, W. J., Murcray, F. H., Brooks, J. N., Van Allen, J., Stocker, R. N., Kosters, J. J., and Barker, D. B. (1974). *Proc. Conf. Clim. Impact Assess. Program, 3rd,* pp. 246–253. Washington, DC.

Murcray, D. G., Barker, D. B., Brooks, J. N., Goldman, A., and Williams, W. J. (1975). *Geophys. Res. Lett.* **2,** 223–225.

Murcray, D. G., Goldman, A., Williams, W. J., Murcray, F. H., Bonomo, F. S., Bradford, C. M., Cook, G. R., Hanst, P. L., and Molina, M. J. (1977). *Geophys. Res. Lett.* **4,** 227–230.

Murcray, D. G., Goldman, A., Bradford, C. M., Cook, G. R., Van Allen, J. W., Bonomo, F. S., and Murcray, F. H. (1978). *Geophys. Res. Lett.* **5,** 527–530.

National Academy of Sciences (1976). "Halocarbons; Effects on Stratospheric Ozone." NAS, Washington, DC.

Park, J. H. (1977). *NASA [Contract. Rep.] CR* **NASA-CR-2925** (Contract NSG-1203).

Patel, C. K. N., Burkhardt, E. G., and Lambert, C. A. (1974). *Science* **184** (June 14), 1173–1176.

Peyton, B. J., Hoell, J., Lange, R. A., Seals, R. K., Savage, M. G., and Allario, F. (1977). *NASA Tech. Memo.* **NASA TMX-73630,** 55–68.

Pollitt, S., Bangham, M. J., Bradsell, R. H., Harries, J. E., Moss, D. G., and Swann, N. R. (1982). To be published.

Rabache, P. (1977). *J. Quant. Spectrosc. Radiat. Transfer* **17,** 673–678.

Raper, O. F., and Farmer, C. B. (1977). *Geophys. Res. Lett.* **4,** 527–529.

Rebours, B., and Rabache, P. (1979). *Infrared Phys.* **19,** 107–113.

Rodgers, C. D. (1976a). *NCAR Tech. Note* **NCAR/TN-116+1A.**

Rodgers, C. D. (1976b). *Rev. Geophys. Space Phys.* **14,** 609–624.

Roesler, F. L. (1974). *In* "Methods of Experimental Physics" (N. Carleton, ed.). Vol **12A,** pp. 531–569. Academic Press, New York.

Rogers, J. W., Stair, A. T., Degges, T. C., Wyatt, C. L., and Baker, D. J. (1977). *Geophys. Res. Lett.* **4,** 366–368.

Rowland, F. S., and Molina, M. J. (1975). *Rev. Geophys. Space Phys.* **13,** 1–35.

Schmeltekopf, A. L., Goldan, P. D., Henderson, W. R., Harrop, W. J., Thompson, T. L., Fehsenfeld, F. C., Schiff, H. I., Crutzen, P. J., Isaksen, I. S. A., and Ferguson, E. E. (1975). *Geophys. Res. Lett.* **2,** 393–396.

Schroeder, D. J. (1974). *In* "Methods of Experimental Physics" (N. Carleton, ed.), Vol. **12A,** pp. 463–489. Academic Press, New York.

Shaw, J. H., and Garing, J. S. (1962). *Infrared Phys.* **2,** 155–173.

Shaw, J. H., Chapmen, R. M., Howard, J. N., and Oxholm, M. L. (1951). *Astrophys. J.* **113,** 268–298.

Smith, S. D., and Pidgeon, C. R. (1964). *Mem. R. Soc. Liege* **9,** 336–349.

Smith, S. D., Holah, G. D., Seeley, J. S., Evans, C., and Hunneman, R. (1972). *In* "Infrared Techniques for Space Research" (V. Manno, and J. Ring, eds.), pp. 199–218. Reidel Publ., Dordrecht, Netherlands.

Stair, A. T., Ulwick, J. C., Baker, D. J., Wyatt, C., and Baker, K. D. (1974). *Geophys. Res. Lett.* **1,** 117–118.

Stone, N. W. B. (1964). Ph.D. Thesis, University of London.

Swann, N. R., Bangham, M. J., and Harries, J. E. (1982). To be published.

Tai, M. H., and Harwit, M. (1976). *Appl. Opt.* **15**, 2664–2666.

Taylor, F. W., Houghton, J. T., Peskett, G. D., Rodgers, C. D., and Williamson, E. J. (1972). *Appl. Opt.* **11**, 135–141.

Toth, R. A., Farmer, C. B., Schindler, R. A., Raper, O. F., and Schaper, P. W. (1973). *Nature (London), Phys. Sci.* **244**, 7–8.

Townes, C. H., and Schawlow, A. L. (1975). "Microwave Spectroscopy." Dover, New York.

Treffers, R. R. (1977). *Appl. Opt.* **16**, 3103–3106.

Twomey, S. (1977). "Introduction to the Mathematics of Inversion in Remote Sensing and Indirect Measurements." Elsevier, Amsterdam.

Vallance-Jones, A. (1971). *Space Sci. Rev.* **11**, 776–826.

Vallance-Jones, A. (1973). *Space Sci. Rev.* **15**, 355–400.

Williams, W. J., Brooks, J. N., Murcray, D. G., Murcray, F. H., Fried, P. M., and Weinman, J. A. (1972). *J. Atmos. Sci.* **29**, 1375–1379.

Williams, W. J., Barker, D. B., Brooks, J. N., Goldman, A., Kosters, J. J., Murcray, F. H., Murcray, D. G., and Snider, D. E. (1976a). *Proc. Soc. Photo-Opt. Instrum. Eng.* Paper 91-03.

Williams, W. J., Kosters, J. J., Goldman, A., and Murcray, D. G. (1976b). *Geophys. Res. Lett.* **3**, 379–382.

Williams, W. J., Kosters, J. J., Goldman, A., and Murcray, D. G. (1976c). *Geophys. Res. Lett.* **3**, 383–385.

Zander, R. (1975). *C. R. Hebd. Seances Acad. Sci.* **281**, 213–214.

Zander, R., Roland, G., and Delbouille, L. (1977). *Geophys. Res. Lett.* **4**, 117–120.

CHAPTER 2

Polarizing (Martin–Puplett) Interferometric Spectrometers for the Near- and Submillimeter Spectra

D. H. Martin

Physics Department
Queen Mary College
University of London
London, United Kingdom

65

I. Introduction

The general spectrometric method known as Fourier transform spectroscopy (FTS) in which spectra are obtained by computational analysis of the interferograms produced by two-beam interferometers is now well established for the near- and submillimeter spectral ranges. The FTS interferometers with which we shall deal in this chapter are of a type first described by Martin and Puplett (1970) and Martin (1972) and offers a number of advantages over more conventional types when high spectral and radiometric resolution is required, when a wide spectral range is to be covered, or when absolute radiometry is necessary (Section II). We shall give a full analysis of this type of interferometer (Section III), a survey of the types of measurement that can be made with it (Section IV), and descriptions of the construction (Section V), alignment (Section VI), and performance (Section VII) of working instruments. Martin–Puplett (MP), or polarizing, interferometers have been used in a number of scientifically important measurements over the last ten years, and these will be referred to when assessing performance.

II. Two-Beam Interferometers and Fourier Transform Spectroscopy

In Fourier transform spectroscopy, the spectrum of an electromagnetic signal is obtained by Fourier-transforming the interferogram produced when the signal is passed through a two-beam interferometer. The principles of FTS have been fully presented in the books by Mertz (1965), Chantry (1971), Bell (1972), Griffiths (1975), and Chamberlain (1979). Genzel and Sakai (1977) have reviewed the development of FTS techniques and given a comprehensive list of references. A short account of the elementary principles of FTS leading to expressions for performance parameters in terms of instrumental design parameters will be given in Section II.A. In Section II.B, we shall examine the four-port character of all two-beam interferometers; this often-neglected aspect is of crucial importance in achieving high performance. The particular advantages of the Martin–Puplett polarizing interferometers will then be examined against this background in Section II.C.

A. PRINCIPLES OF FOURIER TRANSFORM SPECTROSCOPY

A two-beam interferometer is illustrated schematically in Fig. 1. When the movable reflector is progressively displaced, to give a varying path difference x, the detector records an intensity $I(x)$. This interferogram, formed by interference between two beams, one of which is retarded by a time cx with respect to the other, is the squared modulus of the autocorrelation function of the amplitude of the electromagnetic field incident on the interferometer. The power spectrum of the incident field can, therefore, be obtained by Fourier transformation, as can be seen directly from what follows.

A monochromatic input of wave number σ will give, with a well-aligned interferometer, an interferogram of cosine form

$$I(x) - \bar{I} = \tfrac{1}{2}I_0 \cos 2\pi\sigma x, \tag{1}$$

where $\bar{I}$ is a constant offset and I_0 the interferometrically modulated inten-

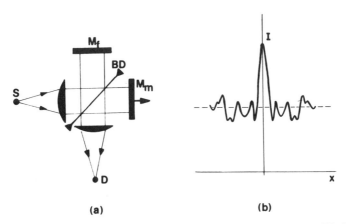

(a)   **(b)**

FIG. 1   (a) A Michelson two-beam interferometer, where S is the source, BD the beam divider, M_f the fixed and M_m the movable reflector, respectively, and D the detector. (b) The interferogram in which the variation of intensity I recorded by the detector as a function of the path difference x.

sity. Because the optical system is linear, an input beam having an arbitrary power spectrum gives an interferogram of the form

$$I(x) - \bar{I} = \int_0^\infty \tfrac{1}{2} S(\sigma) \cos 2\pi\sigma x \, d\sigma, \tag{2}$$

where $S(\sigma) \, d\sigma$ is the interferometrically modulated power in the spectral interval $d\sigma$. We have assumed that the path difference at a given mirror setting is the same for all wave numbers so that $I(x)$ is an even function of x. By Fourier inversion, we obtain

$$S(\sigma) = 4 \int_{-\infty}^{+\infty} \{I(x) - \bar{I}\} \cos 2\pi\sigma x \, dx. \tag{3}$$

In practice, of course, an interferogram is not recorded to an infinite path difference nor, usually, is it sampled continuously. It is sampled discretely, at path differences $x_n = n \, \Delta x$, where Δx is a constant path increment and n denotes the integers, including zero, up to a maximum value $|n| = N$. If n includes both negative and positive values, we have a "double-sided" interferogram, whereas only positive (or negative) values would give a "single-sided" interferogram. We shall consider the former case here.

Suppose, first, that the interferogram had been recorded continuously, but only over the range $x_N > x > -x_N$. Over this finite range, it could be formally fitted by a discrete Fourier series

$$I(x) - \bar{I} = \sum_{m=1}^\infty \frac{a(\sigma_m)}{2} \cos 2\pi\sigma_m x, \tag{4}$$

where $\sigma_m = m \, \Delta\sigma$, $\Delta\sigma = 1/2x_N$, and m is a positive integer and where the coefficients $a(\sigma_m)$ are given by

$$a(\sigma_m) = (2/x_N) \int_{-x_N}^{x_N} (I(x) - \bar{I}) \cos 2\pi\sigma_m x \, dx. \tag{5}$$

Turning now to a discretely sampled interferogram, we can take the lead given by Eq. (5) in deciding what quantity to compute from the recorded data. From the $(2N + 1)$ recorded data words, the following $2N$ linearly independent data words, which we denote as $S'(\sigma_m)$, can be computed:

$$S'(\sigma_m) = (2/x_N) \sum_{n=-(N-1)}^{N} [I(x) - \bar{I}] \cos 2\pi\sigma_m x_n \, \Delta x. \tag{6}$$

Note that we choose not to include $n = -N$ in this summation; in a sense it belongs to the contiguous interval of $2N$ sampling points. We now have to determine the precise relationship between $S'(\sigma_m)$ and $S(\sigma)$. To do this, the full expression for $I(x) - \bar{I}$ given in Eq. (2) is inserted into Eq. (6). After some manipulation, we obtain

$$S'(\sigma_m) = (1/2N)S(\sigma) * C(\sigma, \sigma_m), \qquad (7)$$

where $*$ denotes a convolution, i.e.,

$$S'(\sigma_m) = (1/2N) \int_0^\infty S(\sigma)C(\sigma, \sigma_m) \, d\sigma,$$

$$C(\sigma, \sigma_m) = (\sin N\alpha/\tan \tfrac{1}{2}\alpha) + (\sin N\alpha'/\tan \tfrac{1}{2}\alpha'), \qquad (8)$$

and $\alpha = 2\pi \, \Delta x(\sigma - \sigma_m)$ and $\alpha' = 2\pi \, \Delta x(\sigma + \sigma_m)$.

The second term on the right in (8) will be negligible for all $m \gg 1$, even at $\sigma = 0$, and it will be neglected. Note that the first term on the right-hand side in (8) is periodic in α, i.e., in σ, and this implies that the incident signal at $\sigma + p/\Delta x$, where p is an integer such that $\sigma + p/\Delta x > 0$, will contribute to the computed $S'(\sigma)$ with equal weight for all p. This is "aliasing." To avoid confusion in the computed spectrum from this effect, the value of Δx must be carefully chosen in relation to the ranges of σ over which $S(\sigma)$ is nonzero. If $S(\sigma)$ is zero for all $\sigma > \sigma'$ (as determined by filters or otherwise), Δx should be such that $\Delta x < 1/2\sigma'$. Then, provided $N \gg 1$, little error is incurred by taking, as is customary, the following function for $C(\sigma, \sigma_m)$:

$$C(\sigma, \sigma_m) \rightarrow (2 \sin N\alpha)/\alpha. \qquad (9)$$

From Eq. (7), it is clear that $C(\sigma, \sigma_m)$ is the "instrument resolution function." It is strongly peaked at σ_m with first zeroes at $\sigma - \sigma_m = \pm(1/2x_N)$ (Fig. 2). The (unapodized) spectral resolution afforded by an interferogram that is recorded to a maximum path difference x_N is thus $\Delta\sigma = 1/2x_N$, consistent with the independence of the values of $S'(\sigma_m)$ at successive values of $\sigma_m = m \, \Delta\sigma$ [their independence is implied by the fact that they are the coefficients in the Fourier series that fits the interferogram over the range $x_N > x > -x_N$, i.e., Eqs. (4)–(6).] The summa-

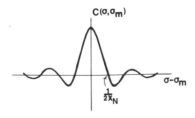

FIG. 2   The instrumental resolution function of a two-beam interferometer.

tion in Eq. (6) could be evaluated for nonintegral values of m to give a more nearly continuous computed spectrum, but the resulting numbers would not be independent of those computed for the integral values and, though this may please the eye, it may also deceive it.

Each point in the discrete spectrum $S'(\sigma_m)$, calculated according to Eq. (6), thus represents the value of $\overline{S(\sigma_m)} \, \Delta\sigma$, where $\overline{S(\sigma_m)}$ is the weighted mean of $S(\sigma)$ about $\sigma = \sigma_m$, i.e.,

$$\overline{S(\sigma_m)} = \int_0^\infty S(\sigma)C(\sigma, \sigma_m) \, d\sigma \bigg/ \int_0^\infty C(\sigma, \sigma_m) \, d\sigma.$$

This follows from Eq. (8) and the fact that the integral in the denominator has the value $1/\Delta x$, provided $m \gg 1$. If $S(\sigma)$ is a slowly varying function of σ, $S'(\sigma_m) = S(\sigma_m) \, \Delta\sigma$.

There will, in practice, be an uncertainty in the value of $S'(\sigma_m)$ at each σ_m because the recorded interferogram $I(x)$ will include variations due to noise from various sources in the system and these will not be distinguished, in taking the Fourier transform, from true interferometric modulation. A component in the noise having real-time frequency ν will be confused with the interferometric modulation produced by the component in the incident beam having spectral frequency $\sigma = \nu/2v$, and the spectral binwidth $\Delta\sigma$ in the computed spectrum will correspond to sampling the noise spectrum over the bandwidth $\Delta\nu = 2v \, \Delta\sigma$, where $v = \frac{1}{2} \Delta x/\Delta t$, Δt being the time interval between sampling points. Thus

$$\nu = 2\sigma v, \qquad \Delta\nu = 2v \, \Delta\sigma = 2v/2x_N = 1/2T, \qquad (10)$$

where T is the full time taken to record one side of the interferogram from $x = 0$ to $x = x_N$.

In a well-designed FTS system, the dominant noise will originate in the detector. Within a narrow frequency band $\Delta\nu$, the rms value of the noise fluctuations, i.e., the variance, will be proportional to $(\Delta\nu)^{1/2}$. It is customary to express the noise in terms of the steady signal power absorbed by the detector, which would give a steady output voltage equal to the observed rms value of the noise voltage in the bandwidth $\Delta\nu$ centered at ν; this equivalent signal power may be written as

$$N(\nu)(\Delta\nu)^{1/2}, \qquad (11)$$

and the quantity $N(\nu)$ so defined is known as the noise-equivalent-power (NEP) of the detector. For "white noise," $N(\nu)$ is independent of ν.

Provided the output from the detector is adequately smoothed or filtered before recording or transforming the interferogram to avoid problems due to aliasing the variance in the value of $S'(\sigma_m)$ will be

$$\varepsilon(\sigma_m) = N(\nu = 2v\sigma_m)/\sqrt{2} \cdot \sqrt{2T} \qquad (12)$$

because, as we have seen, $1/2T$ is the bandwidth corresponding to one spectral resolution bin. The factor $\sqrt{2}$ is required if the true $I(x)$ is known to be an even function, so that only the cosine transform need be taken and the noise is thereby effectively detected coherently.

The foregoing basic relations can now be used to formulate relationships between the anticipated performance levels of an FTS instrument and the design parameters of the system. We shall do this by reference to the signal-to-noise ratios obtained with sources having continuum, not narrow-line, spectra. The signal power available in each spectral bin, if $S(\sigma)$ does not vary much over several resolution intervals, is equal to $S(\sigma) \, \Delta\sigma$, as we have noted earlier. With a source of brightness $B(\sigma)$,

$$S(\sigma) \, \Delta\sigma = B(\sigma) \, \Delta\sigma e(\sigma) \, A\Omega, \tag{13}$$

where $A\Omega$ is the light-grasp, or étendue, of the system and $e(\sigma)$ a throughput efficiency, which includes losses due to modulation and filtering as well as optical throughput losses (see Section VII.F).

Taking Eq. (13) together with (12) leads us to

$$(S/N)/\Delta\sigma \sqrt{T} = 2B(\sigma)e(\sigma)A\Omega/N(\sigma), \tag{14}$$

where the signal-to-noise ratio S/N is defined as the ratio of the recorded signal power in a spectral resolution bin to the noise variance in the bin, and $N(\sigma)$ means $N(\nu = 2v\sigma)$. The quantities on the left-hand side of this relation are measures of performance and those on the right are instrument parameters.

It would appear from this relation that $A\Omega$ should be as large as possible to maximize performance. There is, however, an upper limit for a specified resolving power because the depth of interferometric modulation would deteriorate if the area of illumination of the detector were to exceed that of the central spot of the interference pattern formed in the condenser's focal plane at the detector. It is well known that this limits Ω, the solid-angular spread of the nominally collimated beams, to $2\pi \, \Delta\sigma/\sigma_M$, where σ_M is the maximum spectral wavevector for which efficient interferometric modulation is required. Hence if this maximum value of Ω obtains,

$$(S/N)/(\Delta\sigma)^2 \sqrt{T} = 4\pi B(\sigma)e(\sigma)A/N(\sigma)\sigma_M, \tag{15}$$

where A is the cross-sectional area of the collimated beams.

If the system is to be used for radiometric purposes, it is more useful to express the source brightness in terms of the Rayleigh–Jeans equivalent temperature

$$T_R = B(\sigma)/2k_B c\sigma^2$$

TABLE I

EXAMPLES OF SYSTEM PERFORMANCE[a]

	$\Delta\sigma = 0.1$ cm^{-1}	$\Delta\sigma = 0.01$ cm^{-1}
$\frac{1}{e}$ S/N	10^7, 10^5	10^5, 10^3
$e\Delta T_R$ (K)	10^{-4}, 10^{-2}	10^{-2}, 1

[a] These performance figures are based on the parameters in the text and a measurement time of 400 s. The first number given in each case is for the cooled detector and the second for the room-temperature detector.

and then to specify the temperature increment ΔT_R that would change the S/N by unity, i.e., the "minimum detectable temperature change":

$$\Delta T_R (\Delta\sigma)^2 \sqrt{T} = N(\sigma)\sigma_M / 8\pi k_B c A e(\sigma)\sigma^2. \qquad (16)$$

To gain quantitative guidance on the significance of these relations, we can use the following values for the system's parameters:

$A = 10^4$ mm^2	A value consistent with operational convenience
$\sigma = 50$ cm^{-1}	At the center of the spectral range of interest
$N = 5 \times 10^{-11}$ W Hz$^{-1/2}$	A common specification for a good room-temperature detector
$N = 5 \times 10^{-13}$ W Hz$^{-1/2}$	A common specification for a good liquid-helium-cooled broad-band detector
$T_R = 2 \times 10^3$ °K	Appropriate for a mercury arc lamp, for example

Table I sets out examples of the spectrometric and radiometric performances that relations (15) and (16) would lead one to expect. The performances obtained in practice correspond to values for e of less than 0.01, sometimes much less. It is in tracking down the reasons for such small values of $e(\sigma)$ that the merits and defects of particular types of two-beam interferometer, and of modes of operation, emerge. To do this requires a more complete view of an interferometric system than is explicit in the simple approach just given.

B. THE FOUR-PORT CHARACTER OF TWO-BEAM INTERFEROMETERS

An interferogram is generated by a two-beam interferometer basically because the interference between the two beams results in a changing distribution of the available power between *two* output ports, as the movable mirror is displaced. The sum of the powers in the two output ports is ideally equal to the incident power, and the interferograms in the two ports are therefore complementary (Fig. 3). This complementarity is of importance in distinguishing between those variations in detected power that are due to the interferometric modulation and those that result from spurious fluctuations due to instabilities of one kind or another in the system but, though this has been appreciated for some time, most FTS systems currently in use detect in one port only.

It is also clear that, because reversal of the beams passing through an interferometer interchanges the roles of input and output ports, the presence of two output ports implies the existence of *two* input ports; this, too, provides opportunity for improving or extending the performance of two-beam interferometers.

A two-beam interferometer is thus a four-port device; in this section, we shall identify the ports and consider their accessibility for several types of Michelson interferometer and for Martin–Puplett interferometers.

Figure 4 illustrates a simple Michelson interferometer of the kind employed in the majority of submillimeter FTS systems to date. The beam divider would usually be a flat, polymeric, dielectric film. Figure 4a identifies the input and output ports, respectively, I_1 and O_1, customarily recognized. Figures 4b–d include the others, I_2 and O_2, and show how they

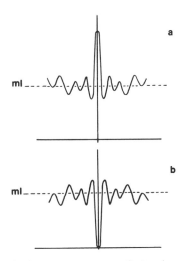

FIG. 3 Interferograms in the two output ports of a two-beam interferometer, where ml indicates the mean level.

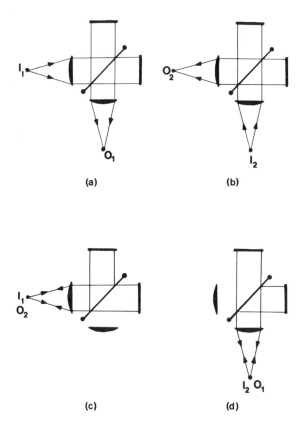

FIG. 4   To illustrate the two input (I_1, I_2) and two output (O_1, O_2) ports of a simple Michelson interferometer. For greater clarity, a single pairing of an input with an output port is shown in each of the diagrams: (a) I_1 and O_1; (b) I_2 and O_2; (c) I_1 and O_2; (d) I_2 and O_1.

can be paired to give rise to interferograms. With a bright source in one input port, the existence of the other input port may not greatly perturb the recorded interferogram (and that is certainly the customary assumption implicitly made), but in precise radiometric measurements, its effect must not be neglected. The fact that, in this configuration, I_1 coincides with O_2 and I_2 with O_1 makes control of the ports difficult and prevents use being made of them to reduce noise, maximize throughput, and extend the range of applications. For this reason, more complex Michelson interferometers have been devised. The ports can be separated by oblique illumination of the beam divider (Fig. 5) or by lateral displacement of the beams as in Fig. 6 (e.g., Burroughs and Chamberlain, 1971; Hanel et al., 1969). This leaves the two output (input) ports at opposite ends of the in-

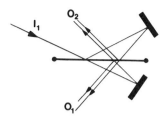

FIG. 5   Port separation by oblique illumination of the beam divider.

strument, which complicates their full utilization. With further elaboration, this can be corrected, as in Fig. 7, or, in order to avoid a varying beam overlap as the mirror is moved, as in Fig. 8 (Chandrasekhar *et al.*, 1976a, b; Genzel and Kuhl, 1978). In such interferometers, whereas the two beams contributing to the interferogram in one output port will have had similar histories—with one reflection from and one transmission through the beam divider—the two that contribute to the interferogram in the other output port will have had different histories—one will have been reflected twice and the other transmitted twice; the interferograms detected in the two output ports may not be strictly complementary, therefore, if the beam divider is absorptive.

Lamellar grating interferometers were important in the early development of FTS. In such a system, the zero order of diffraction from the grating represents one output port and the remaining orders, collectively, the other. This technique is not easily adapted for high resolution or for precise radiometric applications.

The basic configuration of the Michelson polarizing, or Martin–Puplett (MP), interferometer is illustrated in Fig. 9 (Martin and Puplett, 1970;

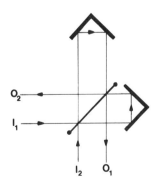

FIG. 6   Port separation by lateral beam displacement.

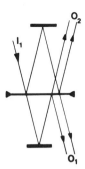

FIG. 7   Separation of input and output ports.

Martin, 1972). Some of its principal advantages derive from the easy accessibility of all four ports in a simple optical configuration. The principle behind the method is most directly understood by reference to Fig. 10, where P_1, P_2 and BD, BC are plane wire-grid polarizers that efficiently transmit one plane of polarization and reflect the orthogonal polarization. A collimated beam A is plane-polarized in the plane at 45° to the normal to the page by transmission through P_1. It is then divided by beam divider BD into a reflected beam a, polarized normally to the paper, and a transmitted orthogonally polarized beam b. After traveling different paths, a and b are recombined at BC, which is so oriented as to reflect a and transmit b. Because a and b have traveled different distances, the recombined beam will be elliptically polarized. The ellipticity and the direction of the major axis is determined by the phase difference of a and b arising from the path difference. The beam then passes through the output polarizer/analyzer P_2 to give, by transmission and reflection, respectively, the two output beams associated with the input beam A, where P_2 is so

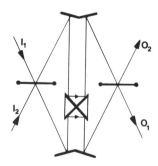

FIG. 8   Separation of input and output ports without a varying beam overlap.

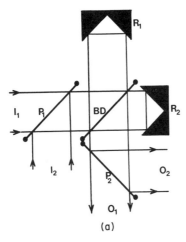

(a)

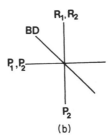

(b)

FIG. 9   (a) Basic configuration of MP interferometer, where P_1, P_2, and BD are the wire grids and R_1 and R_2 rooftop reflectors, one fixed and one movable. (b) The relative orientations of the wires in the grids P_1, P_2, and BD and of the roof edges in R_1 and R_2, as projected onto the planes transverse to the optic axis. Alternative orientations are shown for P_2 "parallel" and "crossed." I_1, I_2 are the two input ports and O_1, O_2 the two output ports.

oriented that it efficiently transmits planes of polarization at 45° to the normal to the page. It may be "crossed" or "parallel" with respect to P_1; changing from one setting to the other simply interchanges the two output beams between the two output ports. For zero path difference, assuming ideal polarizing grids, the incident power will clearly pass wholly into one output port or wholly into the other, depending on whether P_1 and P_2 are crossed or parallel. If the input beam is monochromatic, the direction of the major axis for the recombined beam will rotate as the path difference x is varied continuously by displacing downward the mirror combination M, and the power in each output beam will vary periodically with period

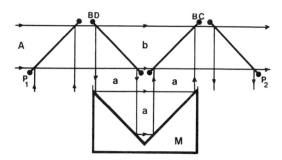

FIG. 10   A polarizing interferometer (see text for discussion).

$x_p = \lambda$, the wavelength of the incident beam. Thus the fraction of the beam power incident on BD that emerges from one output port is

$$f = \tfrac{1}{2}(1 \pm \cos \Delta), \tag{17}$$

where $\Delta = 2\pi x/\lambda$ and the sign alternatives refer, respectively, to the two output ports; the sum of the two is, as it must be, unity at all Δ values. This expression also applies, of course, to a Michelson interferometer, and the basic treatment of FTS in Section II.A (compare Eqs. (1) and (17)) will apply to the MP interferometer as well as to the Michelson interferometer.

The two input ports are now readily identified. They admit the beams that are, respectively, transmitted by and reflected from the input polarizer P_1. The two input beams are therefore orthogonally polarized. At a given path difference, therefore, the division between the two output ports of the power incident in one input port will be the reverse of that of the power incident in the other. If f_{ij} is the fraction of the power incident through the input port i ($i = 1$ or 2) that is transmitted to the output port j ($j = 1$ or 2), we can write

$$f_{ij} = \tfrac{1}{2}(1 \pm \delta'_{ij} \cos \Delta), \tag{18}$$

where δ'_{ij} is equal to $+1$ for $i = j$ and -1 for $i \neq j$ and where the sign alternatives are, respectively, for parallel and crossed settings of P_1 and P_2.

The simple configuration for an MP interferometer shown in Fig. 10 is not satisfactory because the position of zero path difference cannot be reached. The configuration that has made the method practicable is that shown in Fig. 9. The metal roof reflectors R_1 and R_2 are set with their roof lines at 45° to the planes of polarization of the beams incident upon them; their effect is then to rotate the plane of polarization of the reflected beam through 90°, relative to that of the incident beam (see Section II.C.4). The beam that is reflected from the beam divider is, on its return, therefore transmitted through it, and the beam that is transmitted through BD is reflected by it on return. Thus BD serves both as beam divider and beam combiner.

There are other possible configurations. For example, that shown in Fig. 11 is simpler, but there is less ready access to all the input and output ports and the two beams within the interferometer suffer different histories at the beam divider. In high-performance applications, the symmetry of the configuration in Fig. 9 is important (Section III). There will be loss of signal power at the first polarizer if the input beams are not plane-polarized in the appropriate directions. The more elaborate configuration of Fig. 12 avoids this loss. The complexity of such an instrument might be justified in special circumstances because a loss by a factor of 2 in signal-to-noise ratio quadruples the time required to record a spectrum to a specified precision and that could be of significance (for example, in astronomical measurements at a large telescope or from a balloon-borne or space platform).

We have now identified the four ports, both for Michelson and for MP interferometers. Equation (18) has been introduced here in the context of the MP interferometer, but it applies equally well to ideal Michelson interferometers (to demonstrate this, it is necessary to take note of the $\frac{1}{2}\pi$ phase difference between the transmitted and reflected beams from the beam divider). Signal power will enter an interferometer in both input ports (intentionally or otherwise), and Eq. (18) indicates that a detector placed in output port $j = 1$ will register the total power

$$f_{11}I_1 + f_{21}I_2 = \tfrac{1}{2}(I_1 + I_2) \pm \tfrac{1}{2}(I_1 - I_2)\cos\Delta, \qquad (19)$$

where I_1 and I_2 denote here the signal powers entering input ports $i = 1$ and 2, respectively, and are assumed here not to be coherently related. The peak-to-peak amplitude of the interferometrically modulated output is thus equal to $I_1 - I_2$, i.e., *a two-beam interferometer records the dif-*

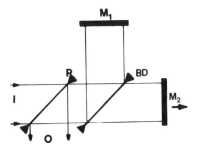

FIG. 11 A simplified polarizing interferometer with input port I and output port O. The second output port and the second input port coincide with I and O, respectively.

FIG. 12 A polarizing interferometer in which *all* the power incident in the two input ports I_1 and I_2 contributes to the interferograms produced at the output ports O_1 and O_2, where M is a moving reflector that serves in both beams.

ference between the powers of its two input beams. The unmodulated component $\frac{1}{2}(I_1 + I_2)$ contains no spectral information but it can carry noise, and it is, for this reason, important to note that this component can be suppressed by recording the difference between the powers in the two output beams, i.e.,

$$(f_{11}I_1 + f_{21}I_2) - (f_{12}I_1 + f_{22}I_2) = (I_1 - I_2) \cos \Delta. \tag{20}$$

C. Advantages of Martin–Puplett Interferometers

The several advantages of Martin–Puplett interferometers over the simplest Michelson interferometers will be examined in this section. These advantages derive first from the accessibility of the four ports and second from the high efficiency of the polarizing beam divider. Some, but not all, of these advantages are also to be found in the more sophisticated Michelson interferometers, as we shall indicate.

1. Output Ports

First we consider the use of the two output ports in the suppression of *signal-carried or signal-related noise.* There are many sources of signal-carried fluctuations. The source brightness may fluctuate (especially in astronomical and atmospheric studies). Variations in detector responsivity, movements of optical components, and varying alignment of the moving mirror will all modulate the signal intensity. Such fluctuations have serious consequences in FTS because the signal power falls on the detector without spectral filtering. It is this that gives the well-known multiplex advantage of interferometric over dispersive methods of spectrometry in *detector-noise*-limited situations, but it is responsible too for a multiplex disadvantage so far as *signal-carried* fluctuations are concerned (Section III.D) because these fluctuations can make themselves felt as variations in the mean level of the interferogram. When both output ports are accessible, this can be, to a large extent, remedied. The interferometric modulations that carry the spectral information are *complementary* in the two outputs, and subtraction of one interferogram from the other therefore enhances these modulations while suppressing the mean level (which carries no spectral information) with its spurious fluctuations, which are *additive* in the two ports [Eq. (20)]. We shall note here two direct and simple ways in which this suppression can be achieved in practice with an MP interferometer.

The first method is simply to use a matched pair of detectors, one in each output port, and connect them in opposition. The second is to rotate either the input polarizer P_1 or the output polarizer P_2 and phase-sensitively detect in either or both output ports at twice the rotation frequency. Such polarization coding will be considered more fully in Section III.B. The more complex Michelson interferometers illustrated in Figs. 6 and 8 would also allow mean-level suppression by the first method, and that in fig. 8 by a method related to the second.

2. *Input Ports*

We shall consider now the advantages of having easy access to the two input ports.

Equations (19) and (20) indicate that any two-beam interferometer records *the difference* between the signal powers entering the two input ports; it is essentially a differential instrument. Few FTS instruments have been designed with this in mind. For high performance, however, it is important to take note of this fact, first in order to avoid some of the errors that may result if the matter is ignored (thermal emission from objects at near-ambient temperature, in the second port, may be of significance at near- and submillimeter wavelengths) and second because there are several ways in which advantage may be taken of it.

Absolute determinations of a signal's power can be made if the second port can be switched (as illustrated in Fig. 13 for an MP interferometer) between two blackbody cavities at different temperatures; the absolute

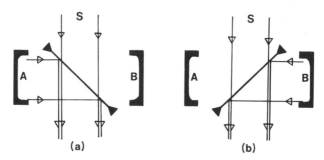

Fig. 13   The alternative settings of the input polarizer that result in the signal S being compared with the signal (a) from reference source A and (b) from reference source B.

value for the signal power is given by interpolation between the two blackbody signals (see Section IV.D).

Explicit use may also be made of an interferometer as a differential instrument if small differences in the spectra of two sources are required, by placing one source in each input port (Section IV.C). Examples are as follows:

(i) detection of the spectrum of an impurity in a host (place impure sample in one input beam and a pure host sample in the other),

(ii) detection of a weak, localized source in the presence of an extended emissive background (e.g., focus a pointlike astronomical source into one port and the surrounding regions of the sky into the other),

(iii) detection of anisotropy in a roughly isotropic field (by detecting the difference between the signals in the two, differently directed, input ports).

3.   Efficient Beam Division

A rather different advantage specific to the MP interferometer is the *high efficiency of wire-grid beam dividers* over a broad and continuous range of spectral frequency.

A wire grid efficiently reflects the plane of polarization having the electric field along the wires because the currents induced by the incident beam flow much as they would in a continuous metal sheet. The reflectivity for that plane of polarization will be comparable, therefore, to that of a metal surface. For the orthogonal plane of polarization, the wires are polarized by the electric field so that the grid behaves more as a thin dielectric film and transmits well. This description breaks down if the wavelength λ is not appreciably greater than the wire spacing d. The results of theoretical studies of the properties of wire grids (see Petit, 1975; Chambers *et al.*, 1980a, b) have been compared with the measured behav-

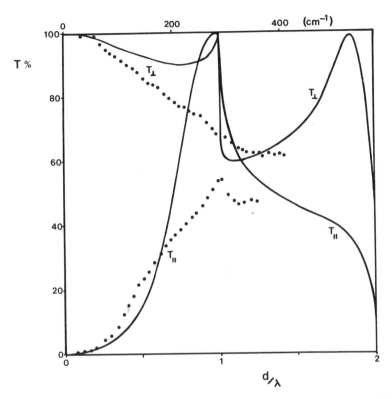

FIG. 14   Theoretical values for the perpendicular and parallel transmission coefficients of a wire grid as dependent on the ratio of wire spacing d to wavelength λ (solid curves). Measured values of the transmission coefficients for a wire grid with $d = 35$ μm are indicated by the dotted curves, and the upper scale applies to this grid. (See Beunen *et al.*, 1981.)

iors of grids wound with fine tungsten wire (Mok *et al.*, 1979; Beunen *et al.*, 1981), and Figs. 14 and 15 illustrate the results. This work indicates that unsupported grids can be wound with spacings of about 25 μm (or more) to give extremely high efficiencies at near-millimeter wavelengths. Figure 39 will present measurements of both transmission and reflection for grids with 12.5- and 25-μm spacings. Good efficiency can be obtained up to 100 cm^{-1} (0.1 mm), and useful efficiency up to at least 300 cm^{-1}.

The dielectric film beam dividers used in Michelson interferometers rely on interference within the film thickness to enhance the reflectance, and there are difficulties in getting high efficiency over a wide spectral range. The efficiency for beam division ($4r^2t^2$, where r and t are the amplitude reflection and transmission coefficients, respectively) varies periodi-

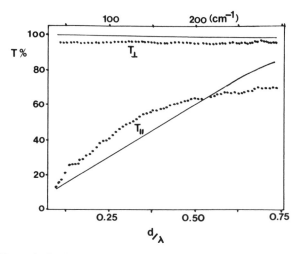

Fig. 15 Theoretical values for the perpendicular and parallel transmission coefficients of a wire grid as dependent on the ratio of wire spacing d to wavelength λ (solid curves). Measured values of the transmission coefficients for a wire grid with $d = 25$ μm are indicated by the dotted curves, and the upper scale applies to this grid. In both the theoretical and measured cases, the wire spacing is five times the wire diameter. (See Mok *et al.*, 1979.)

cally with spectral frequency, the period being determined by the film thickness. The values of efficiency at the maxima and minima are determined by the optical constants for the dielectric material; it is difficult to improve significantly on the values given by Mylar, which are about 75 and 5%, respectively (see Fig. 16). The film thickness can be chosen to give one or more maxima in the spectral range of interest, but it is not possible to maintain a high efficiency over more than about an octave of spectral frequency and this is especially a handicap for measurements at near-millimeter frequencies.

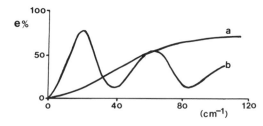

Fig. 16 Efficiencies of dielectric film (Mylar) beam dividers: for a thickness of 12.5 μm (curve a) and a thickness of 76 μm (curve b).

4. *Polarization Rotators*

Wavelength-independent rotators of the planes of polarization are required for the advantageous MP configuration in Fig. 9. Simple metallic rooftop reflectors can be used to this end (Martin and Puplett, 1970). The basis of the idea is illustrated in Fig. 17. The incident beam *i* in Fig. 17a is reflected from a point *O* on the surface of a metallic, or metallized, mirror *M*; if the plane of polarization of the beam is as indicated by the arrow in the flag attaching to it, that of the reflected beam *r* must be as shown in its flag because the superposition at *O* of the *E* fields in the incident and reflected beams must give a resultant that is *normal* to the surface in order to satisfy the boundary conditions for a metallic surface. If the beam then suffers a second reflection at a point *P* on a second metallic surface, to return in the direction of incidence as illustrated by the dashes lines in the figure, the polarization of the emergent beam *e* will be as shown in its flag. The net result is as illustrated in Fig. 17b in which *pp* denotes the plane of incidence and reflection, and the plane of the paper is perpendicular to the direction of incidence. A rotation of the plane of polarization through 2θ has occurred, where θ is the angle between the plane of polarization of the

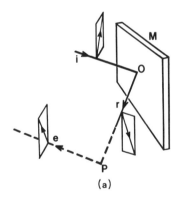

(a)

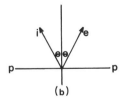

(b)

Fig. 17   Rooftop reflector for the MP interferometer (see text for details).

incident beam and the normal to the plane of reflection. If θ is 45°, as is the case for the roof reflectors R_1 and R_2 in the MP interferometer illustrated in Fig. 9, the net rotation of the plane of polarization is through 90°, as required.

Real metals are not ideal conductors and some degree of ellipticity in the reflected beam must be present, but this will be negligible except at high (optical) frequencies. We have made measurements with a roof reflector having aluminized surfaces, using a plane-polarized beam from a CO_2 laser at a wavelength near 10 μm, and found the reflected beam to be plane polarized to at least 99.9%.

III.   General Theory of Martin–Puplett Interferometers

The simple explanation in Section II.B of the performance of an MP interferometer applies to an ideal system having perfect polarizing components, perfectly aligned. In this section, we shall present a more general and complete analysis of the performance of an MP interferometer system. We shall discuss the basic optical functions of such a system and its insensitivity to the imperfections of construction and alignment that must be present in any real system, analyze two methods of polarization coding or modulation, and trace the effects of signal-carried noise and the methods of noise suppression based on polarization coding. Some aspects of the theory of MP interferometers have been presented by Lambert and Richards (1978), Chambers *et al.* (1980a, b), and Carli and Mencaraglia (1981).

A.   General Theory

1.   *The Two-Beam Unit*

It is best first to consider separately the two-beam unit of the interferometer as illustrated in Fig. 18, comprising the wire-grid polarizing beam divider BD, and the fixed and moving rooftop reflectors M_f and M_m. We shall use r and t to denote the complex amplitude reflection and transmission coefficients of the BD and add a subscript t to indicate that plane of

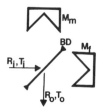

FIG. 18   Two-beam unit (see text for details).

polarization of the incident beam for which the E field is perpendicular to the grid wires and a subscript r for the orthogonal polarization, i.e., that for which the *component* of the E field in the plane of the BD is parallel to the wires. An ideal wire-grid polarizer would therefore have $|r_r| = |t_t| = 1$ and $|r_t| = |t_r| = 0$, but we shall be admitting the possibility of departures from these ideal values and include arbitrary phase angles. The angles of incidence of beams crossing at the BD will be the same (usually, but not necesarily, $\pi/4$), and the r and t coefficients are taken to be the same for incidence on the two faces of the BD.

We use R and T to denote complex amplitudes for plane-parallel monochromatic beams passing through the two-beam unit, R denoting a beam of r-polarization relative to the BD and T one with t-polarization, where R_i, T_i and R_o, T_o denote, respectively, the amplitudes at the input and the output of the two-beam unit (Fig. 18).

The rooftop reflectors serve to rotate the plane of polarization of an incident beam through $\pi/2$, provided the line of the roof edge is at $\pi/4$ to the incident plane of polarization (Section II.C.4). We shall allow for the possibility of imperfect orientation of the rotator.

We thus intend here to recognize two kinds of departure from ideality; first, that due to misorientation of the rotators and second, that due to a polarizer being imperfect in the sense that, for an incident R beam, it will not only reflect (strongly) but will also transmit (weakly), and vice versa for an incident T beam (note, however, that both reflected and transmitted beams are R beams in the first case and T beams in the second, i.e., we assign no rotary properties to the grid, which is assumed to be symmetrical about the line of the grid wires). We shall treat separately

(a)   a perfectly aligned system with imperfect polarizers and
(b)   a misoriented system with perfect polarizers.

In each case, an incident R beam leads to a pure T beam at the output, and a T beam to an R beam; this is so even for the case of a misoriented rotator because the contaminating beam arising from the misorientation will be returned to the source, rather than to the output, if the beam divider is ideal. We can, therefore, write

$$\begin{bmatrix} R_o \\ T_o \end{bmatrix} = b \begin{bmatrix} 0 & d_m + \delta d_f \\ d_f + \delta d_m & 0 \end{bmatrix} \begin{bmatrix} R_i \\ -T_i \end{bmatrix}, \qquad (21)$$

where R_i, T_i and R_o, T_o are the incident and outgoing orthogonal wave amplitudes, respectively, (Fig. 18), b the beam divider efficiency, which is equal to $r_r t_t$, δ the beam-divider defect parameter equal to $(r_t t_r / r_r t_t)$, and d_f and d_m the complex propagation coefficients for the paths via the fixed and movable reflectors, respectively. The absolute values $|d_f|$ and $|d_m|$

FIG. 19 Relative orientations of the axes R, T and S, P and of the wires of the BD projected onto the plane perpendicular to the direction of beam propagation.

measure the loss of amplitude in passing through the unit, excluding loss at the beam divider. The phase angles of d_f and d_m represent the net phase shift along the path so that the ratio of d_m to d_f has phase angle $\Delta = 2\pi x/\lambda$, where x is the path difference. Each off-diagonal term in the matrix in Eq. (21) can be seen to include a main beam passing via the moving (fixed) reflector and the leakage beam passing via the fixed (moving) reflector. For case (a) δ will be zero, but $|d_m|$ and $|d_f|$ will include the loss of amplitude due to imperfect orientation of the rotators.

It will prove useful to transform Eq. (21) so as to refer to wave-amplitude components S and P along axes at $\frac{1}{2}\pi$ to those of R and T (Fig. 19). Noting that the input and output beams are rotated, thus

$$\begin{bmatrix} R_i \\ T_i \end{bmatrix} = \frac{1}{\sqrt{2}} \begin{bmatrix} 1 & -1 \\ 1 & 1 \end{bmatrix} \begin{bmatrix} S_i \\ P_i \end{bmatrix}, \tag{22}$$

$$\begin{bmatrix} S_o \\ P_o \end{bmatrix} = \frac{1}{\sqrt{2}} \begin{bmatrix} 1 & 1 \\ -1 & 1 \end{bmatrix} \begin{bmatrix} R_o \\ T_o \end{bmatrix}, \tag{23}$$

we have the following relation between S_o, P_o and S_i, P_i:

$$\begin{bmatrix} S_o \\ P_o \end{bmatrix} = \tfrac{1}{2} b \begin{bmatrix} d_- & -d_+ \\ d_+ & -d_- \end{bmatrix} \begin{bmatrix} S_i \\ P_i \end{bmatrix}, \tag{24}$$

where,

$$d_+ = (d_f + d_m)(1 + \delta) \quad \text{and} \quad d_- = (d_f - d_m)(1 - \delta). \tag{25}$$

Equation (24) is a statement of the operational function of the two-beam unit.

2. The MP Interferometer as a Polarimeter

The addition of an output wire-grid polarizer to the two-beam unit (Fig. 20) gives a polarimeter. The beam transmitted through the output polarizer (OP) and the beam reflected from it constitute two output ports, A and B. We wish to have expressions for the output powers corresponding to a particular orientation of OP around the propagation axis. We denote the orientation by the angle θ, which is the angle between the axis S in Fig. 19 and the lines of the wires of OP projected onto the plane perpen-

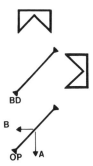

FIG. 20 The MP interferometer as a polarimeter.

dicular to the axis of propagation (Fig. 21). For most purposes, $\theta = 0$ or $\frac{1}{2}\pi$ would be optimum (see the discussion later in this section).

The transmitted output components, referred to the axes of OP, are

$$\begin{bmatrix} A_r \\ A_t \end{bmatrix} = \begin{bmatrix} t_r^o & 0 \\ 0 & t_t^o \end{bmatrix} \begin{bmatrix} \cos\theta & \sin\theta \\ -\sin\theta & \cos\theta \end{bmatrix} \begin{bmatrix} S_o \\ P_o \end{bmatrix}, \qquad (26)$$

where the superscripts on t_t^o and t_r^o are used to denote the coefficients of the OP grid. Using Eq. (24), we can relate A_r, A_t to the inputs S_i, P_i. Multiplying the matrices out gives the following expression for the total intensity in output A:

$$
\begin{aligned}
A_r A_r^* + A_t A_t^* = \tfrac{1}{8}(t_t^o)^2 b\{ & [1 + (\tau^o)^2](d_+^2 + d_-^2)(S_i^2 + P_i^2) \\
& + [1 - (\tau^o)^2](d_-^2 - d_+^2)\cos 2\theta(S_i^2 - P_i^2) \\
& - [1 - (\tau^o)^2](d_+ d_-^* + d_+^* d_-)\sin 2\theta(S_i^2 + P_i^2) \\
& - [1 + (\tau^o)^2](d_+ d_-^* + d_+^* d_-)(S_i P_i^* + S_i^* P_i) \\
& + [1 - (\tau^o)^2](d_-^2 + d_+^2)\sin 2\theta(S_i P_i^* + S_i^* P_i) \\
& - [1 - (\tau^o)^2](d_+ d_-^* - d_+^* d_-)\cos 2\theta(S_i P_i^* - S_i^* P_i)\}. \quad (27)
\end{aligned}
$$

We have used τ^o to denote the (small) quantity $|t_r^o|/|t_t^o|$. Squared terms denote squared moduli.

The corresponding expression for the total power in the output B (that given by reflection from OP, rather than by transmission) is obtained directly from Eq. (27) by replacing t_t^o and τ^o by r_r^o and $\rho^o \equiv |r_t^o|/|r_r^o|$, respec-

FIG. 21 Orientation of OP relative to the S axis.

tively, and by changing the signs of those terms containing the factor $[1 - (\tau^0)^2]$.

It may be seen that each term contains one of the four Stokes polarization parameters of the input beam, namely,

$$
\begin{aligned}
(S_i^2 + P_i^2), && (S_i P_i^* + S_i^* P_i), \\
(S_i^2 - P_i^2), && i(S_i P_i^* - S_i^* P_i),
\end{aligned}
\tag{28}
$$

but to separate them in a measurement will require study of the variations in the output intensity that are consequent upon changes in θ and/or in the d-dependent terms (changes in the path difference). From the definitions of d_+ and d_- [Eq. (25)], it follows that

$$
\begin{aligned}
d_+^2 + d_-^2 &= 2(d_f^2 + d_m^2) + 8\delta'|d_f||d_m| \cos \Delta, \\
d_+^2 - d_-^2 &= 4|d_f||d_m| \cos \Delta + 8\delta'|d_f||d_m|, \\
d_+ d_-^* + d_+^* d_- &= 2(d_f^2 - d_m^2) - 8\delta''|d_f||d_m| \sin \Delta, \\
d_+ d_-^* - d_+^* d_- &= 4i|d_f||d_m| \sin \Delta,
\end{aligned}
\tag{29}
$$

where δ' and δ'' are the real and imaginary parts of δ, respectively, and we have neglected δ^2 compared with unity (note that δ^2 is a product of *four* grid defect parameters).

Let us first see what Eq. (27) gives for perfect polarizers ($b = 1$, $\delta = 0$, $t_t^0 = r_r^0 = 1$, $t_t^0 = r_t^0 = 0$) and perfect alignment ($|d_f| = |d_m| = 1$ and $\theta = 0$ or $\tfrac{1}{2}\pi$). This expression for the output intensity and the corresponding expression for the reflection output port both reduce to (C denotes either A or B)

$$
C_r C_r^* + C_t C_t^* = \tfrac{1}{2}(S_i^2 + P_i^2) \pm \tfrac{1}{2}(\cos \Delta)(S_i^2 - P_i^2) \\
\mp \tfrac{1}{2}i(\sin \Delta)(S_i P_i^* - S_i^* P_i) \tag{30}
$$

where the $\pm$ alternatives are determined by which output port is involved and which θ, either 0 or $\tfrac{1}{2}\pi$. Thus, for a monochromatic input of known λ, three of the Stokes parameters for the incident beam could be determined if measurements were made at three or more values of the path difference. (The fourth, undetermined, Stokes parameter $(S_i P_i^* + S_i^* P_i)$ adds to this information only the quadrant to which the phase angle between S_i and P_i belongs.)

Now let us first take the case of imperfect polarizers with perfect alignment, that is, $\theta = 0$ or $\tfrac{1}{2}\pi$. From Eqs. (27) and (29), we have

$$
\begin{aligned}
A_r A_r^* + A_t A_t^* = \frac{(t_t^0)^2 b}{4} \{ & [1 + (\tau^0)^2](d_f^2 + d_m^2 + 4\delta'|d_f||d_m| \cos \Delta)(S_i^2 + P_i^2) \\
\pm\ & [1 - (\tau^0)^2](2|d_f||d_m| \cos \Delta + 4\delta'|d_f||d_m|)(S_i^2 - P_i^2) \\
-\ & [1 + (\tau^0)^2](d_f^2 - d_m^2 - 4\delta''|d_f||d_m| \sin \Delta)(S_i P_i^* + S_i^* P_i) \\
\mp\ & [1 - (\tau^0)^2](2i|d_f||d_m| \sin \Delta)(S_i P_i^* - S_i^* P_i) \}.
\end{aligned}
\tag{31}
$$

It can be seen immediately that if we have imperfect polarizers, we could not, for a single OP setting, rigorously separate $(S_i^2 + P_i^2)$ from $(S_i^2 - P_i^2)$ (because both contribute to the $\cos \Delta$ variation) nor $(S_i P_i^* + S_i^* P_i)$ from $(S_i P_i^* - S_i^* P_i)$ (because both contribute to the $\sin \Delta$ variation). However, the contribution of $(S_i^2 + P_i^2)$ to the $\cos \Delta$ variation and that of $(S_i P_i^* + S_i^* P_i)$ to the $\sin \Delta$ variation are small, being approximately as $2\delta'$ and $2\delta''$, respectively, compared with unity (note that δ is itself a product of two defect terms, r_t and t_r). Furthermore, there is scope for separating these contributions in that a change in the orientation of OP between the two settings ($\theta = 0, \frac{1}{2}\pi$) changes the sign of some of the terms, as indicated in Eq. (31). Taking the sum and the difference of the two interferograms obtained by transmission through OP in its two settings, we obtain

Sum:
$$\frac{1}{2}\{(t_t^0)^2 b[1 + (\tau^0)^2]\}\{(d_f^2 + d_m^2)(S_i^2 + P_i^2)$$
$$- (d_f^2 - d_m^2)(S_i P_i^* + S_i^* P_i)$$
$$+ 4\delta'|d_f||d_m| \cos \Delta(S_i^2 + P_i^2)$$
$$- 4\delta''|d_f||d_m| \sin \Delta(S_i P_i^* + S_i^* P_i)\}. \qquad (32)$$

Difference:
$$\frac{1}{2}\{(t_t^0)^2 b[1 - (\tau^0)^2]\}\{2|d_f||d_m| \cos \Delta(S_i^2 - P_i^2)$$
$$- 2i|d_f||d_m| \sin \Delta(S_i P_i^* - S_i^* P_i)$$
$$+ 4\delta'|d_f||d_m|(S_i^2 - P_i^2)\}. \qquad (33)$$

If care is taken in setting up the interferometer so that the two beam paths have equal throughputs, i.e., $|d_m| = |d_f|$ (as might be checked by measurements with a bright test source and alternate blocking of the two beams) measurements of the *sum* and *difference* at several Δ values serve to determine, for a monochromatic beam, the values of

$$(S_i^2 - P_i^2)/(S_i^2 + P_i^2) \quad \text{and} \quad (S_i P_i^* - S_i^* P_i)/(S_i^2 + P_i^2), \qquad (34)$$

the first being given by the ratio of the amplitude of the $\cos \Delta$ term in the *difference* to the Δ-independent term in the *sum,* and the second by the corresponding ratio for the $\sin \Delta$ term; recorded data actually give the values of these quantities multiplied by $(1 + 2\tau^0)$, but τ^0 is very small for a good grid and could be separately determined by direct measurements with a bright source.

Turning now to the case of perfect polarizers with imperfect orientations of rotators/reflectors and grids, we obtain, from Eqs. (27) and (29),

$$C_r C_r^* + C_t C_t^* = \frac{1}{4}[(d_m^2 + d_f^2) \pm (d_m^2 - d_f^2) \sin 2\theta](S_i^2 + P_i^2)$$
$$+ \frac{1}{4}[(d_m^2 - d_f^2) \pm (d_m^2 + d_f^2) \sin 2\theta](S_i P_i^* + S_i^* P_i)$$
$$\pm \frac{1}{2}|d_m||d_f| \cos 2\theta(\cos \Delta)(S_i^2 - P_i^2)$$
$$\mp \frac{1}{2}|d_m||d_f| \cos 2\theta(\sin \Delta)i(S_i P_i^* - S_i^* P_i), \qquad (35)$$

where the sign alternatives refer to the two output ports (i.e., $C = A$ and

$C = B$, respectively). These is, here, no admixing of the several Stokes parameters in the cos Δ and sin Δ terms.

Taking the sum and difference of the signal powers in the two output ports gives

Sum: $\quad \frac{1}{2}(d_m^2 + d_f^2)(S_i^2 + P_i^2) + \frac{1}{2}(d_m^2 - d_f^2)(S_i P_i^* + S_i^* P_i).$ $\quad$ (36)

Difference: $\quad \frac{1}{2}(d_m^2 + d_f^2) \sin 2\theta (S_i P_i^* + S_i^* P_i)$
$$+ \frac{1}{2}(d_m^2 - d_f^2) \sin 2\theta (S_i^2 + P_i^2)$$
$$+ |d_m||d_f| \cos 2\theta (\cos \Delta)(S_i^2 - P_i^2)$$
$$- |d_m||d_f| \cos 2\theta (\sin \Delta) i (S_i P_i^* - S_i^* P_i). \quad (37)$$

From this it can be seen that again, provided the two beams are balanced so that $|d_m| = |d_f|$, the ratios of the amplitudes of the cos Δ and sin Δ terms in the difference, respectively, to the sum serve to measure

$$(S_i^2 - P_i^2)/(S_i^2 + P_i^2) \quad \text{and} \quad (S_i P_i^* - S_i^* P_i)/(S_i^2 + P_i^2), \quad (38)$$

in this case to second order in θ. (It should be possible to keep this correction to ~ 1 part in 10^3 straightforwardly, which is likely to be better than the balancing of the beams.)

3. *The MP Interferometer as a Spectrometer*

To complete an instrument for general use as a spectrometer, an input polarizer must be added to the polarimeter described in the preceding section. The overall structure is then as illustrated in Fig. 22. The signal, of which the spectrum is to be determined, is passed through the input polarizer IP before entering the polarimeter. The grid wires of the IP are

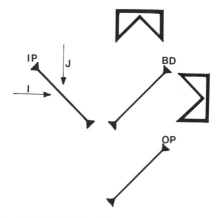

FIG. 22   The MP interferometer as a spectrometer.

oriented so that the emergent beam is plane-polarized symmetrically with respect to the grid axes of the beam divider in the polarimeter. There are two such orientations (orthogonal) and the input signal will contribute mainly to the beam denoted by S_i in the analysis of preceding sections for one orientation and to that denoted by P_i for the other. We shall assume the former.

Account must be taken of any power that enters the polarimeter by reflection from IP rather than by transmission through it. We shall denote the amplitudes of the two signal inputs by I and J (Fig. 22), and to begin, we shall consider them to be monochromatic.

We wish to include in our analysis imperfect polarization by IP and also angular misalignment of IP with respect to the beam divider (BD) in the polarimeter. We denote by I_s and I_p the amplitudes of the plane-polarized orthogonal components of I that are, respectively, ideally passed and blocked by IP in transmission; correspondingly J_s and J_p denote the components of J that are, respectively, ideally blocked and passed by IP in reflection.

The relationships between the outputs from the polarimeter and its inputs S_i and P_i were treated in Section III.2. We need, therefore, to establish the relationships between I, J and S_i, P_i. First, we note

$$\begin{bmatrix} S_i \\ P_i \end{bmatrix} = \begin{bmatrix} \cos\psi & \sin\psi \\ -\sin\psi & \cos\psi \end{bmatrix} \begin{bmatrix} S_i' \\ P_i' \end{bmatrix}, \tag{39}$$

where S_i', P_i' denote components along the axes of IP and ψ the angular misalignment of IP with respect to BD. Second,

$$\begin{bmatrix} S_i' \\ P_i' \end{bmatrix} = \begin{bmatrix} t_t^i & 0 \\ 0 & t_r^i \end{bmatrix} \begin{bmatrix} I_s \\ I_p \end{bmatrix} + \begin{bmatrix} r_t^i & 0 \\ 0 & r_r^i \end{bmatrix} \begin{bmatrix} J_s \\ J_p \end{bmatrix}, \tag{40}$$

where the transmission and reflection coefficients for the wire grid of IP are defined as previously for the grids of BD and OP (the superscript i indicates IP).

Using these, it is a straightforward matter to obtain the relations between the Stokes parameters and the inputs I and J, provided I and J are incoherently related, as will be the case in most applications.

First, for *ideal* polarizers and a misalignment angle ψ, we have

$$S_i^2 + P_i^2 = I_s^2 + J_p^2,$$

$$S_i^2 - P_i^2 = \cos 2\psi (I_s^2 - J_p^2),$$

$$S_i P_i^* - S_i^* P_i = 0, \tag{41}$$

$$S_i P_i^* + S_i^* P_i = -\sin 2\psi (I_s^2 - J_p^2).$$

Second, for perfect alignment but imperfect polarizers:

$$S_i^2 + P_i^2 = (\mathscr{I}_s^2 + \mathscr{I}_p^2) + (\mathscr{J}_s^2 + \mathscr{J}_p^2),$$

$$S_i^2 - P_i^2 = (\mathscr{I}_s^2 - \mathscr{I}_p^2) - (\mathscr{J}_p^2 - \mathscr{J}_s^2),$$

$$S_i P_i^* - S_i^* P_i = (\mathscr{I}_s \mathscr{I}_p^* - \mathscr{I}_s^* \mathscr{I}_p) + (\mathscr{J}_s \mathscr{J}_p^* - \mathscr{J}_s^* \mathscr{J}_p)$$

$$S_i P_i^* + S_i^* P_i = (\mathscr{I}_s \mathscr{I}_p^* + \mathscr{I}_s^* \mathscr{I}_p) + (\mathscr{J}_s \mathscr{J}_p^* + \mathscr{J}_s^* \mathscr{J}_p),$$

(42)

where $\mathscr{I}_s = t_t^i I_s$ and $\mathscr{I}_p = t_t^i I_p$, $\mathscr{J}_s = r_t^i J_s$ and $\mathscr{J}_p = r_t^i J_p$. We can now relate the outputs of an overall spectrometer system to its inputs.

First, we should consider a spectrometer with ideal polarizers and perfect alignment. For this we obtain [from Eqs. (30) and (41) with $\psi = 0$]

$$C_r C_r^* + C_t C_t^* = \tfrac{1}{2}(I_s^2 + J_p^2) \pm \tfrac{1}{2}(I_s^2 - J_p^2) \cos \Delta. \quad (43)$$

Next, we take the case with ideal polarizers but angular misalignments. Taking Eq. (41) together with Eq. (35) gives the following expression for the output power (the sign alternatives referring to the two output ports):

$$C_r C_r^* + C_t C_t^* = \tfrac{1}{4}[(d_m^2 + d_f^2) \pm (d_m^2 - d_f^2) \sin 2\theta](I_s^2 + J_p^2)$$
$$- \tfrac{1}{4}[(d_m^2 - d_f^2) \pm (d_m^2 + d_f^2) \sin 2\theta]\sin 2\psi(I_s^2 - J_p^2)$$
$$\pm \tfrac{1}{2}|d_m| |d_f| \cos 2\theta(\cos 2\psi)(I_s^2 - J_p^2)\cos \Delta \quad (44)$$

This equation takes more expressive form if we use, instead of $|d_m|$ and $|d_f|$, the throughput parameters

$$\beta_m \equiv 1 - |d_m|, \qquad \beta_f \equiv 1 - |d_f|,$$

which go to zero for an ideal system, and if we drop all terms higher than second order in θ, ψ, and β:

$$C_r C_r^* + C_t C_t^* = \tfrac{1}{2}(1 - \beta_m - \beta_f + \tfrac{1}{2}\beta_m^2 + \tfrac{1}{2}\beta_f^2 \pm 2\beta_f\theta \mp 2\beta_m\theta)(I_s^2 + J_p^2)$$
$$- (\beta_f\psi - \beta_m\psi \pm 2\theta\psi)(I_s^2 - J_p^2)$$
$$\pm \tfrac{1}{2}(1 - \beta_m - \beta_f + \beta_m\beta_f - 2\theta^2 - 2\psi^2)(I_s^2 - J_p^2)\cos \Delta$$

(45)

All the first-order correction terms here involve the βs (not θ, ψ) and in practice it is likely that β will exceed θ and ψ; dropping the second-order terms gives

$$C_r C_r^* + C_t C_t^* = \tfrac{1}{2}(1 - \beta_m - \beta_f)(I_s^2 + J_p^2)$$
$$\pm \tfrac{1}{2}(1 - \beta_m - \beta_f)(I_s^2 - J_p^2)\cos \Delta. \quad (46)$$

Equation (45) is to be compared with Eq. (19), which was the basis for the discussion of the four-port character of interferometers in Section II.B. Whereas for perfect alignment the difference between the powers in the two output ports has no Δ-independent term, there are residual Δ-

independent terms in the presence of misalignment; they have, however, second-order (or higher) amplitudes. Equation (46) shows that, to first order in the corrections, the system is ideal, but for a throughput efficiency factor $(1 - \beta_m - \beta_f)$.

Turning now to the case of perfect alignment but imperfect wire grids, we have simply to combine Eq. (42) with Eq. (31). There is some simplification of the resulting expressions if it is assumed that I_s and I_p are equal and incoherently related, and similarly for J_s and J_p. This would be the case for unpolarized sources. More complicated cases can be treated with the preceding equations, but we shall take only the unpolarized case further. Equation (42) reduces to

$$S_i^2 + P_i^2 = [1 + (\tau^i)^2](t_t^i)^2 I_s^2 + [1 + (\rho^i)^2](r_r^i)^2 J_p^2,$$

$$S_i^2 - P_i^2 = [1 - (\tau^i)^2](t_t^i)^2 I_s^2 + [1 - (\rho^i)^2](r_r^i)^2 J_p^2, \qquad (47)$$

$$S_i P_i^* + S_i^* P_i = \tfrac{1}{2}|t_t^i|\tau^i I_s^2 + \tfrac{1}{2}|r_r^i|\rho^i J_p^2.$$

Substitution into Eq. (31) gives, after some reduction (the sign alternatives refer to the OP settings $\theta = 0, \tfrac{1}{2}\pi$),

$$
\begin{aligned}
A_r A_r^* + A_t A_t^* = {}& \tfrac{1}{2}(1 - \rho_r - \tau_t - \tau_t^i + \tau_r^i - \tau_t^o + \tau_r^o \pm 2\delta')I_s^2 \\
& + \tfrac{1}{2}(1 - \rho_r - \tau_t - \rho_r^i + \rho_t^i - \tau_t^o + \tau_r^o \mp 2\delta')J_p^2 \\
& \pm \tfrac{1}{2}(1 - \rho_r - \tau_t - \tau_t^i - \tau_r^i - \tau_t^o - \tau_r^o \pm 2\delta')I_s^2 \cos\Delta \\
& \mp \tfrac{1}{2}(1 - \rho_r - \tau_t - \rho_r^i - \rho_t^i - \tau_t^o - \tau_r^o \mp 2\delta')J_p^2 \cos\Delta.
\end{aligned}
$$
$$(48)$$

In reducing the expression to this form, we have defined wire-grid defect parameters (which approach zero for ideal grids) as follows (these should not be confused with the τ without subscript used previously):

$$\tau_t \equiv 1 - t_t^2, \qquad \tau_r \equiv t_r^2, \qquad \rho_t \equiv r_t^2, \qquad \rho_r \equiv 1 - r_r^2,$$

and δ' is the real part of $\delta \equiv r_t t_r / r_r t_t$. In arriving at this form for Eq. (48), terms higher than first order in the defect parameters have been neglected. The superscript i indicates a parameter for the input grid IP, superscript o indicates a parameter for the output polarizer OP, and the absence of a superscript indicates the wire grid of the beam divider BD.

To obtain the corresponding expression for the output beam that is reflected from, rather than transmitted through, OP, we should replace ρ_r^o, ρ_t^o with τ_t^o, τ_r^o, respectively.

Equation (48) is to be compared with Eq. (19), on which the discussion in Section II.B was based; Eq. (19) can be written as

$$\tfrac{1}{2}I_1 + \tfrac{1}{2}I_2 \pm \tfrac{1}{2}I_1 \cos\Delta \mp \tfrac{1}{2}I_2 \cos\Delta.$$

We can identify I_1, I_2 with I_s^2, J_p^2, respectively (I_1 and I_2 are intensities and

I_s, J_p amplitudes), but must note the introduction of coefficients that are readily interpreted because each ρ_t or τ_r represents a leakage and each ρ_r or τ_t a loss of power for a particular encounter of the relevant beam with a grid on passing through the instrument from input to output, and δ' reflects the effect of phase slippage at the beam divider.

On taking the difference of the output powers for the two settings of OP, the Δ-independent term vanishes for ideal polarizers; it persists for imperfect polarizers but is small, equal to $2\delta'(I_s^2 - J_p^2)$ (and its small magnitude is important in practice; see Sections II.C., and III.D).

Up to this point, we have been concerned with the passage of monochromatic signals through an MP interferometer, whereas the intention is to find the basis for the use of the interferometer for spectroscopy. The optical processes within the interferometer are linear and each spectral component of a signal with an arbitrary spectrum will propagate through the system independently. The signal power is finally nonlinearly detected, but the detection system will have a low-pass filter so that the detected power will be the sum of the powers of the spectral components taken separately.

We can now make explicit the connection between the analyses of this section and the theory of FTS in Section II.A. Equations (43), (45), (46), and (48) contain Δ-independent terms and terms varying as cos Δ, which correspond (with $\Delta \equiv 2\pi\sigma x$) to Eq. (1) on which the description of FTS was based. The "interferometrically modulated intensity I_0" of Eq. (1), whose spectrum is determined by Fourier transformation, can be seen to be the difference of the input powers I_s and J_p, reduced by the throughput coefficients detailed in Eqs. (45)–(48). (Note, however, the relevance of the modulation techniques described in the following section.) These coefficients have magnitudes that may well vary with spectral frequency but they are independent of the signal amplitude at a given spectral frequency and contribute to the instrumental throughput efficiency $e(\sigma)$ of Eq. (13). A spectrometric instrument is used, of course, to determine *relative* signal powers (with and without the sample) or absolute power by comparison with a standard source; in this way, $e(\sigma)$ is eliminated except in that the smaller $e(\sigma)$, the lower will be the signal-to-noise ratio. Other matters bearing on $e(\sigma)$ will be considered in Sections III.B–D.

B.   SIGNAL MODULATION AND POLARIZATION CODING

In a spectrometric instrument, it is necessary to separate the detector's response to the signal input from its response to the thermal power from the surroundings. Almost all methods of doing this are based on modulation coding of the intensity of the signal from the source followed by frequency-selective amplification or demodulation of the detector response. The frequencies of modulation should be in a range where the spectrum of any background fluctuations, and of the intrinsic detector noise, is low.

In most instruments, the modulation is achieved by interposing a chopper, that is, a rotating disk with alternately transparent and opaque sectors, at the source or at an optically formed image of the source. In that way it is possible to distinguish the signal from, in particular, the power that enters the interferometer by its second input port and the output then measures, not the difference between the signal powers in the two input ports, but the difference between the signal proper, which passes through the modulator, and the thermal signal, which is emitted by and/or reflected from the opaque sectors of the chopper. For such a system, Eqs. (13)–(48) need modification. First, J_p^2 should be equated to zero (because it is unmodulated), and then I_s^2 should be replaced by $I_s^2 - K_s^2$), where $2K_s^2$ is the intensity of the signal from the opaque sectors of the chopper. In Eqs. (19) and (20), I_2 should be taken to be this chopper signal.

With a chopper, the true signal power is detected for only one-half of the available measurement time and that necessarily reduces the attainable signal-to-noise ratio by $\sqrt{2}$. For this reason, some spectrometer systems have been developed to exploit the fact that the interferometric modulation itself leads to a variation of the signal in time that should distinguish it from the general background (though not from the signal in the second input port). The most direct approach is to displace the moving mirror at a sufficient speed to ensure that, for the spectral range under study, the resulting frequencies of interferometric modulation are above the frequencies at which the background may significantly vary; the amplifier is then provided with band-isolating filters to cover this range. This "fast-scan" method has been employed to good effect in the infrared. There are basic difficulties in using it for longer wavelengths. Detector noise and background fluctuations usually have strongly rising amplitudes with decreasing frequency below 100 Hz or so. To interferometrically modulate a submillimeter beam, say, 20 cm^{-1}, at 100 Hz requires a mirror speed of 25 mm/sec. To reach the path difference required for a resolution of 0.1 cm^{-1} would take only 2 sec. In such a time it is unlikely that an acceptable signal-to-noise ratio would be reached and a sophisticated data acquisition system, to accumulate data over a longer period, with precise superposition of successive interferograms, is required. It is not easy, moreover, to build a detector with a flat response over the range of modulation frequencies required to scan a broad spectrum.

The second method of exploiting the interferometric modulation is known as phase modulation (Chamberlain, 1971). In the normal operation of an interferometer, one mirror is fixed; in a phase-modulation system, this mirror is given a small-amplitude reciprocating movement along the direction of the incident beam so that the progressive change in path difference given by the normally moving mirror has superimposed on it a small periodic variation. The effect of this, if the detector system is tuned to the modulation frequency, is to differentiate the interferogram. There is no problem in recovering a spectrum from a differentiated interferogram except that the system's efficiency now varies with spectral frequency, being at a maximum when the wavelength is an even multiple of the amplitude of the periodic mirror movement and zero when the wavelength is an odd multiple. In the near-infrared, where the swept spectral range will usually be much less than an octave, this will be no problem, but in the submillimeter spectrum, it is usual to scan several octaves and it would then be necessary to record for at least two values of the amplitude of the periodic mirror movement. Moreover, for high-performance spectroscopy, it would be necessary to maintain constancy of the amplitude of the vibrating mirror to an extremely high precision.

Finally, there is polarization modulation that is specific to MP interferometers. The two orthogonal settings of the output polarizer of an MP interferometer generate complementary interferograms, as illustrated in Fig. 3, and by periodically switching the orientation of the polariser, the difference between them will be recorded if the detector system is tuned to the switching frequency. The unpolarized thermal background will not be modulated and will therefore be undetected.

We have used two methods of polarization modulation. The first is a vibrating reed carrying two small patches of polarizer, with mutually perpendicular polarization axes, set side-by-side so that first one and then the other is placed in the beam, at a focus, as the reed vibrates (Fig. 23). It is possible, but not easy, to design a vibrating reed with polarizers that will oscillate at a frequency above, say, 100 Hz, sweep an area of at least 100 mm^2, and give trouble-free operation for hundreds of hours. They

FIG. 23   Vibrating reed polarization modulator.

must be placed at good foci because they sweep over limited areas.

A simpler robust modulator, which gives a much larger swept area but which requires a larger grid area, is a continuously rotating polarizer; if the angular rotation rate is ω, the demodulation should be at 2ω.

We can readily determine the effects of polarization modulation for a spectrometer with ideal grids by referring to Eq. (44). For a vibrating reed modulator, we require the change in the expression in Eq. (44) on changing θ from 0 to $\frac{1}{2}\pi$ i.e.,

$$|d_m|\,|d_f|(I_s^2 - J_p^2)\cos \Delta. \tag{49}$$

Thus FTS determines the spectrum of $I_s^2 - J_p^2$, the difference in the intensities of the (unpolarized) signals entering the two input ports, where $|d_m|\,|d_f|$ is a throughput efficiency factor. If the settings of the modulator are in error by $\delta\theta$ and that of the input polarizer by $\delta\psi$, we have instead

$$\frac{1}{4}(d_m^2 - d_f^2)\delta\theta(I_s^2 + J_p^2)$$
$$- \frac{1}{4}(d_m^2 + d_f^2)\delta\theta\,\delta\psi(I_s^2 - J_p^2)$$
$$+ |d_m|\,|d_f|[1 - 2(2\delta\theta)^2 - 2(2\delta\psi)^2](I_s^2 - J_p^2)\cos \Delta. \tag{50}$$

The Δ-dependent term is reduced in amplitude, but only by second-order corrections. The Δ-independent terms now present have only second-order amplitudes, and this is just as well because, though they do not distort the spectrum, they could carry noise. For a rotating polarizer modulator, Eq. (44) shows an in-phase (cos 2θ) term

$$|d_m|\,|d_f|(I_s^2 - J_p^2)\cos \Delta \tag{51}$$

(as for the vibrating reed modulator) together with a quadrature (sin 2θ) term

$$\frac{1}{2}(d_m^2 - d_f^2)(I_s^2 + J_p^2) - \frac{1}{2}(d_m^2 + d_f^2)\sin 2\psi(I_s^2 - J_p^2). \tag{52}$$

These quadrature terms are Δ-independent and do not distort the spectrum. They could carry noise, however, and it is as well that they are small (depending on $d_m^2 - d_f^2$ or on sin 2ψ) and that they can be rejected by phase-sensitive detection. The correct phase for detection can be rather easily determined as follows. If one of the two beams in the interferometer is blocked with an absorbing plate, the d coefficient (d_f or d_m) for that beam becomes zero. The inphase term becomes zero [Eq. (51)] and the quadrature term, proportional to the full intensity ($I_s^2 + J_p^2$) is strongly enhanced. The correct phasing of the phase-sensitive detector is thus that which gives zero output when one beam in the interferometer is blocked.

Reference to Eq. (44) shows the near-equivalence of using the input polarizer as modulator rather than the output polarizer (ψ and θ appear similarly in the larger terms), provided the signal and reference sources (I and J) are unpolarized. If the input beams are wholly or partially polarized, the situation is more complicated (the symbols I_s and J_p denote components along the axes of the input polarizer). In such circumstances, it would usually be best not to use the input polarizer for modulation. Correspondingly, if the detector system should be polarization-sensitive, note must be taken of the fact that use of the output polarizer as modulator results in the output beam varying in polarization. With a vibrating reed modulator, this would result in some detection of the unwanted Δ-independent terms and some loss of amplitude of the Δ-dependent term but no distortion of the spectrum. With a rotating polarizer as modulator, it would additionally be necessary for the phase-sensitive detection system to reject well at twice the modulation frequency. It is, however, usually not difficult to make the detector insensitive to polarization: use of a light-pipe condenser achieves this for most detectors.

We have been referring here to interferometers with ideal polarizers. For imperfect polarizers, reference should be made to Eq. (48) rather than Eq. (44). The sign alternatives there refer to settings with $\theta = 0$, $\frac{1}{2}\pi$ and are directly relevant to a vibrating reed modulator. Correct phasing of the phase-sensitive detection system would be necessary when using a rotating modulator with imperfect grids.

C. Beam Spreading

In the analyses of the preceding section, it was assumed implicitly that the beams propagating through the system were perfectly plane-parallel. It is possible to form such a beam only from a point source, and with any other source there will be some beam spreading, which will lead to a decreasing depth of interferometric modulation with increasing path difference and a consequent limit to the attainable spectral resolution. Martin–Puplett interferometers will be no less subject to this limit than other two-beam interferometers. The analysis of this limitation for an incoherent source is well known (see the books cited in the introduction to Section II) and need not be repeated here; we should, however, refer to the analysis we have made (Martin and Lesurf, 1978; Lesurf, 1981) for *coherent* beams of finite cross sections because MP interferometers are particularly advantageous in the long-wave millimeter and submillimeter ranges and an assumption of spatial incoherence across a source whose size is comparable with the wavelength may be untenable.

An electromagnetic wave with a spherical wave front may propagate in free space paraxially along the z axis, with power localized in the transverse planes as a superposition of orthonormal Gaussian beam modes (see the books by Arnaud, 1976; Marcuse, 1972). The fundamental is of the form illustrated in Fig. 24. It is characterized by a Gaussian envelope for the field amplitude in each transverse plane, the width of which varies hyperbolically along the direction of propagation. The beam's wave front remains spherical as it propagates but the radius of curvature depends non-linearly on z, i.e., the center of curvature is not fixed. Correspondingly, the variation of phase angle along the z axis is not purely linear. The complex field amplitude of this fundamental mode, normalized to unit power, can be written as

$$^0\psi = \left(\frac{2}{\pi w^2}\right)^{1/2} \exp\left(-\frac{r^2}{w^2}\right) \exp\left\{-i\left[\frac{kr^2}{2R} + {}^0\phi + k(z - z_0) + \omega t\right]\right\}, \quad (53)$$

where r is the radial coordinate in the transverse plane and ω and k are, respectively, the angular frequency and wavevector of the wave ($k \equiv \omega/c$). The parameters w (the Gaussian width), R (the radius of curvature of the wave front), and $^0\phi$ (the increment to the phase angle associated with the varying center of curvature of the wave front) vary with z in the following ways:

$$w^2(z) = w_0^2(1 + \hat{z}^2), \qquad R(z) = \tfrac{1}{2}kw_0^2\hat{z}(1 + \hat{z}^{-2}), \qquad {}^0\phi(z) = \tan^{-1}\hat{z}, \quad (54)$$

where

$$\hat{z} \equiv 2(z - z_0)/kw_0^2 \quad (55)$$

and w_0 is the minimum value taken by w (the "beam waist"), which occurs at $z = z_0$. Clearly, the departures from a simple spherical wave with fixed center of curvature are significant for $\hat{z} \lesssim 1$.

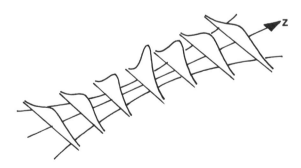

FIG. 24 In a fundamental Gaussian mode, the power distribution in the plane perpendicular to the direction of propagation z has a Gaussian form with varying width as shown.

The complete set of orthonormal modes into which any axially symmetric beam can be decomposed is

$$^p\psi = [^pL(2r^2/w^2)\exp(-i^p\phi)]^\circ\psi, \tag{56}$$

where $p = 0, 1, 2, \ldots$ is the mode number and $^pL(x)$ the pth axially symmetric Laguerre polynomial. The parameter w depends on z as indicated in Eq. (54), independently of the mode number, and $^p\phi$ is given by

$$^p\phi(z) = 2p\tan^{-1}\hat{z} \tag{57}$$

We have not yet made explicit reference to polarization. The forms of Gaussian beam modes are usually established in the literature using a scalar formulation in which the question of polarization is avoided. The first derivations by Goubau and Schwering (1961) are vectorial, however, and it is possible to use their results to determine possible states of polarization. Their first equation can be used to show that, for an axially symmetric mode (in their notation, $m \pm 1 = 0$), the radial and azimuthal components of the electric field in a transverse plane correspond to a plane and uniform polarization across the beam.

A normalized plane-polarized mode is therefore fully characterized by its values for the parameters w_0, z_0, p, and ω (or k) and we denote any such mode by $^n\psi_i$, where the subscript i implies the values w_{0i}, z_{0i}, and ω_i and the superscript n indicates the mode number.

Now, in a two-beam interferometer, we have a situation where two beams are superimposed for detection after suffering differing path delays. We have to evaluate the power in the combined beam and to do that we are involved with evaluating a coupling integral for two beams each of which may be reduced to a linear combination of Laguerre–Gaussian components but with a position for its beam waist that is displaced along the axis with respect to the other. The depth of interferometric modulation will thus be determined by coupling integrals for pairs of Laguerre–Gaussian modes, i.e.,

$$\langle n, i \mid m, j \rangle \equiv \int {}^n\psi_i{}^m\psi_j^* \, d\sigma, \tag{58}$$

where $d\sigma$ is an element of area in the transverse plane at z and the integral extends over this plane. We can obtain, for two co-axial Gaussian modes with beam waists at z_{0i} and z_{0j}, respectively,

$$\langle n, i \mid m, j \rangle = {}^{nm}C_{ij}(z_i, z_j)\exp\{-i[k(z_i - z_j) + {}^{nm}\phi_{ij}]\}, \tag{59}$$

where $z_i = z - z_{0i}$ and $z_j = z - z_{0j}$, i.e., the distances of the plane z for which the integral is evaluated from the two beam waists.

In this expression, the *amplitude* is

$$^{nm}C_{ij}(z_i, z_j) \equiv \int \frac{2}{w_i w_j} {}^n L\left(\frac{2r^2}{w_i}\right) \cdot {}^m L\left(\frac{2r^2}{w_j}\right) \exp(-Q_{ij}r^2)\, d\sigma \qquad (60)$$

in which Q_{ij} is the r-independent quantity

$$Q_{ij} = \frac{1}{w_{0i}^2(i + i\hat{z}_i)} + \frac{1}{w_{0j}^2(1 - i\hat{z}_j)} \qquad (61)$$

and the *phase angle* is

$$^{nm}\phi_{ij}(z) \equiv {}^n\phi_i(z_i) - {}^m\phi_j(z_j). \qquad (62)$$

For two *fundamental* modes, the amplitude $^{nm}C_{ij}$ reduces to

$$^{00}C_{ij} = 2/w_i w_j Q_{ij} \qquad (63)$$

and taking this together with Eqs. (61) and (62), we obtain, after some manipulation,

$$^{00}C_{ij} \exp(-i{}^{00}\phi_{ij}) = \left[\hat{w}_0 + \frac{i}{2k}\left(\frac{z_i - z_j}{w_{0i} w_{0j}}\right)\right]^{-1}, \qquad (64)$$

where $\hat{w}_0 \equiv \frac{1}{2}[(w_{0i}/w_{0j}) + (w_{0j}/w_{0i})]$, which is close to unity except when w_{0i} and w_{0j} are very different in magnitude, in which case it exceeds unity. Hence we have, for two fundamental modes, the general result

$$\langle 0, i | 0, j \rangle = \exp\{-i[k(z_i - z_j) + {}^{00}\theta_{ij}]\} \Big/ \left(\hat{w}_0^2 + \frac{1}{4}\frac{(z_i - z_j)^2}{k^2 w_{0i}^2 w_{0j}^2}\right)^{1/2} \qquad (65)$$

where $\tan {}^{00}\theta_{ij} \equiv (z_i - z_j)/2w_0^2 k$ and $\overline{w_0^2} \equiv \frac{1}{2}(w_{0i}^2 + w_{0j}^2)$.

Now, for normalized plane waves, we would have simply

$$\exp -ik(z_i - z_j), \qquad (66)$$

so the Gaussian form leads to a reduction in the amplitude of the depth of interferometric modulation which increases with the path difference. This sets a limit to the attainable resolution, which depends on the sizes of the beam waists, i.e., on the convergence of the beams; correspondingly, a specification of the required resolution would set lower limits to the cross-sectional dimensions of the beams used. For pure fundamental modes, these limits are not severe (see Lesurf, 1981), but a pure Gaussian beam corresponds to a relatively low equivalent throughput or étendue, $A\Omega \simeq \lambda^2/2$. With superimposed higher order modes (to fill out the beam cross section more uniformly), it would be necessary to evaluate the coefficients $^{nm}C_{ij}$ previously discussed. These coefficients have a close connection with those fully investigated by Kogelnik (1964), and their values can be derived in the same way. We should not take this matter further here; we have been concerned simply with indicating how the efficiency of interferometric modulation for coherent beams could be calculated for any particular case.

There is a further consideration to be taken into account, when going beyond the simple approach based on plane-parallel beams, which may assume importance at long (near-millimeter) wavelengths and in precision radiometry. This is the fact that thermal power from the environment or enclosure of the interferometer can be *diffracted* into the beams and reach the detector. Of such inputs, only those that are diffracted from the input polarizer would be detected in an MP interferometer. Their effect can be assessed directly by the following general argument. If an interferometer were completely enclosed at a uniform temperature, blackbody power would fill the system and movement of the interferometer's reflector would produce no net interferometric modulation. Now, if an aperture is opened in the enclosure to reveal an external, extended (thermal) source to the interferometer, the net interferometric modulation would be calculable by supposing that there is a source coincident with the aperture at the temperature of the background, i.e., all diffraction effects would be covered by considering the radiative antenna pattern of the aperture alone. The input deriving from any localized source in the field of view would also be diffracted at the aperture, of course, and a similar calculation would be required for that.

D.   Signal-Related Fluctuations and Noise

The signal emerging from an interferometer is a superposition of all spectral components, each coded by its interferometric modulation. This "multiplex" aspect of the FTS method leads to a signal-to-noise ratio that depends very much on whether or not the level of the dominant source of noise varies with the signal level at the detector. The well-known "multiplex advantage" (relative to dispersive methods of spectrometry) accrues if the dominant noise is independent of signal strength. It is less well-known that the multiplexing may be *dis*advantageous if the dominant noise is signal-carried or signal-related. It has been remarked in Section II.C.1 that FTS instruments that suppress the mean level of an interferogram, including MP interferometers, avoid or significantly reduce this disadvantage; the purpose of this section will be to examine this question further. We note that there will be signal-related noise if

(a)   the incident signal itself is subject to fluctuations,

(b)   there are movements or vibrations of optical components (especially reflecting components) that lead to a fluctuating misalignment of the system, or

(c)   there are variations in the detector responsivity.

These variations modulate the total intensity, represented by the mean level of the interferogram, and if there is no mean-level suppression, this is mistaken for interferometric modulation in computing the Fourier transform, giving serious distortion, or noise, in the measured spectrum.

If, when the path difference is x, the recorded signal is reduced by a factor $B(x)$ relative to that which would have been detected but for the fluctuations, the quantity that will replace $I(x) - \bar{I}$ in taking the truncated Fourier transform [Eq. (6)] will be

$$B(x_n)I(x_n) = \begin{cases} B(x_n)\{I(x_n) - \bar{I}\} + \bar{I}B(x_n) & \text{without mean-level suppression,} \\ B(x_n)\{I(x_n) - \bar{I}\} & \text{with mean-level suppression.} \end{cases}$$

The purpose of mean-level suppression is, therefore, to remove from the measured spectrum the false contribution given by the truncated Fourier transform of $\bar{I}B(x_n)$, i.e.,

$$\bar{I}B(\sigma) * C(\sigma, \sigma_m), \tag{67}$$

where $B(\sigma)$ is the Fourier transform of $B(x)$ and $C(\sigma, \sigma_m)$ the instrument resolution function [Eqs. (7) and (8)]. This false, noisy contribution to the measured spectrum can be large because it has $\bar{I}$ as a factor, which is the mean intensity over the total spectrum of the incident signal. (This would clearly be a multiplex *dis*advantage because in dispersive methods of spectrometry the multiplying factor for $B(\sigma)$ would be the intensity in *one* spectral resolution interval.) Suppression of the mean level before detection is for this reason an important requirement for high-performance FTS (Section II.C.1). It must, of course, be recognized that if the method of suppression involves a switching process fluctuations at the switching frequency will not be eliminated.

Turning now to the term that persists even when there is complete suppression of the mean level, we can use the convolution theorem on Fourier transforms to note that

$$\{I(x) - \bar{I}\}B(x) \equiv \mathcal{F}S(\sigma)\mathcal{F}B(\sigma) = \mathcal{F}[S(\sigma) * B(\sigma)] \tag{68}$$

(where $\mathcal{F}$ denotes the operation of taking the Fourier transform and the asterisk denotes the convolution) so that taking the truncated Fourier transform of the recorded interferogram yields

$$S(\sigma) * B(\sigma) * C(\sigma, \sigma_m) \tag{69}$$

rather than $S(\sigma) * C(\sigma, \sigma_m)$ [Eq. (7)]. That is to say, it is as if the spectrum of the incident signal were $S(\sigma) * B(\sigma)$ rather than $S(\sigma)$. If we write $B(\sigma) \equiv 1 - b(\sigma)$, we have

$$S(\sigma) * B(\sigma) = S(\sigma) - S(\sigma) * b(\sigma),$$

and $S(\sigma) * b(\sigma)$ can be seen to be the distortion, or noise, added to the true spectrum.

For a spectrum comprising sharp lines, for example, the consequence of this is that each line will be flanked with noise signals, which fall off away from the line, to each side, as the power spectrum of $b(\sigma)$. For high-performance FTS, the consequences of such signal-carried noise would have to be investigated carefully in each case.

There is a general consideration of importance here arising from the fact that nonlinear processes are present when there is signal-dependent noise. Transmission and reflection spectra of laboratory samples are usually obtained by taking the ratio of the spectrum obtained with the sample to that of a "background" without the sample. The background spectrum can often be recorded many times, therefore with low-noise, and stored for use in normalizing the spectra of individual samples. The noise in the resultant ratio spectrum is therefore determined mainly by that in the spectrum obtained with the sample. If an interferometer can be used to record directly the *difference* spectrum (and the MP interferometer can be so used; see Section IV.B.1), there may be a considerable improvement in the signal-to-noise ratio. This is because the difference spectrum $S_s(\sigma) - S_B(\sigma)$ will frequently be much weaker than either the sample spectrum $S_s(\sigma)$ or the background spectrum $S_B(\sigma)$. Consequently, the noise, or distortion, spectrum $\{S_s(\sigma) - S_B(\sigma)\} * b(\sigma)$ would be much smaller than $S_s(\sigma) * b(\sigma)$. The same consideration applies in astronomical applicatiions in which it is often necessary to subtract the sky emission

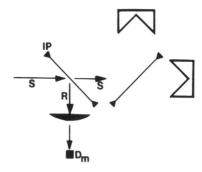

FIG. 25   The portion R of an incoming beam S that is reflected from the input polarizer IP can be detected by a detector D_m to monitor the variations in S.

when recording the spectrum of a weak astronomical source (see Section IV.B.2).

If the noise is carried into the system by fluctuations of the incident signal itself (an example of this is the fluctuating attenuation of an astronomical signal by the earth's atmosphere) and if these fluctuations are suffered equally by all spectral components of the incident signal, their effect on the form of the spectrum can be eliminated or reduced by monitoring the *total* intensity of the signal and correcting the recorded interferogram in the computer before taking the Fourier transform. A straightforward way to do this with an MP interferometer is illustrated in Fig. 25. It would be important, however, to be certain that the signal fluctuations were equally and simultaneously carried by all the spectral components, otherwise this procedure would itself distort the spectrum.

IV.   Measurement Techniques Incorporating MP Interferometers: Modular Configurations

An MP interferometer can be used not only for spectrometry of conventional kinds but also for differential (double-beam) spectrometry, absolute radiometry, and polarimetry. It is possible, moreover, to devise a basic set of modular units that allows the several kinds of measurements to be made by assembling the modules in appropriate configurations. This approach is described in this section with particular reference to the systems developed in our laboratory—Fig. 26 shows a modular system configured for radiometric measurements.

A.   MODULAR APPROACH

Figure 27 lists the several modular units required and shows the symbol for each used in the subsequent figures to illustrate the configurations. Each module is a rectangular box with an optical unit mounted within it and with a circular aperture in one or more of its faces. When incorporated in a configuration, each box abuts face to face with its neighbors and fixing bolts, in standardized locations in the abutting faces, give a strong, well-aligned, structural frame. The modules can be readily reassembled in a new configuration with immediate alignment of the apertures. The optical units can then be adjusted for precise alignment through removable top covers. The boxes are designed to support a vacuum if required and O-rings round the apertures provide vacuum seals between the boxes.

1.   The Beam-Divider Module

This module is a wire-grid polarizer stretched over a support ring, which can be adjusted so that it is symmetrically placed with respect to

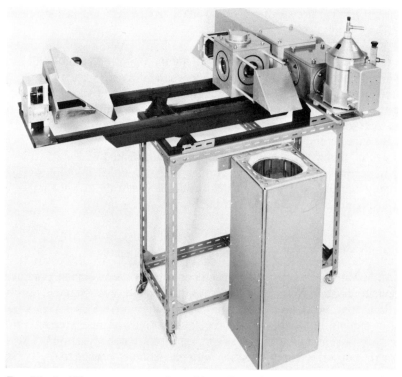

FIG. 26 An MP spectrometer configured for radiometric measurements of atmospheric emission. The glass Dewar in the right foreground contains a blackbody cavity for calibration, cooled by liquid nitrogen; a second blackbody cavity at another temperature would be placed in the corresponding position behind the radiometer. The mirror on the left is set to select a beam at the required elevation and to direct it into the instrument. The helium cryostat containing the detector is on the right.

the four apertures in the vertical faces of the cubical box in which it is mounted, i.e., at 45° to the beam axes (Fig. 28).

2. The Fixed-Reflector Module

This module is a rooftop reflector on a universal mount set normally to the axis of the single aperture of the box in which it is mounted (Fig. 28).

3. The Movable-Reflector Module

This module has a rooftop reflector mounted on a carriage which moves along guide rods (Fig. 28). The carriage is driven by a nut running on a precision screw thread, which is turned by an integral motor and shaft

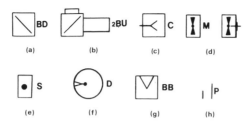

FIG. 27 Symbols used to represent the following modular units in Figs. 33–34: (a) beam divider; (b) two-beam unit; (c) condenser/collimator; (d) modulator, and with light pipe; (e) source; (f) detector; (g) blackbody; (h) polarizers.

encoder. The reflector has its optical axis parallel to the movement. This unit is an exception to the rule just given in that it is supported, not by the enclosing box, but from an endplate. This plate, which has an aperture for the beam, abuts to the face of the beam-divider module to which the unit is invariably to be coupled. The enclosing box, without an end face, then slides without contact over the reflector unit and abuts, in the usual way, with the same face of the beam-divider module.

4. The Two-Beam Module

The two-beam module, which is at the center of every configuration, is composed of a beam-divider module, a fixed-reflector module, and a movable-reflector module (Fig. 28).

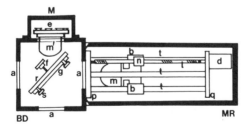

FIG. 28 Two-beam module comprising the fixed-reflector module M (containing the fixed rooftop reflector m' on a universal adjustable mount e), the movable-reflector module MR (containing the rooftop reflector m, which is supported by bearings b on three guide rods t and moves along the rods when driven by the nut n on the precision screw thread l which is turned by the stepping motor d; the end plates p and q complete the frame, which is supported by p at the face of the beam-divider module), and the beam-divider module BD (which contains the grid, mounted on a ring g and stretched over the support ring r, which is on a three-point mount f; three bolts s hold the divider ring over the support ring). The beams enter and leave the module by the apertures a.

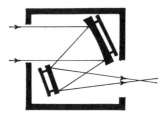

FIG. 29   Condenser/collimator module with mirrors.

5.   The Condenser/Collimator Module

This has apertures in two of the vertical faces of the enclosing box and mirrors or lenses arranged so that a collimated beam entering one aperture is brought to a focus just outside the other (Figs. 29 and 30).

6.   The Modulator Module

This module contains a polarization modulator, either a rotating polarizing disk (Fig. 31b) or a vibrating reed (Fig. 31a). The modulator is normally to be near a focus to keep it as small as possible; it is therefore set inside the modulator box at a distance from the aperture equal to that between the (external) focus provided by the condenser/collimator unit and its aperture. After passing through the modulator, the beam may diverge, into a collimator module, say, or it may be picked up by a light-pipe centered in the exit aperture (Fig. 32).

7.   The Source Module

This module contains a water-cooled, stabilized, high-pressure mercury arc lamp at such a position, on the axis of the aperture in the enclosing box, as would place it at the focus of a neighboring collimator module.

8.   The Calibration-Source Blackbody Module

This module contains a blackbody cavity or surface at a precisely known temperature and is used for radiometric calibration. We have used

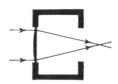

FIG. 30   Condenser/collimator module with lens.

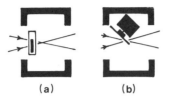

FIG. 31   Modulator module with (a) vibrating reed or (b) rotating polarizing disk.

three types of blackbody. The simplest is a sheet of Eccosorb ES-AN72 (available from Emerson and Cuming, Inc.) attached to the inside wall of an expanded polystyrene box filled with liquid air or nitrogen, or iced water. This box can be slid into a standard modular box with its top removed. The signal from the source is admitted to the interferometer through the $\sim$ 20-mm-thick expanded polystyrene wall to which the Eccosorb is attached. Below 25 cm^{-1}, the absorption and scattering in the expanded polystyrene are low (a few percent) and a correction to allow for this can be made. For frequencies above 25 cm^{-1}, the correction required is too large for the purpose of absolute calibration, but we have made frequent use of such a source for general test and measurement purposes up to 50 cm^{-1} because it is simple and fills the spectrometer's beam without collimator or alignment.

We have also used a calibration-source module in which a mirror is used to direct the beam downward into an open-top Dewar vessel containing refrigerant and an Eccosorb cone as absorber. A small correction to the effective temperature to allow for reflection at the free surface of the liquid nitrogen or air is required (about 5 °K). Absorption by water vapor in the air above the refrigerant restricts the useful spectral range to frequencies below about 25 cm^{-1}.

We have also constructed a liquid-nitrogen cryostat with a blackbody cavity in its vacuum space and a Mylar side window (25 μm thick), for which a very small reflection correction is required. Whereas Eccosorb can be used as an absorber when immersed in a liquid refrigerant, its use in a vacuum space would not be reliable because its temperature would be nonuniform and not accurately measurable. We have used a metallic cav-

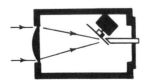

FIG. 32   Modulator module with light-pipe.

ity with a very thin layer of dielectric and resistance-matched metal film on the inner surfaces such that interference effects enhance the absorption in the film (Carli, 1974). The cavity is carefully designed to fill the beam of the interferometer. This calibration source can be used throughout the millimeter and submillimeter spectrum.

9. *The Detector Module*

We have used germanium and InSb bolometers in compact liquid-helium cryostats as modular units. The cryostat, with a small (25-mm) window, abuts to a condenser module or to the light-pipe output from a modulator module. A Golay or pyroelectric detector can be incorporated in a modular box with lens, cone, or mirror condenser and used instead of a cooled detector for many applications.

10. *Wire-Grid Polarizers*

In addition to the modular units, it is useful to have wire-grid polarizers, which may be inserted into an aperture of a module or at the focus of a condenser/collimator unit.

B. Configurations for Basic Spectroscopic Measurements

For laboratory measurements on materials under a wide variety of conditions, it is necessary to have suitable modules to house the samples and introduce them into the configuration. Such a module should pick up the beam at an input focus, direct it to interact appropriately with the sample, and finally recondense it to an output focus at which the next module in the configuration will accept it. An important advantage of MP interferometry is the variety of dispositions of the sampling and modulating functions it allows (Fig. 33). This flexibility of choice springs from the fact that the beam is polarization-coded at the input and decoded at the output so that, provided there is no *polarized* emission from any part of the system between the input and output polarizers, only power that has passed through *both* polarizers will be detected. This is in contrast with the situation in which a sample and its mounting are placed before a Michelson interferometer with a chopper (a shutter that periodically blocks the beam) after it, before the detector. This would result in the thermal emission from the sample and its environment being modulated and detected. If the sample were at a raised or reduced temperature (relative to that of the chopper) the contamination of the true transmitted or reflected signal would be large.

The output beam from an MP interferometer is polarized, and if the polarization modulator is at the output end, the output polarization alter-

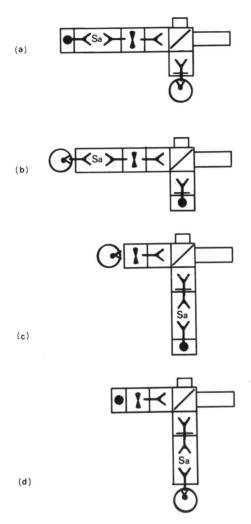

FIG. 33 Alternative configurations (see Fig. 27 for an explanation of the symbols). Sa denotes a module containing the sample under study. (a) Fore sampling and fore modulation; (b) aft sampling and aft modulation; (c) fore sampling and aft modulation; (d) aft sampling and fore modulation.

nates. This has to be borne in mind when deciding on a configuration. Fore modulation with aft sampling (Fig. 33d) gives a fixed polarization at the sample, and the orthogonal polarization is straightforwardly obtained by rotating the output polarizer through 90°. This is usually the most convenient configuration for a wide variety of transmission and reflection measurements on gases, liquids, and solids. The configuration involving aft modulation with aft sampling would give a periodically varying direction of polarization of the beam through the sample unless steps were taken to suppress the polarization, such as passing the beam along a length of light-pipe with circular cross section between modulator and sample. For emission studies, one of the fore sampling configurations of Fig. 33 would be used, of course. For a remote source (e.g., astronomical or atmospheric), aft modulation allows direct input of the already collimated incident beam to the two-beam unit. Provided consideration is given to the matter of polarization, there is considerable flexibility of configuration choice, and one should seek the most convenient disposition for each experiment, in which the sample could be at low or high temperature, in a magnetic field, at high pressure, etc., or might need to be long (gaseous samples), easily changed for another, and so on.

C. DOUBLE-BEAM MEASUREMENTS (DIFFERENCE SPECTRA; SKY SUBTRACTION)

In a conventional *single*-beam determination of the attenuation on transmission through, or reflection from, a sample, a measurement is made of the source's power transmitted to the detector, first in the absence and then in the presence of the sample in the beam. If the fractional reduction in the beam intensity attributable to the sample is $\mu(\nu)$, the two recorded spectra are, respectively, $I_0(\nu)$ and $I_0(\nu)\mu(\nu)$ and the ratio gives $\mu(\nu)$. A *double*-beam determination of $\mu(\nu)$ is possible when a configuration can be found that permits the direct measurement of the difference $I_0(\nu) - I_0(\nu)\mu(\nu)$ followed by measurement of $I_0(\nu)$. Such configurations are possible with interferometers that give access to both input ports. For both single- and double-beam cases, two spectra must be recorded, but there is an advantage in the double-beam method in that the consequences of changes in the source intensity $I_0(\nu)$ are less than for the single-beam case. If $I_0(\nu)$ remains constant during the recording of each interferogram but changes by an (unknown) fractional amount δ between the two, the fractional error incurred by neglecting the change when calculating the value of $\mu(\nu)$ is $\delta[1 - \mu(\nu)]$ for the single-beam method and $\delta\mu(\nu)$ for the double-beam. For small $\mu(\nu)$, the error is thus much reduced by double-beaming. Moreover, double-beaming brings advantages also when the source intensity changes *during* the recording of an interferogram (see Section III.D).

Similar considerations apply when the interest in a sample centers on the measurement of the spectrum of one component in it, e.g., an impurity or contaminant. If a sample of the host material, without the component of interest, is available, there is advantage in recording directly the difference spectrum of sample and host and then the spectrum of host alone.

A third situation in which similar considerations apply arises in astronomy. In a submillimeter-wave astronomical observation, it is necessary to separate the signal that derives from the astronomical source from that (usually much larger) due to emission from the atmosphere in the line of sight of the source. The usual basis for making the separation is the assumption that the signal due to the atmosphere will not change when the telescope is directed slightly away from the astronomical source. "Sky subtraction" is given by recording the difference between the signals so recorded.

Most ways of recording difference spectra or sky subtraction data have involved an oscillating mirror to switch the beam received by the detector between sample and reference channels. The two input ports of an interferometer, in particular of an MP interferometer, allow a *static* method that should be much less prone to mechanical stability and microphonic problems. This is illustrated in Figs. 34 and 35. An input system is required to overlay the beam deriving from the first channel with an orthogonally polarized beam deriving from the second; the interferometer then directly measures the difference. The system in Fig. 34 is used for two-beam measurements on gases. Figure 35 shows two systems for sky subtraction: that in Fig. 35a being based on standard modules; that in Fig. 35b is more compact.

D. RADIOMETRIC MEASUREMENTS AND ABSOLUTE CALIBRATION

Any two-beam interferometer records the difference of the power spectra of the beams entering its two input ports. If precise measurements

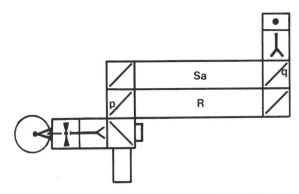

FIG. 34 Configuration that allows the direct recording of the difference spectrum of a gas (in Sa) and reference (in R). The components p and q are wire grids.

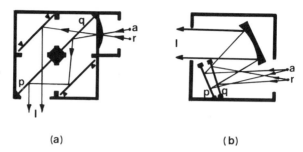

(a)   (b)

Fig. 35 Input systems that produce a collimated beam *I* in which are overlaid orthogonally polarized beams from the images of an astronomical source a and of a neighboring sky point r. The components p and q are wire grids. (a) Based on standard modules; (b) based on more compact units.

of the signal entering the first input port are to be made, it is essential that the signal entering the second port at least be under control and constant. The necessity for this control has often been overlooked. Moreover, advantage can be taken of the second port for absolute radiometric calibration. The MP interferometer is particularly well suited to precision absolute radiometric measurements because both input ports are readily accessible.

The simplest procedure is to have in the reference port a blackbody cavity of known and steady temperature and to calibrate the system by placing a second blackbody cavity of known but different temperature in the signal port. A system that may be more readily brought under automatic control is illustrated in Fig. 36 (see also Fig. 26). The two reference blackbody cavities are at different temperatures and fill the collection aperture of the spectrometer. The input polarizer has alternative settings: in the first, it reflects power from one cavity into the spectrometer; in the second, power from the other. A signal from the source under study

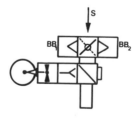

Fig. 36 A radiometer, where BB_1 and BB_2 are blackbody sources at different temperatures. The signal S enters by transmission through the input polarizer, which has alternative settings (solid and dashed lines).

enters the spectrometer by transmission through the input polarizer in both settings. When the polarizer is in the first setting, therefore, the spectrometer records the difference of the signals from the source and from the first reference cavity; for the second setting, it measures the difference between the source signal and that from the second cavity. Spectra are recorded for both settings of the polarizer.

If the source is extended and fills the aperture of the spectrometer, its absolute brightness can be determined by interpolation between the known brightness of the blackbody cavities. For the first setting, the power spectrum recorded is

$$S_1(\sigma) = R(\sigma)[A(\sigma) - B(\sigma, T_1)],$$

and for the second setting, it is

$$S_2(\sigma) = R(\sigma)[A(\sigma) - B(\sigma, T_2)],$$

where $R(\sigma)$ is the optical throughput efficiency of the system as a function of wave number σ, $A(\sigma)$ the brightness of the source, and $B(\sigma, T_1)$, $B(\sigma, T_2)$ those of blackbodies at temperatures T_1 and T_2. Therefore $A(\sigma)$ can be determined absolutely from

$$A(\sigma) = \left(\frac{S_2(\sigma)}{S_2(\sigma) - S_1(\sigma)}\right) B(\sigma, T_1) - \left(\frac{S_1(\sigma)}{S_2(\sigma) - S_1(\sigma)}\right) B(\sigma, T_2).$$

If the brightness of the source is expressed in terms of an equivalent, frequency-dependent, temperature $T(\sigma)$, then

$$T(\sigma) = \left(\frac{S_2(\sigma)}{S_2(\sigma) - S_1(\sigma)}\right) T_1 - \left(\frac{S_1(\sigma)}{S_2(\sigma) - S_1(\sigma)}\right) T_2.$$

The use of collection optics between source and spectrometer would not change this approach so long as power from the source uniformly fills, or overfills, the collection aperture of the spectrometer and corrections are made for loss at, and emission from, the mirrors. If it cannot be arranged that signal power from the source fill the aperture, the simplicity of this procedure as a method of making absolute measurements is reduced. Steps would have to be taken to determine the ratio of the collection apertures for the source and for the calibration cavities and to correct for any signal deriving from the extended background to the source.

When working at millimeter wavelengths, the effects of diffraction may not be negligible in precision measurements; that is to say, with aperture diameters no more than 50–100 times the wavelength, the existence of the subsidiary lobes of the antenna pattern may have to be recognized and care exercised in reducing them, correcting for them, or filling them with the source.

E. Polarimetric Measurements

In Section III.A.2 we made explicit the fact that the two-beam unit of an MP interferometer is essentially a polarimeter because it allows one to measure Stokes parameters of an input beam (Fig. 37). For a strong monochromatic input beam, three of the four parameters can be measured, viz., $S_i^2 + P_i^2$, $S_i^2 - P_i^2$, and $S_i P_i \cos \theta$, where S_i and P_i are the amplitudes of the orthogonal plane-polarized components of the input beam and θ the phase angle between them. (The fourth parameter, $S_i P_i \sin \theta$, would be needed if the quadrant of θ were required, but the simple MP interferometer will not give this.) To determine these parameters, it is simply necessary to measure cosine and sine components of the sinusoidal interferogram and the mean level on which it is superposed.

If the input beam is not monochromatic, the spectra of the parameters $S_i^2 - P_i^2$ and $S_i P_i \cos \theta$ can be determined by cosine and sine Fourier transformations of the interferogram, but the spectrum of the intensity $S_i^2 + P_i^2$ cannot be extracted from the dc level. To find the spectrum of $S_i^2 + P_i^2$, the spectra of S_i^2 and P_i^2 can be separately measured by introducing an input polarizer, appropriately orientated for each case. The same measurements could also be used to give $S_i^2 - P_i^2$, of course, but this difference will be more precisely determined by the direct method (compare the discussion of double-beam measurements in Section IV.C). In order to separate reliably the cosine and sine transforms, it would be necessary to know the point of zero path difference with precision and also any phase shifts in the optical and electronic systems; these would be determined when using the input polarizer to measure S_i^2 and P_i^2.

F. Measurement of Complex Refractive Index Spectra

Two-beam interferometry can be used to measure the complex refractive index spectra of liquids and solid materials available as large homogeneous samples (see Birch and Parker, 1979; Birch, 1978). Interferograms are recorded with and without a sample placed in one of the two beams of the two-beam unit and phase and amplitude spectra are determined from complex Fourier transforms of the interferograms; from these the spectra of the real and imaginary parts of the refractive index are computed. An

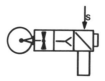

Fig. 37　A polarimeter (i.e., spectrometer without input polarizer).

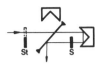

FIG. 38   A refractometer, where S is the sample and St a stop having alternative positions (solid and dashed lines).

MP interferometer can be used for this purpose. A transmitting sample in one arm of an MP interferometer passes the beam twice, once before and once after reflection from the roof-edge reflector. The plane of polarization rotates by 90° between the two passes. A single pass is obtained if the sample is placed in one-half of the beam only, with half of the input beam blocked, as illustrated in Fig. 38. By moving the stop into the other half of the input beam, measurements can be made for the orthogonal polarization.

For liquids or highly absorbing solids, a thin sample is usually supported on the surface of the plane reflector of a Michelson interferometer, or a thicker sample is used in lieu of the reflector. This can be done with an MP interferometer, provided that 45° incidence, rather than normal incidence, is acceptable. If normal incidence were necessary, the simpler configuration of Fig. 11 could be used, albeit with loss of mean-level suppression in the interferogram.

G. MEASUREMENTS OF OPTICAL TRANSFER FUNCTIONS

A major application of interferometry in the optical spectrum is the testing of optical components; in particular, the determination of optical transfer functions. The potential of an MP interferometer for this purpose has been studied by Lambert and Richards (1978).

H. TWO-BEAM INTERFEROMETERS AS DIPLEXERS

The use of an MP interferometer as a high-efficiency coupler to overlay the signal and local oscillator beams in a millimeter-wave heterodyne receiver has been reported by Wrixon and Kelly (1978), Martin and Lesurf (1978), and Lesurf (1981).

V.   Construction of Special Components for MP Interferometers

In this section we shall consider aspects of the construction of components peculiar to the MP technique of interferometry. We shall not deal with components that differ little from the corresponding components of Michelson interferometers.

A. Polarizing Grids

Theoretical treatments of the polarizing properties of conducting wire grids (Section II.C.3) give a very high efficiency for wavelengths exceeding four times the center-to-center spacing of the wires. The wires' cross-sectional shape, the mark-to-space ratio, and the ohmic dissipation have secondary effects in this range. It would be expected, however, that nonuniformity of the wire spacing and, especially, lack of flatness of a grid could lead, in practice, to some degradation of performance, perhaps particularly in reflection. It is therefore necessary to find ways of winding and mounting fine wire grids in such a way that they be flat to better than $\lambda/4$ for the shortest wavelength of concern over an area of at least 80-mm diameter. We found it possible fairly straightforwardly to wind free-standing grids of fine tungsten wire with spacing of 10 μm or more and to flatten them by stretching, as a drum skin, over a flat-ground ring support. A specially adapted coil winder is used to wind the 5- or 10-μm tungsten wire onto a rectangular plate, which is recessed to carry the stainless steel ring frame to which the wires are to be fixed. When the grid has been wound, a solvent adhesive is painted over the faces of the ring frame to secure the wires. The wires are then cut around the outer rim of the ring to release it from the rectangular frame, a second ring is laid over the first, and the two are bolted together to form a sandwich with the grid and glue as filling. To secure the wires, cold-setting epoxy resin may be used to seal the cut ends of the wires protruding from the outer edges of the rings. We have also used gold-coated tungsten wire, fixed the grid to a gold-plated ring with colloidally applied solder, and heated it in an oven. Grids with a spacing of 25 μm on ring frames with diameters up to 160 mm have been wound in this way. They can be stretched over support rings without spoiling. The methods for winding grids of this type have been described by Costley et al. (1977b) and Ade et al. (1979).[1]

Performance data for such grids are shown in Fig. 39, where R_p and T_p denote the fractional specular power reflection and transmission, respectively, for the plane of polarization with E field parallel to the wires, and R_t and T_t those for the transverse polarization.

In view of the difficulties in achieving a flatness ~ 5 μm or better with an unsupported grid, methods for producing a grid photolithographically in an evaporated film of aluminium on a 2-μm-thick sheet of Mylar have been explored. Such grids can be obtained commercially with diameters up to 30 mm. We, and others (Challener et al., 1980), have made larger

---

[1] Tungsten wire grids are offered by Analytical Accessories Ltd. (SPECAC), Orpington, Kent, United Kingdom.

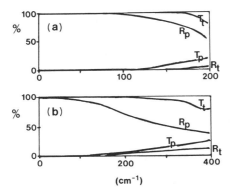

FIG. 39   Measured reflection and transmission coefficients for parallel and transversely polarized beams incident normally on tungsten wire grids with a spacing of 25 μm and wire diameter of 10 μm (a) and a spacing of 12.5 μm with wire diameter of 5 μm (b). (See Ade *et al.*, 1979.)

grids in this way; it is difficult to ensure the necessary uniformity over the film of aluminium thickness, photoresist thickness, UV exposure, development rate, and etching rate, but grids with good efficiencies up to, say, 200 cm^{-1} can be made.

The polarizing grids are required, of course, not only for beam dividers and input and output polarizers but also for the polarization modulators (Section V.D).

B. ROOFTOP REFLECTORS

Each of the two beams in the basic MP interferometer encounters a rooftop reflector (Fig. 40), the main purpose of which is to produce a rotation of the plane of polarization of the reflected beam by 90° relative to that of the incident beam (Section II.C.4). The rotation function does not impose constraints on the angle between the faces of the reflector, but the need to overlay the interferometer's two beams, so as to maximize the in-

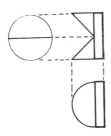

FIG. 40   Rooftop reflector.

terferometric modulation, does. How closely equal must the roof angles of the two rooftop reflectors be and how nearly must the angle be to 90°? The angle through which a beam is reflected by a rooftop reflector is $\pi - 2\alpha$, where ($\frac{1}{2}\pi + \alpha$) is the roof angle, if it is incident in the plane perpendicular to the roof line. Each beam in the interferometer is made up of two half-beams, as illustrated in Fig. 41, and these half-beams are separated by an angle 4α after reflection. If α_f and α_m (for the fixed and moving reflectors, respectively) are not zero, we have four half-beams, two from each reflector, propagating in slightly different directions through the interferometer, even when it is optimally aligned. Each produces an Airy ring pattern in the focal plane of the condenser, i.e., at the detector, and the central spot of each subtends an angle at the condenser equal approximately to λ/D, where D is the beamwidth. To have full interferometric modulation between all four half-beams, the roof-mirror angles must be such that the Airy ring patterns coincide within their central spots, i.e., the roof-mirror angles must be well within $\lambda/4D$ of each other and well within $\lambda/4D$ of 90°,

$$\alpha_f, \ \alpha_m \ll \lambda/4D.$$

For example, for $\lambda = 1$ mm and a beamwidth of 100 mm,

$$\alpha_f, \ \alpha_m \ll 0.003 \quad \text{rad}.$$

Thus, for an interferometer to work in the millimeter and submillimeter range (up to 100 cm^{-1}), we need roof reflectors whose roof angles are within about 1 arcmin of 90°. Such an accuracy is not difficult to achieve with conventional methods of glass working.

A similar precision is required for the relative alignment of the two roof-

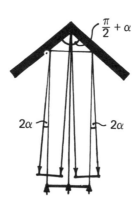

Fig. 41   Half-beams with roof angle $\pi/2 + \alpha$.

top reflectors with respect to rotation about the transverse axes perpendicular to the roof-edges, and that must be provided by the reflectors' mounts (perhaps by a coarse adjustment for the moving reflector and a fine adjustment for the fixed).

It is not usually necessary, however, to provide high precision rotational adjustments around the other, orthogonal, axes nor high precision transverse lateral adjustments. This is because the coherence length across the beam will usually be many wavelengths (achieved by restricting the size of source and/or detector) in order that a good spectral resolution be possible. In fact, if the beam-spread angle due to source or detector size is δ, the coherence length across the beam will be on the order of λ/δ. The maximum resolving power R is also determined by δ, being equal to $8/\delta^2$, and the coherence length is, therefore, of order $\lambda\sqrt{R}/3$. For a system capable of a resolving power of, say, 10^3, the coherence length would be about 10λ. Rotational and lateral adjustments to superpose the beams to within 10λ accuracy are therefore required. The procedures for alignment of an MP interferometer will be dealt with in Section VI.A.

C. Polarization Modulators

Polarization coding, or modulation, was discussed in Section III.C. We have used polarization modulators of three types. The simplest is that illustrated in Fig. 42b. A tungsten wire grid is wound (Section V.A) onto a light-weight circular rim with four thin spokes and a central boss, and this polarizing disk is mounted on the shaft of a suitable electric motor. The unit is placed at a focus of the beam, which is polarization-modulated by transmission through the rotating grid; it is set obliquely to the beam so that power reflected from the grid into the ongoing beam comes from a part of the system that can be kept stable and uniform in temperature (wobbling motion of the rotating disk could otherwise introduce noise, which can be significant in a high-performance radiometric system). Beams having widths up to 30 mm can be straightforwardly polarization-modulated in this way at frequencies up to 120 Hz. As with chopper modulators, an electrical signal synchronous with the modulation to drive the phase-sensitive detector (psd) in the signal amplifiction channel is provided by a solid-state lamp and photodiode unit. This can be mounted so that light is reflected to the diode from the rotating boss, alternate quadrants of which are painted black.

The unwanted modulation of the beam by the spokes is small if the spokes are thin, and its consequences are suppressed by the psd because, with four spokes, it occurs at a higher frequency than the polarization modulation and because the spokes are so oriented with respect to the

FIG. 42   Polarization modulators. (See the discussion in the text.)

grid wires that it will be in antiphase with the relevant harmonic of the modulation.

A more compact, spokeless modulator of the rotating type can be constructed using a torque motor. Motors of this type are expensive, but it would be possible in this way to obtain higher modulation frequencies such as might be required for transient or fluctuating signals.

A vibrating-reed modulator, which gives nearly true orthogonal polarization modulation, is shown in Fig. 42a. A light-weight frame comprising two circular apertures, each spanned by a wire-grid polarizer, moves by twisting a phosphor-bronze strip to which it is attached. The modulator operates at the natural resonant frequency of this twist motion. The moving arm carries two small permanent magnets, each near a coil on the base to provide a drive to maintain the movement and a sensor for a servo-loop control of the amplitude. The same circuit provides the signal for the reference channel for the psd. A beamwidth up to 10-mm diameter can be modulated in this way at frequencies up to 80 Hz, and because the grids required are of small area, high-quality photolithographic grids can be used to give good polarization modulation for spectral frequencies up to at least 500 cm^{-1}. The movement is resonant and the power dissipation is consequently very low.

D. LENSES AND MIRRORS

We have used two types of condenser and collimator, the first based on mirrors and the second on lenses (Figs. 29 and 30). Suitable mirrors are widely available but lenses are not, and for this reason we should give some information about their fabrication. The material used is polyethylene of high molecular weight and high density in the form of extruded rods of 80-mm diameter. This material machines well and has low spectral absorption losses in the range below 40 cm^{-1} (Birch *et al.*, 1981). At 10 cm^{-1}, the absorption coefficient is less than 0.03 cm^{-1} and increases monotonically to 0.3 cm^{-1} at 38 cm^{-1}; a slab of 1-cm thickness will attenuate a normally incident beam by less than 3% at 10 cm^{-1} and by 26% at 38 cm^{-1} by absorption, and by about 4% at each surface by reflection. The aperture ratios required for the lenses are usually large and this makes aspherical surfaces necessary. For a focal length of 125 mm and $f/1.5$, the required thickness at the center of the lens is about 15 mm. Such lenses are cut from blanks by an appropriate lathe tool guided by a shaped form.

VI. Alignment of MP Interferometers

We shall not dwell on the general procedures used to align any two-beam interferometer, but there are several additional steps peculiar to MP interferometers and these we shall deal with here.

Most of the alignment of an MP interferometer is best done with a Mylar film rather than a wire grid stretched over the beam-divider support ring. The two-beam unit, i.e., the beam-divider module together with the fixed- and movable-reflector modules, is taken first so that the rooftop reflectors can be viewed from the entrance and exit ports, both directly and by reflection from the Mylar film.

The movable reflector is brought approximately to the position of zero path difference and is set with its roof edge in whatever orientation is preferred (usually perpendicular to, at 45° to, or in the plane of incidence, but other directions could be chosen, perhaps so as to provide a particular polarization of the beam passing through the sample). The fixed reflector is then rotated around its axis until its roof edge is parallel to that of the movable reflector as seen by reflection in the Mylar film to within about 1°. One or other of the well-known procedures for aligning a Michelson interferometer should then be followed in order to bring into parallelism the line of movement of the movable reflector, its optical axis, and the optical axis of the fixed reflector as seen by reflection in the Mylar film. Because the roof reflectors' included angles are 90°, it is not necessary to make the rotational adjustments about the axes parallel to the roof edges with precision. The precision required for the rotational settings around the axes perpendicular to both the roof edges and optical axes is determined by the considerations given in Section V.B.

When this has been done and with the movable reflector approximately in the position of zero path difference, the fixed reflector is brought, by displacement transverse to the optical axis, into coincidence with the image of the movable reflector seen by reflection in the Mylar film (and if a small rotational adjustment around the optical axis is also required to reach good coincidence, it should be made). Superposition of the reflectors' images should be within about 10λ if a resolving power of 10^3 is required and within λ if source and detector sizes as large as would be allowed if a resolving power of 10 is sufficient are used (Section V.B).

At this point, the Mylar film is to be removed and replaced by the wire-grid polarizer, but it is first necessary to determine the correct orientation for the wires. A simple procedure for this is to replace the Mylar film with another on which a clear line, through its center, has been penned. When this line is viewed from the output port, it will be seen against its images (coincident) in the roof reflectors. The bolts that hold the rim on which the film is mounted over the support ring are loosened and the rim is rotated until the penned line is perpendicular to its images. The beam-divider support ring is then marked at the point where the line reaches it. The Mylar beam divider is removed and replaced with the wire-grid set so that the wires passing through its center reach the support ring at the point previously marked on it. The wire grid can be removed and replaced without going through the alignment procedure again.

The input or output polarizer is mounted next, in the two-beam unit, and the wires set parallel to or perpendicular to the line of the roof edges.

The rest of the required configuration is then assembled and aligned in straightforward ways. There is one further adjustment peculiar to MP instruments, i.e., setting the phase of the phase-sensitive detector in the detector's amplifier train, as described in Section III.B. With a bright source in the entrance aperture, one beam is stopped inside the two-beam unit; the phase shift in the reference channel from the polarization modulator is then adjusted to give zero output from the psd.

An interferometer that has been aligned by these procedures should satisfy two criteria relating to the form of the interferogram. First, the interferogram for an unpolarized source should be symmetrical about the zero path-difference position. If it is not, the optical alignment procedures must be repeated; an asymmetric interferogram is not attributable to incorrect orientations of the polarizers [Eq. (44)]. Second, the output signal should be near zero for large path differences, provided the source has no sharp spectral features and the spectral band is limited by filters to wavelengths at which the polarizers are efficient. If it is not and if the interferogram *is* symmetrical, the fault may lie in incorrect orientations of the wire grids or roof reflectors, and these should be checked, or in an inadequate superposition of the roof reflectors, which reduces the depth of interferometric modulation, or in an optical obstruction that causes more power to be admitted through one beam than through the other. Removal of the input (or output) polarizer should, of course, lead to zero recorded signal at all positions of the moving reflector; failure to find this would indicate some polarization of the source (Section III.A.2).

VII. Measurements Made with MP Interferometers and Performance Tests

The special advantages of Martin–Puplett interferometers have been recognized in several laboratories, and such instruments are now in use in a number of near- and submillimeter-wave measurement programs. These include studies of solids and liquids in transmission and reflection, radiometric measurements of emission from the stratosphere, from high-temperature (thermonuclear) plasmas, and from the cosmic-background. Some examples of spectra obtained with MP systems will be given in this section and an assessment will be made of performances relative to expectation. The instruments employed are those in the authors' laboratory, if not otherwise indicated.

A. LABORATORY TESTS

Figures 43 and 44 illustrate the low-frequency performance of an MP instrument in a configuration involving a mercury-arc lamp as source, a liquid-helium-cooled InSb detector, and a rotating polarization modulator. The main dimensions of the modules in this instrument are similar to those of the modules in Fig. 26.

Figure 43 shows the background spectrum at a linearly apodized resolution of 0.05 cm^{-1} with the instrument evacuated. Above 30 cm^{-1}, the signal level falls away as a result of the decreasing detectivity of the InSb detector. The decreasing signal level to low frequencies is due to decreasing brightness of the source. Most of the structure in this spectrum is due to interference effects in the detector or the cryostat windows, but the several sharp dips where water vapor has strong absorption lines betray residual traces of water vapor in the instrument and confirm the resolution. The spectrum covers more than two octaves, but there are no regions of low efficiency such as would be given by a dielectric beam divider.

The best way to determine the signal-to-noise ratio to be expected in a

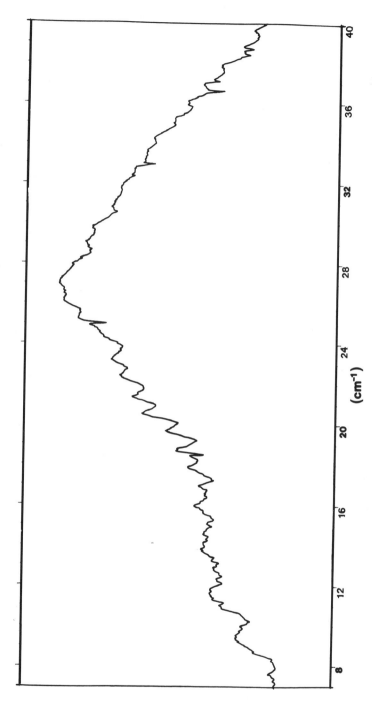

Fig. 43 Background spectrum at 0.05-cm⁻¹ resolution.

(cm⁻¹)

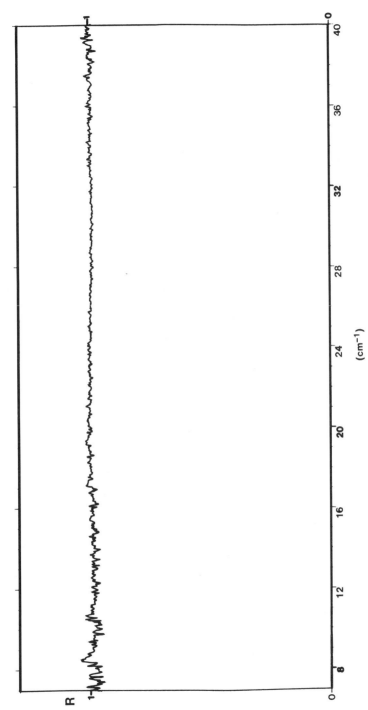

FIG. 44 Ratio of background spectrum (see text discussion).

129

spectrum of a sample is to take the ratio of two such backgrounds. An example of this is shown in Fig. 44. Over the range 16–34 cm^{-1}, the signal-to-noise ratio is about 100 or better, falling to about 50 at 10 and 40 cm^{-1}. The noise appears to be not quite white because, whereas the signal levels at 8 and 40 cm^{-1} are about the same (Fig. 43), the signal-to-noise ratio at 40 cm^{-1} is about twice that at 8 cm^{-1}.

The interferograms from which these spectra were obtained by Fourier transformation were sampled at 64-μm increments of path difference, and about 14 min were taken to run to the path difference of 200 mm required for a linearly apodized resolution of 0.05 cm^{-1} (which corresponds to about $\frac{1}{2}$ mm in Figs. 43 and 44).

Figure 45 shows a transmission spectrum for water vapor obtained with this instrument at a linearly apodized resolution of 0.1 cm^{-1}. A 3-m light-pipe was used to contain the specimen, and the figure presents the ratio of the spectrum obtained with water vapor at 15 torr in the pipe to that given with the cell evacuated. The sampling interval was 80 μm and the recording time was 40 min. Most of the visible structure in the range 15–40 cm^{-1} is reproducible. For example, the lines marked by arrows at 21.96 and 28.68 cm^{-1} are attributable to the 1_1-1_{-1} and 2_0-2_{-2} transitions and that at 37.9 cm^{-1} to the 1_0-0_0 transition of H_2O^{18}. The former involve the weakly populated, first-vibrational excited state; at this level of intensity, there are many transitions of water vapor that could give the many weak features in the spectrum.

The spectrum of water vapor shown in Fig. 46 was recorded with an uncooled Golay detector. Most of the visible structure is reproducible and

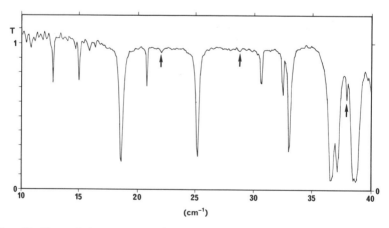

FIG. 45 Transmission spectrum of water vapor obtained with an MP interferometer using an InSb detector.

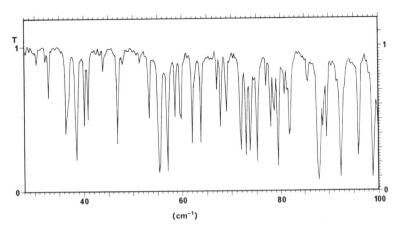

FIG. 46   Transmission spectrum of water vapor obtained with an uncooled Golay detector.

assignable. The linearly apodized resolution was 0.25 cm^{-1}, the sampling interval was 16 μm, and the recording time was 35 min (the full recorded spectrum ran to twice the spectral range accommodated in Fig. 46).

The spectrum of water vapor is too rich in structure at low levels to provide a clear demonstration of the effectiveness of an instrument in recording weak features in the presence of strong. We have selected the spectrum shown in Fig. 47 to do this. It shows a portion of the transmission spectrum of a sample of erbium oxide at 8 °K recorded to a linearly apodized resolution of 0.2 cm^{-1} with a germanium bolometer. There are strong absorption lines at 40 and 75 cm^{-1} because of the erbium oxide, but there was a trace of water vapor in the spectrometer when the interferogram was recorded and the sharp weak-absorption lines at the water vapor frequencies (indicated by the bars along the top of the figure) serve to show that the signal-to-noise ratio is at least 300. Note that the sample was in a helium cryostat with attendant losses of signal level.

B.   STUDIES OF SOLIDS AND LIQUIDS

The potential of MP interferometry for near-millimeter and submillimeter studies of solids and liquids has been well demonstrated by recent measurements on magnetic, polymeric, molecular, and ionic solids and on dielectric liquids, in many cases down to low temperatures and in some cases at high magnetic field strengths or at high pressure (see, as examples, the studies of magnetic crystals reported by Bloor and Dean (1972) and Bloor *et al.* (1976), of molecular crystals by Keeler and Batchelder (1972) and by Burton and Akimoto (1980) and of the complex

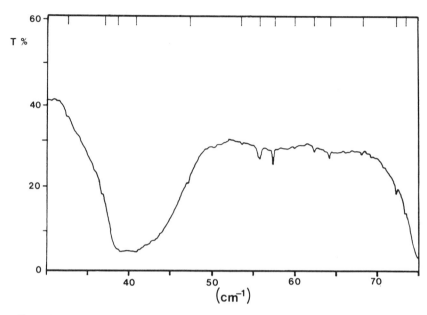

FIG. 47 Transmission spectrum of a sample of ErO with sharp weak-absorption lines, due to residual water vapor, superimposed.

refractive indices of dielectric solids and liquids by Afsar and Chantry (1981), Afsar et al. (1976), Ledsham et al. (1976, 1977), and Afsar and Hasted (1978).

Figure 48 shows, by way of illustration, the results of measurements of transmission through a single crystal of praseodymium magnesium nitrate at 1.6 °K (Bloor et al., 1976). These data belong to an extensive study of the electronic transitions of rare-earth ions that interact with the crystal fields of neighboring ions and, by exchange, with other rare-earth ions (Bloor and Copland, 1972). Spectra were recorded over the range 10–100 cm⁻¹ as functions of magnetic field up to 5 T and of crystal orientation. For this work, the high throughput efficiency of the MP interferometer and its insensitivity to signal-carried fluctuations were essential because there was, unavoidably, some significant loss of signal power in reaching the small crystal in a helium cryostat and superconducting magnet and because a large number of spectra had to be obtained to fill the full range of magnetic field and orientation.

Figure 49 illustrates the data obtained in another extensive study of a solid (Bloor and Kennedy, 1980), a chain-extended, well-ordered, crystalline-conjugated polymer. Of note here is the wide spectral range of four octaves and the series of temperatures for which data were taken.

The data in Fig. 50 are illustrative of the precision measurements of the

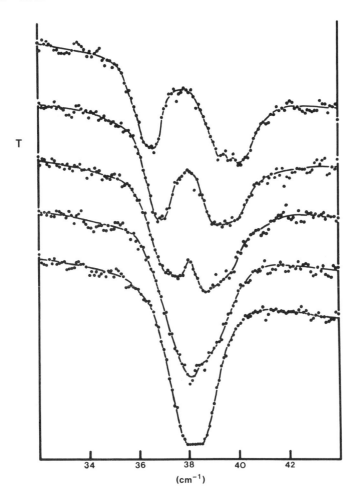

T

34   36   38   40   42

(cm⁻¹)

FIG. 48   Transmission spectra for a crystal of praseodymium magnesium nitrate showing the splitting of an electronic transition of the Pr ions on applying a magnetic field. The several spectra are for magnetic fields from 0.5 to 4.5 T in 1-T steps and they are displaced from one another along the transmission axis for greater clarity; the transmission is in arbitrary units. (See Bloor *et al.*, 1976).

complex refractive index of solids and liquids that can be made in the millimeter and submillimeter materials testing facility at the National Physical Laboratory, Teddington, U.K. (Afsar and Chantry, 1981).

C. ATMOSPHERIC EMISSION STUDIES

The effectiveness of MP instruments in giving data at high spectral and radiometric precision in difficult operational circumstances has been dem-

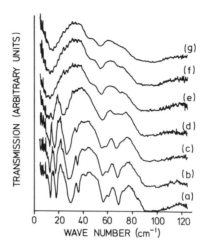

FIG. 49 Transmission spectra of a single crystal of the conjugated polymer *bis*(*p*-toluene sulphonate) diacetylene at temperatures (a) 4.2 °K; (b) 90 °K; (c) 125 °K; (d) 175 °K; (e) 200 °K; (f) 250 °K; and (g) 277 °K. The incident beam was polarized parallel to the polymer chain axis. The several spectra were recorded at a resolution of 0.2 cm⁻¹; they are relatively displaced along the transmission axis for greater clarity (see Bloor and Kennedy, 1980).

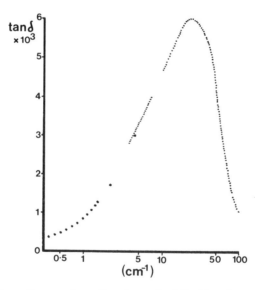

FIG. 50 The absorption spectrum (loss tangent) of liquid carbon tetrachloride. The smaller circles denote values obtained using MP interferometers (by Afsar and Chantry) and the larger circles denote values obtained by microwave measurements (by Dagg and Reesor and by Van Loon and Finsy). (See Afsar and Chantry, 1981.)

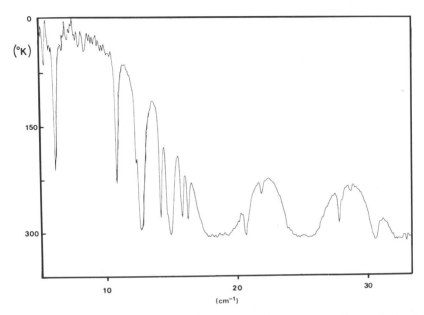

FIG. 51   Spectrum of atmospheric emission measured at mountain altitude, absolutely calibrated and expressed in terms of the effective temperature. (After Hills *et al.*, 1978.)

onstrated in several programs of study of near- and submillimeter emission from the stratosphere involving measurements from mountain sites, aircraft, and balloon-borne platforms.

Figure 51 shows the atmospheric emission spectrum measured at a mountain site (Teide, Tenerife, Canary Islands) with an apodized resolution of 0.1 cm^{-1}. This spectrum is calibrated absolutely, by the radiometric method described in section IV.D. A helium-cooled germanium bolometer was used and the recording time was 4 min for each of the two interferograms required. The system used is that shown in Fig. 26. The strong emission lines are due to water vapor and oxygen; much of the small-scale structure in the spectrum below 10 cm^{-1} is probably noise, but some of the structure near 8 cm^{-1} may well be attributable to ozone. The full study from which this spectrum was taken required that frequent and regular measurements be made at several elevation angles over an extended period, as part of a site-testing program for the U.K. millimeter-wave telescope (Hills *et al.*, 1978;.

Measurements have also been made at a mountain top of the near-millimeter solar spectrum as modified by transmission through the atmosphere. Because the angular size of the sun was small compared with the angular spread of the radiometer's antenna pattern, sky subtraction (Section IV.C) was essential. By removing the input polarizer, the radiometer

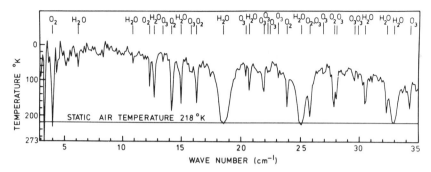

Fig. 52 Spectrum of atmospheric emission measured from an aircraft, absolutely calibrated and expressed in terms of the effective temperature.

was also used as a polarimeter (Section IV.E), and an upper limit of 0.5% was set to the polarization of the near-millimeter emission of the sun (Carli and Mencaraglia, 1977).

Figure 52 shows the absolutely calibrated spectrum of atmospheric emission measured with an MP radiometer on a CV990 aircraft above California at an altitude of 39,000 ft for a line of sight at 14° elevation. The apodized resolution is 0.1 cm^{-1} and the interferogram was recorded, with a germanium bolometer cooled by liquid helium, in 4 min. The absolute calibration was by reference to a cavity cooled by liquid nitrogen (Section IV.D). Most of the emission lines can be assigned to H_2O, O_2, and O_3 as indicated. Much of the structure below 10 cm^{-1} may be system noise (the aircraft environment was vibrationally hostile and the fact that the noise in the spectrum is no more than about four times that expected from detector noise testifies to the effectiveness of the use of the two output ports to suppress signal-carried noise; see Section III.D) but that near 8 cm^{-1} may be attributable to ozone.

Figure 53 shows a small portion of one of the many spectra recorded on the aircraft at higher resolution, 0.03 cm^{-1} apodized. Each interferogram gave a spectrum from 5 to 50 cm^{-1} and took 7 min to record (Carli *et al.*, 1975, and 1977). The stronger lines are attributable to H_2O and many of the numerous weaker narrow lines to ozone as can be seen by comparison with the calculated spectrum in the figure (Bangham, 1978).

Following from these measurements, a larger MP instrument, giving a resolution of 0.003 cm^{-1} unapodized, was designed and constructed at the University and CNR-IROE, Firenze, and was carried by balloon to 40-km altitude above Texas in a program to study minor constituents of the stratosphere. The radiometer was used as a differential system (Section IV.D) to compare the emission at low elevation with that at zenith (where

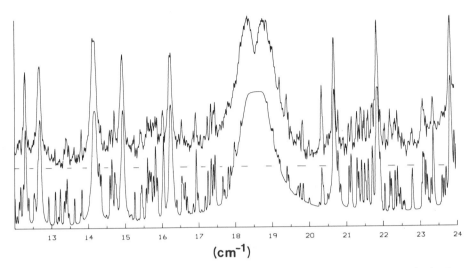

(cm^{-1})

FIG. 53   Upper curve: part of a high-resolution atmospheric emission spectrum obtained from an aircraft (Carli *et al.*, 1975, 1977). Lower curve: a computed spectrum based on oxygen, ozone, and water vapor (Bangham, 1978), displaced downward so as to allow comparison with the measured spectrum (Crown Copyright, National Physical Laboratory).

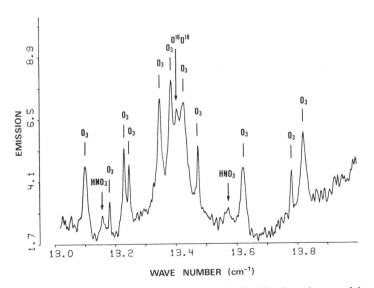

WAVE NUMBER (cm^{-1})

FIG. 54   Atmospheric emission (arbitrary units) in the $13-14$-cm^{-1} spectral interval, measured from 39.8-km altitude with zenith angle 94.5°, field of view 1.5°, spectral resolution 0.0033 cm^{-1} unapodized. Each spectrum is an average of four spectra, each taken in 4 min. The species responsible for the clear lines are indicated. (After Carli *et al.*, 1980.)

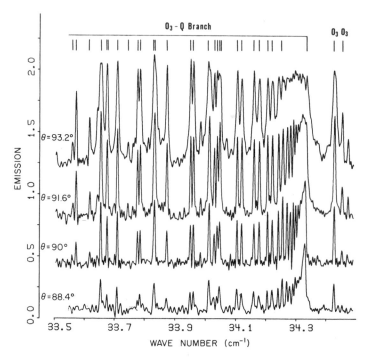

FIG. 55 Atmospheric emission (arbitrary units) in the 33.5–34.5-cm⁻¹ spectral interval, measured from 39.8-km altitude at four different zenith angles θ; field of view 1.5°, spectral resolution 0.0033 cm⁻¹ unapodized. Each spectrum is the average of three measurements, each taking 4 min to record. (After Carli *et al.*, 1980.)

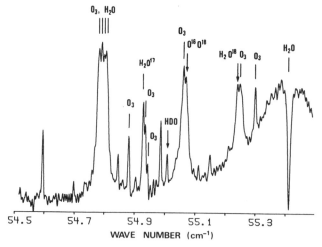

FIG. 56 Atmospheric emission (arbitrary units) in the 54.5–55.5-cm⁻¹ spectral interval, measured from 39.8-km altitude with zenith angle 93.2°, field of view 1.5°, resolution 0.004 cm⁻¹. Spectrum taken in 4 min. The species responsible for most of the clear lines are indicated, but some lines are still unassigned. (After Carli *et al.*, 1980.)

the attenuation is so low that this effectively gives an absolute calibration). Figures 53–56 show short sections from four of the large number of very impressive high-resolution spectra obtained (Carli *et al.*, 1980); the operational details are given with the figures. Many lines attributable to molecular species important in stratospheric chemistry are found in the spectra, and a number of clear lines in the spectra are not yet assigned.

D. PLASMA DIAGNOSTICS

In several laboratories around the world, there are large-scale installations in which, in a pulsed operation, plasmas are created with densities and temperatures approaching those required for controlled thermonuclear fusion. Most of these involve a toroidal hydrogen plasma held in place by a magnetic field of a few tesla for up to 1 sec and reaching electron temperatures of a few kiloelectronvolts at a density of 10^{13} cm^{-3} or more. The electrons will follow helical trajectories in the magnetic field at a frequency equal to $B/1.07$ (cm^{-1}), where B (in teslas) is the magnetic field strength, i.e., in the near-millimeter range. This cyclotron motion is important because it is a mechanism by which energy is lost by radiation from the plasma and, conversely, by which energy could be pumped into the plasma and also because measurement of the intensity of emission of cyclotron radiation serves to determine the electron temperature and changes in electron density. Fast-scanning polarizing MP interferometers are now in use at most of these installations for such diagnostic work; it is the high throughput efficiency of the MP instruments over several octaves at low frequencies that is particularly required (Costley *et al.*, 1974, 1977a; Hutchinson and Komm, 1977; Hutchinson, 1978; Bartlett *et al.*, 1978). Figure 57 shows measured cyclotron emission signals from the

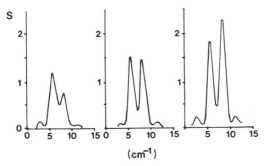

FIG. 57   Emission spectra recorded during a pulse of the Princeton tokomak. Each took 20 msec to record, the first at 250 msec into the pulse, the second at 370 msec, and the third at 465 msec. (See Stauffer and Boyd, 1978.)

Princeton tokomak (Stauffer and Boyd, 1978) as an illustration of such measurements. The fundamental and three harmonics of the cycltron frequency are seen (the plasma is optically thin for the fundamental), and there are clear and informative changes in relative intensities of the harmonics in the course of the pulse.

E. Cosmic Background Studies

The most striking application of MP instruments to date has been the detection of the near- and submillimeter radiation incident isotropically on the earth from outside the galaxy, which is believed to be a relic of radiation filling the universe after the big bang that initiated universal expansion. This radiation has a continuous spectrum and was first detected at 7.35-cm wavelength by Penzias and Wilson in 1964. Subsequent measurements down to wavelengths of a few millimeters established a ν^2-dependence characteristic of a blackbody spectrum in the long-wave limit and an absolute value corresponding to about 3 °K. The cosmological big bang origins of the radiation would only be firmly established, however, if the spectrum were shown to have blackbody character in the mid- and short-wave regions too. To demonstrate that requires measurements at near- and submillimeter wavelengths from a platform above the absorbing and emitting lower atmosphere. A precision radiometer with high throughput efficiency and coolable to liquid-helium temperatures is necessary for this and a thorough analysis of weak side lobes in the antenna pattern of the radiometer is essential (to exclude the possibility of contamination by radiation from the earth). Three high-altitude balloon-borne experiments to determine the submillimeter spectrum have been made and each incorporated an MP interferometer. The first was that of Robson *et al.* (1974), and the others were by Woody *et al.* (1975) and Woody and Richards (1979). The spectrum obtained by Robson *et al.* (1974) is shown in Fig. 58. It includes the emission from the thin polyethylene window covering the liquid-helium cryostat, which contained the radiometer, and, superimposed on this, atmospheric emission lines attributable to oxygen, ozone, and water vapor. Robson *et al.* (1974) used the window emission for intensity calibration and concluded that the small peak in the recorded spectrum at ~ 5 cm^{-1} was the cosmic background spectrum with a near blackbody form corresponding to a temperature of 2.96 °K. With this calibration, however, the atmospheric lines appear to be weaker than expected for an altitude of 40 km. If the atmospheric lines are themselves used for calibration, the power recorded in the range 6–16 cm^{-1} is decidedly higher than that of a 3 °K blackbody but well below what would be indicated by extrapolation of the ν^2-dependence found at longer wavelengths.

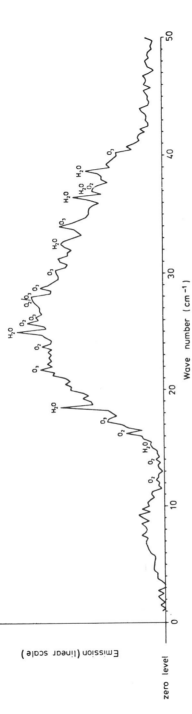

Emission (linear scale)

zero level

Wave number (cm⁻¹)

FIG. 58   The spectrum recorded by an MP radiometer, which was lifted on a balloon-borne platform to an altitude of 40 km, cooled to 1.7 °K, and directed away from the earth at elevation angle 60°. (After Robson et $al.$, 1974; Clegg, 1978.) Four hours were required to record this spectrum to a resolution of 1 cm⁻¹, using two InSb detectors at 1.7 °K, one in each output port. As presented here, the spectrum is uncorrected for the variation of instrumental throughput and responsivity with frequency. The broad peak from 16 to 40 cm⁻¹ is attributable to the very weak thermal emission from the thin polyethylene window covering the aperture in the liquid-helium cryostat. The superimposed lines are attributable to emission from residual atmospheric oxygen, ozone, and water vapor, as indicated. The secondary peak from 5 to 11 cm⁻¹ may be the cosmic background spectrum; in any case, the extremely low signal level below 14 cm⁻¹ sets tight upper limits to the background intensity (see text).

The spectra obtained in the experiments of Woody *et al.* (1975) and Woody and Richards (1979) were free of contamination by window emission (it was found that the liquid helium cryostat could be operated at altitude without a cover) and the intensities of the atmospheric emission lines (scaled using the absolute calibration conducted in the laboratory after the flight) were consistent with the best estimates of stratospheric concentrations of ozone, oxygen, and water vapor. This lends confidence to the subtraction from the measured spectrum of a contribution attributable to the atmosphere, leaving a residual spectrum attributed to the cosmic background. This residual spectrum is close to a blackbody spectrum for about 2.9 °K, as shown in Fig. 59. These impressive measurements are examined in detail by Woody and Richards (1981).

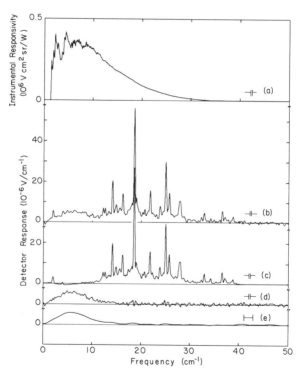

FIG. 59   (a) Spectrum recorded by an MP radiometer on a balloon-borne platform at 43-km altitude. The data took 45 min to record at an apodized resolution of 0.25 cm⁻¹ and at elevation angle 65°. The detector was a germanium bolometer cooled by liquid ³He to 0.36 °K. (b) Calculated emission spectrum for the atmosphere under the conditions of (a), including the effect of the 1.67 °K reference temperature. (c) The difference between (a) and (b), i.e., the spectrum attributed to the cosmic background. (d) Same as (c), but smoothed to a resolution of 1.8 cm⁻¹. (After Woody and Richards, 1979.)

F. Assessment of Performance

The basic treatment of FTS given in Section II.A led to Eq. (14), which related the performance parameters (signal-to-noise ratio S/N, resolution $\Delta\sigma$, and recording time T) to the system design parameters [source brightness $B(\sigma)$, light-grasp $A\Omega$, and detector noise-equivalent-power $N(\sigma)$], and the throughput efficiency $e(\sigma)$:

$$S/N/\Delta\sigma \sqrt{T} = 2B(\sigma)e(\sigma)A\Omega/N(\sigma). \tag{70}$$

It was not possible there to make the several contributions to $e(\sigma)$ explicit but we can now do so. The following factors that do not enter a simple formulation need to be taken into account.

1. *Interferometric Modulation.* In FTS, each spectral component of a signal is cosine-modulated, by interference, in order to allow demodulation by Fourier transformation. This results in a reduction of signal-to-noise ratio relative to a hypothetical situation in which the spectral components are in some way distinguished from each other without coding by interference (as would be implied by putting $e(\sigma) = 1$). Carli (1972) first showed that the contribution to $e(\sigma)$ associated with interferometric modulation and demodulation is $\sqrt{\frac{1}{8}}$ when a single detector in one of the two output ports is used and $\sqrt{2} \times \sqrt{\frac{1}{8}}$ when there is a detector in each output port.

Polarization modulation (Section III.B) is a way of switching the two output beams alternately to a single detector. Because the interferograms in the two output ports are complementary, this doubles the modulated signal, giving, overall, a contribution to $e(\sigma)$ of $1/\sqrt{2}$ for a single detector and 1 for two. With polarization modulation and two detectors, therefore, the ideal value, 1, is obtained.

In addition, the introduction of polarization modulation has the consequence that the relevant detector noise-equivalent-power is not that which corresponds to the real-time frequency of interferometric modulation for spectral frequency σ [see Eq. (10)] but that corresponding to the modulation frequency, which can more readily be lifted above the range where low-frequency ("$1/f$") noise occurs in the detector system.

2. *Chopper Modulation.* Some systems use a chopper modulator (rather than fast-scanning or polarization modulation; see Section III.B). If a chopper is used in transmission only, it reduces the time for which the signal is detected. Carli and Natale (1979) show that square-wave modulation followed by cosine demodulation in a phase-sensitive detector gives a factor $\sqrt{2}/\pi$ in $e(\sigma)$. If the chopper is used in reflection as well as transmission, with two detectors, this factor becomes $2/\pi$. Taking these together with the factor for interferometric modulation gives $1/2\pi$ and

$1/\sqrt{2\pi}$ for one and two detectors, respectively. The comment in paragraph 1 on the value to be taken for $N(\sigma)$ also applies here, of course.

3. *Time Efficiency.* Allowance must also be made in $e(\sigma)$ for the fact that the system noise bandwidth is not usually set by true integration over the sampling interval T/n, where T is the total recording time and n the number of independent points or bins in the derived spectrum. Usually, the noise is smoothed by an RC network. The S/N given by RC smoothing would be the same as that given by true integration if the RC time constant were made equal to half the sampling interval; this would, however, give some loss of independence of the signal levels computed for neighboring points or bins of the spectrum. Decreasing the time constant would strengthen the independence but would reduce the S/N by a factor $(2n\tau/T)^{1/2}$.

There might also be other practical reasons for T being greater than $n\tau$, for example, the introduction of dead time to allow decay of microphonic disturbance following a stepped movement of the reflector in the interferometer. That would introduce a factor $(n\tau/T)^{1/2}$ in $e(\sigma)$. We should, therefore, recognize that a factor on the order of $1/\sqrt{2}$ might frequently enter $e(\sigma)$ to allow for such time-efficiency factors.

4. *Position of Zero Path Difference.* In setting up Eq. (14), it was assumed that a cosine transform would be used to derive the spectrum from the interferogram. This implies that the position of zero path difference is independently known and that the interferogram is symmetrical about that position. If it is necessary to find zero path difference from the interferogram itself, by taking both sine and cosine transforms and combining them, noise will enter both transforms and a factor $1/\sqrt{2}$ will have to be included in $e(\sigma)$ to allow for that.

5. *Throughput Losses.* There are then the throughput losses arising from optical imperfections in the interferometer, such as imperfect grids and imperfect alignment (Section III.A) and the loss of interferometric modulation due to beam spreading (Section III.C). These would usually be unimportant, in a well-designed interferometer, compared with losses of signal power at windows, lenses, and other optical components involved when a laboratory sample has to be at reduced or elevated temperature, at high pressure, in a magnetic field, etc., or when the sample is small so that the beam has to be condensed and then recollimated. These losses will depend, of course, on the details of the experiment; they could be as low as one-half, say, or as high as four-fifths, in typical experiments, and higher still in special cases.

6. *Losses at Detector.* There may be similar losses to those referred to in the paragraph above in that part of the system where the signal power is condensed on to the detector. Such losses would best be included when taking a value for the detector noise-equivalent-power, but it has to be

stressed that data quoted in the literature for detector NEPs often refer, not to the power incident upon the window in the detector enclosure, but to that incident upon the sensitive element of the detector following transmission through any windows and condensing optics.

7. *Losses at Source.* With an MP interferometer of the simplest kind there will be a loss of signal power if the incident beam is not plane-polarized. For an unpolarized beam, that loss will be one-half, and this is another factor for $e(\sigma)$. If it is required to investigate the polarized properties of a source or sample, this factor can be discounted, of course.

Thus, a system of low efficiency (chopper, single detector, RC smoothing, losses at sample and detector) may have an $e(\sigma)$ as low as

$$\frac{1}{2\pi} \times \frac{1}{\sqrt{2}} \times \frac{1}{\sqrt{2}} \times \frac{1}{5} \times \frac{1}{2} \approx \frac{1}{120},$$

which is lower than 1%. A system of high efficiency (polarization modulation, two detectors, time integration, no loss at sample, polarized input) would have

$$e(\sigma) = 1 \times \frac{1}{\sqrt{2}} \times \frac{1}{\sqrt{2}} = \frac{1}{2}.$$

The factor 60 separating these efficiencies would correspond to a factor $(60)^2 = 3600$ in the time required to record a spectrum to specified resolutions in frequency and intensity. Moreover, in going from Eq. (14) to Eqs. (15), (16), and Table I, the value taken for Ω is the maximum value consistent with the specified resolving power $\sigma/\Delta\sigma$. The areas of available detectors and/or samples may be less than would be allowed if only low resolving power is required.

Detailed examination of the spectra shown in this section reveal performances that are within an order of magnitude of those that would be expected on the basis of the equations of Section II and the considerations just examined. This is a stronger statement than may appear at first sight because these expectations are based on the assumption that the *detector* is the dominant source of noise. Thus the claim, finally, for MP interferometry is that *signal-related* noise can be sufficiently suppressed in such an interferometer (Section III.D) that performances approaching the limits set by the extremely low noise levels of liquid-helium-cooled detectors can be realized in practice.

ACKNOWLEDGMENTS

Mr. E. Puplett played a key part in every aspect of the development of the MP method of interferometry at Queen Mary College, from the inception, through the design and testing of a succession of instruments, to the use of them in atmospheric studies. I should also like to acknowledge the contributions made by Dr. B. Carli in the analysis of the method and our observational work, by Drs. S. El Atawy and P. A. R. Ade in developing cooled bolometric detectors, by Mr. W. Duncan and Dr. S. Fonti in the use of MP interferometers for absolute radiometry, and by Mr. C. Spratling in writing software for data acquisition and analysis. I am indebted to colleagues in the workshop and technical services in the Physics Department at Queen Mary College, under Mr. F. Hands and Mr. K. Nickels, for their work in constructing our instruments and detectors. A grant from the Paul Instrument Fund of the Royal Society supported this work. I thank Mr. S. A. Vickers for providing me with information on his vibrating-reed modulator (Fig. 42); J. A. Beunen, A. E. Costley, G. F. Neill, C. L. Mok, T. J. Parker, G. Tait, W. G. Chambers, P. A. R. Ade, and C. T. Cunningham, whose data were used in Figs. 14, 15, and 39; D. Bloor, J. R. Dean, V. Sells, and the publishers of *Infrared Physics* for permission to use Fig. 48; D. Bloor and R. J. Kennedy and the publishers of *Chemical Physics* for permission to use Fig. 49; M. N. Afsar, G. W. Chantry, I. R. Dugg, G. E. Reesor, R. Finsy, and R. Van Loon, whose data are included in Fig. 50; M. J. Bangham and the publishers of *Infrared Physics* for permission to use Fig. 53; B. Carli, F. Mencaraglia, A. Bonnetti, and the publishers of the *International Journal of Infrared and Millimeter Waves* for permission to use Figs. 54–56, F. J. Stauffer and D. A. Boyd, whose data were used for Fig. 57; P. E. Clegg, for permission to use fig. 58; and P. L. Richards for permission to use Fig. 59.

References

Ade, P. A. R., Costley, A. E., Cunningham, C. T., Mok, C. L., Neill, G. F., and Parker, T. J. (1979). *Infrared Phys.* **19**, 599.

Afsar, M. N., and Chantry, G. W. (1981). *Int. J. Infrared Millimeter Waves* **2**, 107.

Afsar, M. N., and Hasted, J. B. (1978). *Infrared Phys.* **18**, 835.

Afsar, M. N., Hasted, J. B., and Chamberlain, J. (1976). *Infrared Phys.* **16**, 301.

Arnaud, J. A. (1976). "Beam and Fiber Optics." Academic Press, New York.

Bangham, M. J. (1978). *Infrared Phys.* **18**, 357, 799.

Bartlett, D. V., Costley, A. E., and Robinson, L. C. (1978). *Infrared Phys.* **18**, 749.

Bell, R. J. (1972). "Introductory Fourier Transform Spectroscopy." Academic Press, New York.

Beunen, J. A., Costley, A. E., Neill, G. F., Mok, C. L., Parker, T. J., and Tait, G. (1981). *J. Opt. Soc. Am.* **71**, 184.

Birch, J. R. (1978). *Infrared Phys.* **18**, 275.

Birch, J. R., and Parker, T. J. (1979). "Infrared and Millimeter Waves" (K. J. Button, ed.), Vol. 2, Chap. 3. Academic Press, New York.

Birch, J. R., Dromey, J. D., and Lesurf, J. (1981). *Infrared Phys.* **21**, 229.

Bloor, D., and Copland, G. M. (1972). *Rep. Prog. Phys.* **35**, 1173.

Bloor, D., and Dean, J. R. (1972). *J. Phys. C* **5**, 1237.

Bloor, D., and Kennedy, R. J. (1980). *Chem. Phys.* **47**, 1.

Bloor, D. Sells, V., and Dean, J. R. (1976). *Infrared Phys.* **16**, 425.

Burroughs, W. J., and Chamberlain, J. (1971). *Infrared Phys.* **11**, 1.

Burton, C. H., and Akimoto, Y. (1980). *Infrared Phys.* **20**, 115.

Carli, B. (1972). *Infrared Phys.* **12**, 251.

Carli, B. (1974). *IEEE Trans. Microwave Theory Tech.* **MTT-22**, 1094.

Carli, B., and Mencaraglia, F. (1977). *Infrared Phys.* **17**, 293.

Carli, B., and Mencaraglia, F. (1981). *Int. J. Infrared Millimeter Waves* **2**, 87.

Carli, B., and Natale, v. (1979). *Appl. Opt.* **18**, 3954.

Carli, B., Martin, D. H., Puplett, E., and Harries, J. E. (1975). *Nature (London)* **257**, 649.

Carli, B., Martin, D. H., Puplett, E., and Harries, J. E. (1977). *J. Opt. Soc. Am.* **67**, 917.
Carli, B., Mencaraglia, F., and Bonetti, A. (1980). *Int. J. Infrared Millimeter Waves* **1**, 263.
Challener, W. A., Richards, P. L., Zilio, S. C., and Garvin, H. L. (1980). *Infrared Phys.* **20**, 215.
Chamberlain, J. (1971). *Infrared Phys.* **11**, 25, 57.
Chamberlain, J. (1979). "Principles of Interferometric Spectroscopy." Wiley, New York.
Chambers, W. G., Mok, C. L., and Parker, T. J. (1980a). *J. Phys. D* **13**, 515.
Chambers, W. G., Mok, C. L., and Parker, T. J. (1980b). *J. Phys. A* **13**, 1433.
Chandrasekhar, H. R., Genzel, L., and Kuhl, J. (1976a). *Opt. Commum.* **17**, 106.
Chandrasekhar, H. R., Genzel, L., and Kuhl, J. (1976b). *Opt. Commun.* **18**, 381.
Chantry, G. W. (1971). "Submillimeter Spectroscopy." Academic Press, New York.
Clegg, P. E. (1978). *In* "Infrared Astronomy" (G. Setti and G. G. Fazio, eds.), pp. 181–198. Reidel Publ., Dordrecht, Netherlands.
Costley, A. E., Hastie, R. J., Paul, J. N. M., and Chamberlain, J. (1974). *Phys. Rev. Lett.* **33**, 758.
Costley, A. E., Hastie, R. J., Paul, J. N. M., and Chamberlain, J. (1977a). *Phys. Rev. Lett.* **38**, 1477.
Costley, A. E., Hursey, K. H., Neill, G. F., and Ward, J. M. (1977b). *J. Opt. Soc. Am.* **67**, 979.
Genzel, L., and Kuhl, J. (1978). *Infrared Phys.* **18**, 113.
Genzel, L., and Sakai, K. (1977). *J. Opt. Soc. Am.* **67**, 871.
Goubau, G., and Schwering, F. (1961). *IRE Trans. Antennas Propag.* **AP9**, 248.
Griffiths, P. R. (1975). "Chemical Infrared Fourier Transform Spectroscopy." Wiley, New York.
Hanel, R., Forman, M., Meilleur, T., Westcott, R., and Pritchard, J. (1969). *App. Opt.* **8**, 2059.
Hills, R. E., Webster, A. S., Alston, D. A., Morse, P. L. R., Zammitt, C. C., Martin, D. H., Rice, D. P., and Robson, E. I. (1978). *Infrared Phys.* **18**, 819.
Hutchinson, I. H. (1978). *Infrared Phys.* **18**, 763.
Hutchinson, I. H., and Komm, D. S. (1977). *Nucl. Fusion* **17**, 1077.
Keeler, G. J., and Batchelder, D. N. (1972). *J. Phys. C* **5**, 3264.
Kogelnik, H. (1964). "Proceedings of Symposium on Quasi-Optics." Polytechnic Press, Brooklyn.
Lambert, D. K., and Richards, P. L. (1978). *J. Opt. Soc. Am.* **68**, 1124.
Ledsham, D. A., Chambers, W. G., and Parker, T. J. (1976). *Infrared Phys.* **16**, 515.
Ledsham, D. A., Chambers, W. G., and Parker, T. D. (1977). *Infrared Phys.* **17**, 165.
Lesurf, J. (1981). *Infrared Phys.* **21**, 383.
Marcuse, D. (1972). "Light Transmission Optics." Van Nostrand-Reinhold, Princeton, New Jersey.
Martin, D. H. (1972). *In* "Infrared Detection Techniques for Space Research" (V. Manno and J. Ring, eds.), p. 267. Reidel Publ., Dordrecht, Netherlands.
Martin, D. H., and Lesurf, J. (1978). *Infrared Phys.* **18**, 405.
Martin, D. H., and Puplett, E. (1970). *Infrared Phys.* **10**, 105.
Mertz, L. (1965). "Transformations in Optics." Wiley, New York.
Mok, C. L., Chambers, W. G., Parker, T. J., and Costley, A. E. (1979). *Infrared Phys.* **19**, 437.
Petit, R. (1975). *Nouv. Rev. Opt.* **6**, 129.
Robson, E. I., Vickers, D. G., Huizinga, J. S., Beckman, J. E., and Clegg, P. E. (1974). *Nature (London)* **251**, 591.
Stauffer, F. J., and Boyd, D. A. (1978). *Infrared Phys.* **18**, 755.

Woody, D. P., and Richards, P. L. (1979). *Phys. Rev. Lett.* **42,** 295.

Woody, D. P., and Richards, P. L. (1981). *Astrophys. J.* **248,** 18.

Woody, D. P., Mather, J. C., Nishioka, N. S., and Richards, P. L. (1975). *Phys. Rev. Lett.* **34,** 1036.

Wrixon, G. T., and Kelly, W. M. (1978). *Infrared Phys.* **18,** 413.

CHAPTER 3

Infrared Detectors for Low-Background Astronomy: Incoherent and Coherent Devices from One Micrometer to One Millimeter

P. L. Richards

Department of Physics
University of California
Berkeley, California

L. T. Greenberg

Space Sciences Laboratory
The Aerospace Corporation
El Segundo, California

149

I. Introduction

The conditions under which infrared measurements are made from cooled space telescopes are radically different from the usual ground-based environment with its intense flux of thermal infrared photons. In order to create the instrumentation necessary to profit fully from the space environment, it is important that both the infrared astronomy community and also the wider device development community have an overview of the status of development of those types of infrared detectors which will be useful for low-background space astronomy. For a number of reasons, this overview has been difficult to obtain. The backgrounds encountered in most earth-, or aircraft-, based measurements are usually large enough that direct experience with low-background systems has not been generally available. Because of security classification or proprietary interests, many of the best low-background detectors have not been described in the open literature. Although the security classification problem has eased considerably in recent years for discrete low-background detectors, much of the best work remains buried in technical reports that are difficult to locate.

In this chapter we shall give a general survey of the present status of the types of incoherent and coherent infrared detectors that are expected to be useful for low-background space astronomy. The discussion will be focused on their physical principle of operation, and examples will be given of the performance currently available. No attempt will be made to give an exhaustive scholarly treatment of any one subject. Our goal is rather to give the reader a context into which further information can be fitted.

The references provided are not intended to be exhaustive. Relatively few original sources are quoted, all of which refer to recent developments. Wherever possible, references are given to recent review articles and sources to facilitate searches of the literature in areas of interest to the reader. Books and reprint collections of general interest include those by Willardson and Beer (1970, 1977), Arams (1973), Hudson and Hudson (1975), Keyes (1977), Kingston (1978), and Wolfe and Zissis (1978). The proceedings of the Meetings of the IRIS Specialty Group on Infrared Detectors is a valuable resource, which appears annually. It was classified as secret prior to 1976, but since then it has been published in two volumes, one of which is unclassified. The distribution is controlled. Copies can be requested from the Office of Naval Research, Branch Office Chicago, 536 South Clark Street, Chicago, Illinois 60605.

The technical information in this chapter will begin in Section II on general principles of infrared detectors and will then continue in Section III on photon detectors, Section IV on bolometers, Section V on coherent detectors, and Section VI on important supporting technologies. A comparison will be made in Appendix I of coherent and incoherent modes of detection and a glossary will be included in Appendix II.

II. General Principles of Infrared Receivers

Astronomical observations usually involve the measurement of the flux of radiant energy emitted by distant objects, At infrared wavelengths it is often convenient to visualize this flux of energy as a stream of photons. The uncertainty principle of quantum mechanics places certain restrictions on the simultaneous measurement of both the number of photons (intensity) and the phase of the photons (that is, the phase of the electromagnetic field). Any detector that determines both the intensity and the phase is called a *coherent detector*. Coherent detectors (except for upconverters) have inherent spontaneous emission of photons. These additional photons are always present along with the signal to be measured. They introduce a source of noise that limits the performance of the coherent detector so that it is consistent with the requirements of the uncertainty principle. The noise introduced by this spontaneous emission is roughly equivalent to the noise produced by the background noise of a blackbody at temperature $T = hc/\lambda k$, where λ is the average wavelength of the receiver's response, h is Planck's constant, c is the speed of light, and k is Boltzmann's constant. A detector that is limited only by this spontaneous emission is called an *ideal coherent detector*.

Direct detectors, which produce an output voltage or current proportional to the signal power, do not measure the phase of the photon's electromagnetic field. Because the phase is not determined by such incoherent measurements, the uncertainty principle places no restriction on the precision of the measurement of the number of photons. Thus, incoherent (direct) detectors can be, in principle, completely noiseless and can detect the arrival of a single signal photon. Such a direct detector (with unit quantum efficiency $\eta = 1$) is called an *ideal photon counter*.

As a general rule, coherent detectors are more useful at the longer infrared wavelengths and for experiments requiring narrow infrared bandwidths. Conversely, incoherent detectors are favored for short-wavelength, broad-band measurements. The tradeoff between these techniques will be discussed in some detail in Appendix I. The infrared portion of the electromagnetic spectrum is a crossover region where the eventual choice between coherent and incoherent detectors is dictated by the astronomical requirements such as spectral range, spectral resolution, and spatial resolution. Thus, it can be expected that both coherent and incoherent infrared detectors will be needed to realize the full astronomical potential of the infrared wavelength range.

In practice, very nearly ideal incoherent photon counters can be constructed at optical wavelengths. In the infrared, however, incoherent detectors fall at least two to three orders of magnitude short of counting single photons. Though nearly ideal coherent detectors exist at a few selected wavelengths in the infrared, generally, coherent detectors for most of this region are also far from approaching the ideal limit.

In nearly all astronomical experiments, unwanted (background) photons reach the detector in addition to the signal. The background level and the scientific requirements of a particular astronomical measurement not only affect the type of detector that should be employed (i.e., coherent or incoherent) but also the degree of perfection required in the receiver system. Most present infrared observations, especially those made from the ground, are made under conditions with a very high background. Although an ideal photon counter could be used under such conditions, it is usually not necessary to measure the arrival of each individual photon. As an example, consider a direct detector used for near-infrared photometry on a large ground-based telescope. Under these conditions, the background would correspond to $\sim 10^{12}$ photons/sec on the detector. The noise fluctuations in such a background due to the Poisson statistics of photon detection will be $\simeq 10^6 \eta^{-1/2}$ photons in 1 sec (referred to the detector input), where η is the quantum efficiency. Thus, any incoherent detector that can detect 10^6 photons in 1 sec will be limited by background noise. A detector with relatively low responsivity, which is amplifier-noise-limited at 10^6 photons/sec, but which has unit quantum efficiency, will perform as well under these conditions as an ideal photon counter. (Responsivity is the ratio of the detector output signal to the incident power and often has units of amperes per watt.) This example provides the insight that under background-limited conditions it is not the responsivity of the detector that should be optimized but rather the quantum efficiency. Under low-background conditions, such as can be obtained with a cooled space telescope, the fluctuations in the background become so small that, because of the effects of amplifier noise, responsivity often becomes more important than quantum efficiency per se.

Except for devices using photocathodes and electron multiplication, all incoherent infrared detectors presently in use (photoconductors, photo-voltaic devices, and bolometers) make use of sophisticated electronic circuitry to measure very weak currents or voltages produced in the device. Though these weak signals are often produced by individual photo events in the detector, fundamental considerations involving the capacitance of the detectors and of solid-state electronics preclude the detection of single photons.

The sensitivity of direct detectors is often measured in terms of the noise equivalent power (NEP). The NEP of a detector is defined to be the radiant power, in watts, that must be incident on the detector to produce a signal equal to the rms noise of the detector (with its associated electronics) within an electrical bandwidth of 1 Hz. Thus, the smaller the NEP, the more sensitive the detector. The NEP of a detector is directly related to the minimum number of photons at a particular wavelength that can be detected in 1 sec. For example, an NEP of 10^{-15} W/Hz$^{1/2}$ would allow the detection at unity signal-to-noise ratio of 3.6×10^3 photons in 1 sec at a wavelength of 1 μm, or 3.6×10^6 photons in 1 sec at a wavelength of 1 mm.

In addition to the NEP, many other characteristics must be considered in the selection of a receiver for astronomical observations. These characteristics are not usually independent parameters, so detector performance must be optimized in a complex tradeoff that depends on the application. Important characteristics include

(1) speed of response, which can often be traded for NEP. Faster detectors often have higher NEPs;

(2) linearity of response, or at least reproducibility, which is essential to allow quantitative measurements of flux;

(3) suitability for incorporation in arrays, because in many cases the time required to make an observation is inversely proportional to the number of detectors that can be used;

(4) operating characteristics such as detector temperature, power dissipation, sensitivity to energetic nuclear particles, and ruggedness, because these characteristics often determine the suitability for a given mission.

For the purpose of this chapter, the term "low background" can be defined to be that measurement regime in which the amplifier noise becomes the limiting factor rather than the statistical noise of the incident photons. This concept can be used for either coherent or incoherent detectors. It is not an absolute definition in any sense. It depends for, example, on the quantum efficiencies and responsivities of available detectors and the noise in available amplifiers, as well as the wavelength, the bandwidth, and the background conditions of the desired measurement. The high- and low-background limits are quite different for coherent and incoherent detectors. This definition has the merit of focusing attention on a specific set of problems that are of great importance for space astronomy.

References in which the general principles of infrared detection are discussed include Emmons *et al.* (1975), Gillett *et al.* (1977), Kruse (1977), and Blaney (1978).

A. Detector Needs for Low-Background Space Astronomy

Several currently contemplated or ongoing NASA scientific programs in infrared space astronomy are described in the following reports:

(1) Report on Space Science 1975, National Academy of Science, 1976.

(2) Shuttle Infrared Telescope Facility (SIRTF), Interim Report, NASA Ames Research Center, April 14, 1978, and SIRTF Design Optimization Study Final Technical Report, Ames Research Center, September 1979

(3) Infrared Astronomical Satellite (IRAS), Mission Definition Report, Goddard Space Flight Center, May 1976.

(4) Spacelab 2, Infrared Telescope Design Study, Final Report, Smithsonian Astrophysical Observatory, April 1978.

(5) Cosmic Background Explorer (COBE), Interim Report, Goddard Space Flight Center, February 1, 1977.

(6) Submillimeter Space Telescope, Jet Propulsion Laboratory, April 10, 1978.

(7) Space Telescope Infrared Photometer, Final Report, Lockheed Missiles and Space Company, March 1976.

(8) Infrared Photometer for the Space Telescope, Phase B Definition Study, Final Report, Ball Brothers Research Corporation, June 1976.

Nearly all of these infrared astronomical observations from space platforms will require detectors capable of achieving optimum performance under conditions of extremely low background. In some cases, orbiting infrared observatories will allow background reductions by factors of 10^6 to 10^8 over those commonly encountered on the ground. In order to make optimum observational use of these platforms, much work will be needed on the problems of low-background infrared measurements.

The most pressing need is for the development of single-element direct detectors and direct-detector arrays that can be background-limited on cryogenic telescopes operated in space. Figure 1 shows the irreducible noise of detectors owing to various sources of infrared background. The figure is directly applicable to moderately broad-band photometry and

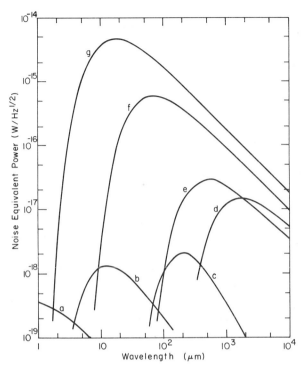

FIG. 1 The theoretical limits to detector NEP set by several background sources relevant for space missions are illustrated as a function of wavelength λ. Curves a–g were all
calculated assuming a spectral bandpass $\Delta\lambda/\lambda = 0.1$, a throughput that accepts 84% of the
energy from a point source ($A\Omega = 3.7\lambda^2$), a quantum efficiency for incident photons $\eta = 1$,
and an optical efficiency $\epsilon = 1$. For other values of bandpass, efficiency, or throughput, the
NEP scales as $[(A\Omega/\epsilon\eta)(\Delta\lambda/\lambda)]^{1/2}$. Where more than one source is present, the NEP values
should be combined in quadrature. Curves a–g show limits set by statistical fluctuations in
the photon rate from the following sources:

(a) zodiacal scattered light at the ecliptic pole,
(b) zodiacal dust emission at the ecliptic pole,
(c) interstellar dust emission at high galactic latitude,
(d) the 3 °K cosmic background radiation,
(e) a 10 °K, 10% emissive telescope,
(f) a 77 °K, 10% emissive telescope, and
(g) a 300 °K, 10% emissive telescope.

These curves were calculated using Poisson statistics because the Bose correction factor is
important only at long wavelengths for efficient systems with high background temperatures
($\lambda \gtrsim 10^3$ μm, for a 10-% efficient system and a 100% emissive 300 °K source).

polarimetry with diffraction-limited apertures and can be scaled in a simple way for other bandwidths or apertures. As the various sections of this chapter will indicate, detectors do not exist for any part of the infrared spectrum that will be limited by the astrophysical background. This is even more strongly the case when detectors are used for narrow-band spectroscopy because the background-limited NEP varies directly as the square root of the infrared bandwidth. Improvements in single-element detectors would pay off handsomely because the integration time to achieve a given signal-to-noise ratio varies as the square of the NEP.

The development of low NEP one- and two-dimensional arrays would also have a substantial impact on infrared space astronomy. As with single detectors, the stress should be on sensitivity (low noise and high quantum efficiency) rather than speed. Photometric mapping is a straight-forward example of the potential use of arrays. The observing time required to map a region of the sky with a fixed ratio of signal to noise is reduced by a factor equal to the number of elements in the array, provided that each element of the array has the same NEP as the single detector used for comparison. This potential gain is rapidly lost, however, if the array elements have significantly larger NEPs than that of the single detector because of the dependence of integration time on the square of the NEP. In low backgrounds, the dominant noise in an array (as well as in a single detector) is often the noise in the readout system. Under these conditions, improved readout electronics are very desirable. There may sometimes be practical advantages to integrating arrays that can store the signal in the form of a charge for a significant time and then read it out quickly.

Spectroscopy has applications for low-noise, single-element detectors as well as for one- and two-dimensional arrays. Grating spectroscopy, which becomes an attractive technique when the detector noise has been reduced to the point where the dominant noise comes from the fluctuations in the arrival rate of the photons from the background or from the observed source, requires low-noise, one-dimensional, and long, narrow, two-dimensional arrays. Fourier transform spectroscopy has use for both single-element detectors of low NEP and moderate speed as well as arrays for spectroscopic imaging. Observations of spectral lines from small objects with Fabry–Perot interferometers require single-element detectors with the highest possible sensitivity. There is a complicated tradeoff among these spectroscopic techniques that depends not only on the scientific mission but also on the properties of available detectors. This tradeoff is discussed in the appendixes to the April 14, 1978, SIRTF Interim Report.

Heterodyne techniques will find increasing application in high-resolution spectroscopy, especially at wavelengths longer than ~10 μm. The full utilization of heterodyne techniques will require the development of tunable local oscillators and broad-band mixers. Broad mixer and IF amplifier bandwidths are needed to facilitate line searches and multi- (frequency) channel operation with a single receiver.

Not all space applications require detectors optimized for low backgrounds. For example, backgrounds in planetary astronomy can be very much higher than the astrophysical backgrounds used to calculate curves a–d in Fig. 1. However, planetary probes (and other multiyear space missions) have severely limited cryogenic capabilities. Under these operating conditions, available detectors for wavelengths $\gtrsim 5$ μm are often not photon-noise-limited, even for the relatively high planetary backgrounds. Thus, considerable benefit could be obtained from improved detectors for long wavelengths that are designed to operate in the temperature range 50 °K $\leq T \leq$ 150 °K, which can be achieved with passive cooling techniques.

III. Photon Detectors

In Sections A–F we shall discuss the operating principles and performance of detectors that respond to the rate at which photons, not power, are incident upon the active area. Several types of photon detectors can be distinguished from one another. In external photodetectors, a current proportional to the incident photon rate is produced in the form of free electrons emitted from a photocathode into a vacuum. These detectors will be discussed in Section A. In subsequent sections, B–F, we shall discuss internal photodetectors, where the current flow occurs inside the detector material. Then, in Section G a brief discussion of an unconventional type of photon detector, known as a quantum counter, will be given.

A. PHOTOCATHODES

Photocathodes have been employed in a large variety of highly successful astronomical detectors for wavelengths less than 1 μm. Photomultiplier tubes, image intensification tubes, and image dissectors are but a few of the devices that make use of the photoemission of electrons from a cathode surface. In large measure, these devices have gained their success because each free photoelectron may be accelerated to a high energy by an electric field and detected as a single event. Such detectors have good linearity and, in many cases, closely approach the performance of an ideal photon counter. Large developmental efforts have been devoted to improving the performance of photocathodes, both in terms of increasing their quantum efficiency and extending their response into the infrared. By far the most significant improvement has been the development of negative-electron-affinity III–V photocathodes (Spicer, 1977). With these cathodes, quantum efficiencies of 20 to 30% in the visible and 10% near 1 μm have been obtained in practical single-photon counting systems.

In present photocathode materials, interfacial potentials of ~0.8 eV limit the longest wavelengths of response to ~1.0 μm. Other material combinations may be discovered that will allow lower interfacial potentials and hence longer wavelength response. For the near term, however, it seems more likely that other techniques may significantly extend the response of photocathodes. An example of perhaps the most straightforward technique is field-assisted photoemission.

In a field-assisted photocathode, a very thin metallic layer is placed within the photocathode structure. This layer, which is transparent to electrons owing to its extreme thinness, allows an external voltage bias to be applied to the cathode. The voltage produces an internal electric field which enhances the negative electron affinity of the cathode. This technique has demonstrated photoresponse out to 1.5 μm in the laboratory. The possibility of response at even longer wavelengths seems to exist, because semiconducting alloys can be tailored to have band gap energies corresponding to nearly any desired near-infrared wavelength. Present work on photocathodes beyond 1 μm deals with the many solid-state physics problems that remain to be solved.

The wide variety of demonstrated techniques for detecting single photoelectrons would certainly allow any reasonably efficient photoemitter in the infrared to be an unexcelled detector at the very low astronomical backgrounds expected in the 2–3-μm spectral region. For example, an IR photocathode combined in an image tube with existing CCD electron detectors would offer an enormous step forward.

B. Photovoltaic and Photoconductive Detectors

Internal photodetectors can be made from either intrinsic or extrinsic semiconductor materials. In the former case, free-charge carriers are generated from the bulk semiconducting material itself. In the latter case, they are generated from impurity atoms deliberately introduced into the bulk material. Figure 2 compares the electronic energy levels in an intrinsic semiconductor with the corresponding levels in both p- and n-type extrinsic semiconductors. All three materials are shown with the same long-wavelength cutoff.

In an *intrinsic* semiconductor, absorption of a photon with energy greater than the band gap simultaneously creates a free hole in the valence

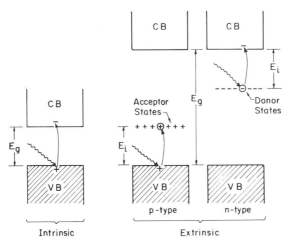

FIG. 2 Electronic energy levels in an intrinsic semiconductor compared with *p*- and *n*-type extrinsic materials with the same long-wavelength cutoff. The electronic transitions, which result from the absorption of a photon and the resulting mobile charges, are shown in each case. Ionized (neutralized) impurity states are shown circled. They are immobile and do not contribute to current flow. In the figure, VB is the valence band, CB the conduction band, E_g the gap energy, and E_i the impurity ionization energy.

band and a free electron in the conduction band. These free carriers are then able to move under the influence of an electric field and thereby produce a current. Thus photoresponse exists short of a cutoff wavelength $\lambda_c = hc/E_g$, where E_g is the semiconductive band gap. Considerable effort has gone into producing detectors from semiconductors with a wide range of band gaps. High-performance intrinsic detectors for low-background conditions are currently available only for the wavelength region $\lambda \lesssim 15 \ \mu$m. Intrinsic response upward of 20 μm has, however, been reported in alloy–semiconductors such as HgCdTe.

In a *p*-type *extrinsic* material, the absorption of a photon with energy larger than the separation E_i between the valence band and the acceptor states (between the conduction band and the donor states in *n*-type material) causes the generation of one mobile hole (electron) and an immobile ionized impurity. The cutoff wavelength $\lambda_c = hc/E_i$ for extrinsic photoconductivity depends on both the properties of the host semiconductor and of the particular dopant. Of special interest are extrinsic detectors in silicon and germanium, denoted generically as Si:XX and Ge:XX (where XX represents any of a number of dopants). These are clearly the best available photon detectors for 15 μm $< \lambda <$ 210 μm.

Figure 3 shows the long-wavelength cutoffs for a variety of intrinsic and extrinsic detectors that are reasonably well developed. Some character-

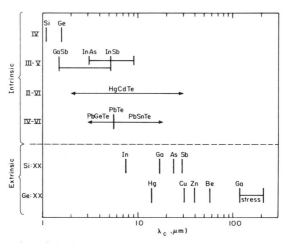

Fıg. 3   Illustration of the long-wavelength cutoff λ_c of various intrinsic and extrinsic photon detectors. Horizontal lines are used in cases where a range of values of λ_c can be obtained by alloying or by the application of a uniaxial stress. An arrow indicates that further extension of the range is possible. In the case of the alloy InAsSb, λ_c has a maximum value of ~ 9 μm for a mixture of Sb and As.

istics of these detectors will be discussed further in Sections C–E. Relevant general discussions of these detectors can be found in Willardson and Beer (1970, 1977) and also in Keyes (1977).

Several properties of intrinsic detectors make them preferable to extrinsic detectors. At high concentrations the impurity states in extrinsic semiconductors overlap in real space and form energy bands. These and other effects severely limit the doping levels that can be used. As a result, the typical absorption lengths are the order of millimeters in extrinsic materials as compared with micrometers in intrinsic materials. Thus, intrinsic detectors are much thinner than extrinsic detectors. This thinness of intrinsic detectors greatly reduces their susceptibility to optical cross talk from internal reflections when used in arrays and to the effects of energetic nuclear particles. Another practical advantage of intrinsic detectors is their operating temperature. For the same long-wavelength spectral cutoff, the operating temperature required to avoid adverse effects from thermal generation of carriers is not as low for intrinsic as it is for extrinsic detectors. The advantage is compounded by the ability of the fabricator of many intrinsic detectors to adjust the energy gap so that the range of sensitivity corresponds to the spectral region of interest. Finally, a number of curious anomalous effects have been observed at low background in many extrinsic silicon and germanium detectors, which so far have not been encountered in most intrinsic detectors.

Internal photodetectors can be operated in two basic modes: photoconductive (PC) and photovoltaic (PV). In photoconductive detectors, the free carriers migrate under an externally applied electric field. Such detectors can be made of either extrinsic or intrinsic materials. In a PV detector (also referred to as a photodiode), there is a separation of the two types of carrier under the influence of an internal electric field (sometimes assisted by an external field). Such detectors are always made from intrinsic semiconductors. The internal field may be due to the presence, for example, of a $p-n$ junction, a heterojunction, a metal–insulator–semiconductor (MIS) structure, or a Schottky (metal–semiconductor) barrier. If the PV detector is operated in the rarely used, open-circuited configuration, then the separation of charge produces a voltage across the detector contacts. This effect gives the photovoltaic device its generic name. Usually, however, a photodiode is operated in an essentially short-circuited configuration and the current produced by the process of charge separation is measured.

The amount of charge transferred per incident photon determines the responsivity of either type of detector. The difference between the two types of charge collection mechanisms, however, has important consequences. In an idealized photoconductor, every free carrier generated in the volume of the device is capable of producing a current. The amount of charge transferred depends on the carrier lifetime and mobility, the applied electric field, and the detector size. A quantity known as the photoconductive gain, usually symbolized as G, is defined to be the ratio of the transferred charge to the charge on a single electron. In the simple case, this is equal to the ratio of the carrier lifetime to the carrier transit time across the detector. Observed values range from less than 10^{-3} to greater than 10^4. Gains greater than 1 require, in addition to long lifetime, what are known as "ohmic contacts." Such contacts allow the free passage of carriers from one electrode into the semiconductor to replenish those carriers removed at the other electrode. Making noiseless ohmic contacts is a substantial part of the art of PC detector fabrication. In terms of G, the current responsivity S can be written as

$$S = G\eta e\lambda/hc,$$

where the responsive quantum efficiency η is equal to the number of free carriers (or carrier pairs) generated divided by the number of photons incident on the active area of the detector. For the sake of conciseness, careful distinctions among the various types of quantum efficiency are not made in this chapter. Some of the issues involved will be discussed in Appendix II.

In a PV device, one unit of charge is transferred per electron–hole pair generated within the small active region near the junction. Thus the equation for the responsivity of a PV detector is similar to that for a PC detector, except that $G = 1$ and η is now the fraction of incident photons that produce carriers that are collected. Typical values of η are >0.6 for both types of intrinsic detectors. By comparison, extrinsic detectors often have values of $\eta < 0.3$.

The control of η in a PV detector is better understood than the control of $G\eta$ in a PC detector. Therefore, PV detectors offer practical advantages in photometric accuracy and in uniformity as array elements. Furthermore, photodiodes typically have much higher impedances than intrinsic photoconductors and do not require bias. Thus, photodiodes offer lower power dissipation and easier coupling to multiplexers, both vitally important characteristics for the feasibility of very large arrays. Strong Department of Defense interest in arrays has therefore focused much of the intrinsic detector development effort on PV devices.

Extrinsic silicon and germanium PC detectors are frequently used for low background applications, particularly for wavelengths longer than 10 μm. The relatively advanced CCD and related MOS technologies in silicon have spurred an intensive effort aimed toward the development of arrays of PC detectors with the associated processing electronics incorporated into monolithic structures.

1. Noise in PV and PC Detectors

The fundamental source of noise in the photodetection process is the statistical fluctuation in the number of incident photons (signal plus background), which is commonly referred to as *photon noise*. This fluctuation in the arrival rate of photons produces a fluctuation in the rate of generation of free carriers, which appears as noise in the detector output current. Because of the determinancy of the charge collection mechanism in PV detectors, there is no additional fluctuation resulting from this process. In photoconductive detectors, however, additional fluctuations, which are equal in magnitude but uncorrelated, are caused by the statistics of carrier recombination. Thus, in photon-noise-limited operation, PV detectors are more sensitive by a factor of $2^{1/2}$ than PC detectors. As the background level is reduced, other sources of noise become important and this advantage disappears. Even at high background, some photoconductors such as HgCdTe can be operated under conditions where space-charge effects (related to the phenomenon known as sweepout) suppress the recombination noise. Thus, these PC detectors have the same photon-noise-limited sensitivity as PV detectors.

A commonly used figure of merit is the noise equivalent power previously defined (in Section II) as the signal power required to produce a signal-to-noise ratio equal to 1, where the noise (rms) is measured in a

1-Hz bandwidth. (Another widely used figure of merit, the D-star (D^*) is not used in this chapter but is defined in Appendix II.) The NEP of a photon detector cannot be less than the limit set by the fluctuations in the background radiation,

$$\text{NEP} \geq (hc/\lambda)(2\dot{N})^{1/2},$$

where $\dot{N}$ is the number of background plus signal photons incident on the detector per second. The difference between the NEP calculated using Poisson statistics as previously assumed and that calculated using the exact Boise–Einstein statistics is usually negligible in the wavelength region where photon detectors are available. The NEP of a real detection system is given by the ratio of the rms current–noise i_n to the responsivity, i.e.,

$$\text{NEP} = i_n/S = (i_{n,d}^2 + i_{n,a}^2)^{1/2}/S,$$

where $i_{n,d}$ is the current noise of the detector and $i_{n,a}$ the noise of the associated amplifier. Amplifier noise contributions will be discussed in Section VI.A.

Let us first consider the sources of noise particular to PV detectors. A photodiode with zero-bias voltage and no photon background can have only Johnson noise because it must be in thermal equilibrium. Excess low-frequency noise (commonly referred to as $1/f$ noise) often appears when a bias is applied. For this reason, photodiodes are usually operated at zero bias, although, theoretically they are more sensitive with reverse bias. The NEP of an unbiased PV detector at temperature T and illuminated by $\dot{N}$ photons per second is

$$\text{NEP}_{\text{PV}} = (hc/\eta e\lambda)[2e^2\eta\dot{N} + (4kT/R_0) + i_{n,a}^2]^{1/2},$$

where the first term in the square brackets is the shot noise in the photocurrent (photon noise), the second term the Johnson noise of the diode resistance at zero bias R_0, and the third term the amplifier noise. At high backgrounds, where the second and third terms are negligible, the NEP of a PV detector is $\eta^{-1/2}$ times larger than the NEP of an ideal photon detector. Because η is typically greater than 0.6, PV detectors are nearly ideal high-background detectors for those wavelengths where they are available.

The usefulness of a PV detector in low background (small $\dot{N}$) improves with increasing values of R_0. Because PV detectors are thin planar devices, R_0 is usually inversely proportional to detector area. A common figure of merit is the R_0A product. This important parameter depends on the particular device structure (e.g., p–n junction, Schottky–barrier junction) and also on a number of semiconductor properties including band structure, minority-carrier lifetimes and mobilities, acceptor and donor densities, surface generation rates, and defect densities. (Because of leakage effects along the periphery of a PV detector, R_0A may not be strictly independent of device area.) The R_0A product is also strongly dependent upon E_g/T. Thus the required operating temperature decreases with increasing λ_c. For the better developed PV detectors, values of R_0A at low temperatures are often large enough that Johnson noise is negligible compared with amplifier noise.

As discussed in Section VI.A, the effective amplifier noise term becomes larger if the detector capacitance C is comparable to or larger than the amplifier input capacitance. Thus, for the lowest NEP, one is restricted to small area detectors. (An important parameter is the detector capacitance-to-area (C/A) ratio, which depends on several material properties, including the dielectric constant, band gap, and carrier concentration.) A particularly severe case is the PbSnTe PV detector, which has very high junction capacitance (~ 1 μF/cm^2). In an effort to reduce the deleterious effect of this high capacitance, geometrical enhancement techniques have been developed to yield ratios of optical area to electrical area to electrical area greater than 30.

Let us now consider the case of PC detectors. The NEP of a PC detector (including again its amplifier) may be written as

$$\text{NEP}_{PC} = (hc/G\eta e\lambda)(4G^2e^2\eta\dot{N} + i_{n,e}^2 + i_{n,a}^2)^{1/2},$$

where the excess noise term $i_{n,e}$ and the gain G are often frequency-, illumination-, bias-, and temperature-dependent. As discussed in an earlier paragraph, the photon-noise-limited NEP is $2^{1/2}$ times higher than for a PV detector with the same value of η. Neglecting the excess noise term, we see that at high backgrounds the only figure of merit is η, and at low backgrounds, the product $G\eta$ should be maximized. In intrinsic PC detectors, η is usually $\gtrsim 0.5$, as is the case with PV detectors. However, in extrinsic detectors, a much lower absorption coefficient may yield substantially lower values of η. To increase the absorption efficiency, extrinsic detectors are sometimes mounted in chambers with reflective walls, called *integrating cavities*.

The high capacitance of PV detectors is the chief limitation to their ultimate low-background sensitivity. For this reason, PC detectors with their low capacitance (the order of 10 pF/cm^2) can be more sensitive, particularly because $G\eta > 1$ is, in principle, possible.

An unconventional, but revealing, way to summarize the status of development of photon detectors is to plot the minimum photon rate that can be detected in a 1-Hz bandwidth ($NE\dot{N}$) as a function of $\dot{N}$. Available data are shown in Fig. 4. The apparent limit at 10^3 photons/sec $\cdot$ Hz$^{1/2}$ is caused by amplifier noise (see Section VI).

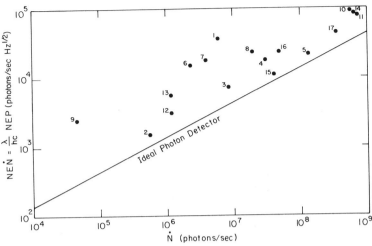

FIG. 4 The measured noise equivalent photon flux NEṄ for a number of low-background detector tests is plotted as a function of the photon flux Ṅ incident on each detector. The sloping line represents the performance of an ideal ($\eta = 1$) photon detector obeying Poisson statistics. Because NEṄ is defined for a unit bandwidth the ideal limit is NEṄ = $(2\dot{N})^{1/2}$. The points that are plotted came from a variety of sources and in many cases do not represent optimal performance for a particular detector or detector type. The listing at each point refers to detector type, test wavelength (not necessarily wavelength of peak performance), detector temperature, comment (if any).

1. PbS (PC), 3 μm, 150 °K,
2. InSb (PV), 2 μm, 4 °K,
3. InSb (PV), 5 μm, 65 °K, (flashed),
4. InSb (CID), 4.2 μm, 77 °K,
5. InSb (CID), 4.2 μm, 77 °K,
6. PbSnTe (PV), 10 μm, 30 °K,
7. InAsSb (PV), 4.3 μm, 80° K,
8. GaInSb (PV), 4 μm, 80 °K,
9. HgCdTe (PC, trapping mode),
   3 μm, 5 °K,

10. HgCdTe (PV), 4 μm, 100 °K,
11. HgCdTe (CCD), 4.3 μm, 77 °K,
12. Si:As, 15 μm, 2.5 °K, (typical IRAS),
13. Si:Sb, 15 μm, 2.5 °K, (typical IRAS),
14. Ge:Cu, 10 μm, 4.2 °K,
15. Ge:Ga, 94 μm, 3 °K, cavity,
16. Ge:Ga, 94 μm, 3 °K, no cavity,
17. Ge:Ga, 150 μm, 2 °K, stressed.

C. INTRINSIC PV AND PC DETECTORS

Infrared detectors have been fabricated from a wide variety of intrinsic semiconductor materials. The first detectors were made from elemental semiconductors such as Ge and Si and simple compound semiconductors such as PbS, PbSe, InAs, and InSb. Interest in longer wavelength response than is available in these simple materials and the desirability of tailorable spectral response has led to the exploration of a number of ternary and occasionally quaternary alloy–semiconductor systems. Currently the most advanced alloy systems for infrared detection are GaInSb, InAsSb, PbSnTe, and HgCdTe.

The first available intrinsic detectors were generally operated in the PC mode. These, however, have been largely superseded by photovoltaic detectors. This trend is true for both high-performance single detectors and array elements. The use of intrinsic detectors in focal planes has been described by Longo *et al.* (1978).

1. *Group IV Semiconductors*

The well-known semiconductors Ge and Si exhibit intrinsic absorption in the near-infrared region less than 1.6 and 1.1 μm, respectively. Silicon photodiodes exhibit high quantum efficiency and are amplifier-noise-limited in nearly all cases. For moderately high backgrounds, where the detectors are photon-noise-limited, their high quantum efficiency makes them better than photomultipliers. Avalanche photodiodes in silicon are reasonably well developed and can reduce the amplifier noise limit somewhat. Recent advances in Ge purification have allowed high-quality Ge photodiodes to be made.

Charge-coupled-device (CCD) and charge-injection-device (CID) imagers are well developed in intrinsic silicon. Extremely low noise CCD arrays with long integration times are being developed by NASA for use on the Space Telescope.

2. *Group III–V Semiconductors*

Visible and infrared photodetectors have been made of many combinations of Al, Ga, and In with P, As, and Sb. The chief infrared materials have been InAs and InSb, with the alloys GaInSb and InAsSb having undergone recent development.

Astronomers have been using InSb PV devices for ground-based observations for several years, with continually improving sensitivity. Values of R_0A greater than 10^7 $\Omega \cdot$ cm^2 at 77 °K have been reported. A technique called "flashing" that uses brief illumination of a cold (~60 °K) detector with strong, near-IR radiation has achieved values higher than 10^9 $\Omega \cdot$ cm^2. Cooling to 5 °K, without flashing, has yielded R_0A products greater than 10^{10} $\Omega \cdot$ cm^2, which easily challenge the best amplifiers. Upon cooling below 30 °K, some InSb detectors show a gradual falloff in responsivity with increasing wavelength. It is possible that this loss could be avoided in a different device geometry. Photodiodes of InSb have typical capacitances of ~50 nF/cm^2, which are comparable to other III–V and II–VI devices. Techniques used to reduce the much higher capacitance of IV–VI diodes could probably be used in III–V and II–VI detectors and would provide a useful gain in sensitivity.

Schottky barrier detectors, CCDs, and CIDs have all been demonstrated in InSb, which is the best-studied and one of the best-behaved compound–semiconductor materials. The InSb CID is used in one of the most successful of the present intrinsic detector arrays. Background-noise-limited operation at 77 °K with $\eta \simeq 0.5$ has been demonstrated at moderate ($\sim 3 = 10^{12}$ photons/cm$^2 \cdot$ sec) background levels. Substantial improvement has been demonstrated for individual CID elements operating at lower temperatures.

The alloy systems GaInSb and InAsSb can be varied to yield devices with λ_c values covering the range from 1.6 to 9 μm. Values of R_0A greater than 10^6 $\Omega \cdot$ cm^2 at 80 °K have been reported for diodes with $\lambda_c \simeq 4.3$ μm in both systems. Self-filtering devices have been demonstrated with a short wavelength cuton produced by a different composition alloy in a backside-illuminated structure.

3. Group IV–VI Semiconductors

Photoconductive detectors of PbS and PbSe for the 1–5-μm region have been available for many years, but their performance at low backgrounds has been surpassed by PV detectors such as InSb. The main IV–VI detector material of recent interest is PbSnTe, although detectors have been made in several other IV–VI ternary alloys, e.g., PbSnSe, PbSSe, PbGeTe. Detectors have been fabricated with response beyond 20 μm, although the main effort has been in detectors for 8 to 14 μm. For this region, PV detectors with very high values of R_0A at low temperature are now available. These are generally amplifier-noise-limited because of the high junction capacitance of PbSnTe diodes. The high thermal expansion of PbSnTe presents a problem for interconnection of large diode arrays with Si multiplexer chips. Because of problems caused by this high capacitance and thermal expansion, further development of lead–salt detectors is not being pushed.

4. Group II–VI Semiconductors

One of the most versatile infrared detector materials is HgCdTe. By varying the alloy composition, detectors with λ_c values from 2 to 30 μm have been fabricated. High-performance PC detectors for intermediate backgrounds have been available for several years, and there is, at present, an intense development effort in PV devices. High values of R_0A ($>10^6$ $\Omega \cdot$ cm^2) have been attained in detectors with $\lambda_c < 5$ μm. It is expected that high values of R_0A will be available in longer-wavelength compositions at low temperatures within a few years.

At the present time, PC HgCdTe detectors are not under as heavy a development as PV HgCdTe devices. Below a transition temperature that depends on composition, HgCdTe PC detectors change from the normal PC mode of operation to what is known as a "trapping mode," where very large values of G ($>10^4$) are observed. The trapping mode mechanism is thought to be caused by the presence of a depletion layer at the photoconductor surface. Models for the enhanced photoconductive response caused by such a depletion layer are being studied. Trapping mode devices are characterized by a strong dependence of responsivity and spectral response on background illumination and by spatial nonuniformity. Although very high sensitivities have been obtained with the large gain, effective use of trapping mode devices may be difficult until these properties are understood and controlled. Also under development are CCDs and CIDs in HgCdTe as well as more conventional hybridization of PV detectors with Si CCD multiplexers.

D. EXTRINSIC PHOTOCONDUCTIVITY IN Si AND Ge

Impurity photoconductivity in Si and Ge has been studied extensively for over 25 years (Bratt, 1977). For several reasons most of the recent development effort has gone into Si rather than Ge; these include the mature MOS and CCD technologies in Si, the possibility of smaller detectors because of stronger absorption in Si, and relatively small Department of Defense interest in the region $\lambda > 30$ μm where Ge:XX must be used.

The long-wavelength limit λ_c to the response of an extrinsic photoconductor depends strongly on both the host material and the dopant and weakly on temperature. For a given host–dopant combination, there is usually a maximum in the photon absorption coefficient at a wavelength typically 0.6–0.95 times λ_c. The detailed shape of a detector's spectral response function depends on a number of fabrication parameters, such as doping concentration, detector thickness, and contact geometry, as well as on the fundamental variation of the absorption coefficient with wavelength. The spectral response can be further modified by the use of antireflection coatings, backsurface reflectors, and integrating cavities. Optimization of the performance is a complex process, which is made more difficult by the presence of the poorly understood anomalous effects to be discussed next. Typically, an extrinsic detector will have quantum efficiency greater than half its peak value over a wavelength range from λ_c/x to λ_c, where x varies from 2 to 6.

Photoconductivity has been studied in extrinsic silicon and germanium with a wide variety of impurities. The choice of dopant is influenced by a number of factors besides the cutoff wavelength λ_c and the absorption cross section. These factors include ease of uniform doping at a controllable level without introducing undesirable impurities and compatibility with other processing steps. Because of these diverse requirements the best dopant for a discrete detector may not, for example, be best for a monolithic array. Values of λ_c for some of the more fully developed dopants in both silicon and germanium appear in Fig. 3.

Doped germanium detectors have been developed to cover a wide wavelength region, but currently those detectors with $\lambda_c < 30$ μm, such as Ge:Au, Ge:Hg, and Ge:Cd, have been supplanted by Si:XX detectors. However, Ge:Cu ($\lambda_c = 31$ μm) still remains important for use as a photomixer for heterodyne receivers. The shallow impurity levels in Ge (produced, for example, by Zn, Be, and Ga) make it possible to cover the region from 30 to 120 μm. By subjecting Ge:Ga to uniaxial stress, the long-wavelength cutoff can be further extended to 210 μm. Thus Ge:Ga is the best developed of these long-wavelength detector materials for low-background applications and is being used in both the IRAS and Spacelab 2 IRT experiments for all spectral bands with wavelengths longer than 30 μm.

Because of the small doping permitted in Ge:Ga, several millimeters of optical path length are needed to produce adequate absorption, especially near 30 μm. Integrating cavities are especially useful under these conditions. Germanium also has an inherent lattice absorption between 30 and 40 μm, which competes with the impurity absorption. A fully developed Ge:Be or Ge:Zn detector should offer an improvement over Ge:Ga in this interval. Germanium:Zinc was one of the first extrinsic detectors developed and could probably be improved by more recently developed processing techniques. Development of the Ge:Be detector is relatively recent and has been too slow for use on the previously mentioned space missions.

Another consideration for low-background space-based observations is sensitivity to energetic nuclear particles. The high doping levels used in extrinsic silicon detectors make the absorption length for photons relatively short ($\lesssim 1$ mm). The small detector dimensions and low atomic number of Si should make these detectors less sensitive to energetic nuclear particles than Ge:XX detectors with their larger volume and atomic number.

1. Responsivity and Anomalous Effects in Si:XX and Ge:XX Detectors

As previously discussed, the responsivity of PC detectors is proportional to the product $G\eta$ and is directly measurable. Although several methods exist, it is more difficult to accurately determine G and η separately. Typical values for η in well-developed materials range from 0.1 to

0.4. Values approaching 1.0 have been reported (usually when an integrating cavity has been used). It is not uncommon for the dc photoconductive gain to be greater than 1 in the well-developed Si:XX detectors and in some short-wavelength Ge:XX detectors. Values as large as 40 have been reported. (High values of G have generally been observed in detectors operating at temperatures near to their maximum operating temperature. For example, the responsivity of S̃i:In has been observed to vary by more than a factor of 10 between 20 and 50 °K.)

Although dc photoconductive gains greater than 1 can be obtained and do give high dc responsivity, the useful responsivity in low-background operation is often lower for two reasons: gain saturation and anomalous effects.

When the infrared signals vary faster than the rate of dielectric relaxation, the photoconductive gain is reduced by space-charge effects (also known as sweepout) and saturates at a value of ~0.5. A theory predicting gain saturation is available and gives qualitative agreement with the observations, though there has been some difficulty in obtaining detailed agreement.

Under conditions of high-bias and low-background levels, extrinsic photoconductors may exhibit a number of peculiar effects. Many of these effects are thought to be related to properties of the semiconductor–contact interface and to the long times needed to establish equilibrium charge and field distributions in very high resistivity detectors, but full explanations are not available. The lack of reproducibility of these effects, even in nearly identical detectors, has hampered systematic understanding or empirical control. Among these anomalies are the "memory effects," in which the amplitude of the response to a small signal may depend on the light level falling on the detector in the preceding few seconds. As the bias and, therefore, the photoconductive gain are decreased, the importance of these memory effects and other long time constant (10–30-sec) anomalies decreases. With present Si:As and Si:Sb detectors, one can expect to sacrifice about a factor of 2 in responsivity in order to reduce the memory effects significantly.

Spatial nonuniformity of both spectral and overall response has also been observed in some detectors with partially illuminated surfaces under conditions of low background and high bias.

At low illumination levels, many extrinsic detectors produce spontaneous current spikes. These are typically very fast pulses (usually limited by the preamplifier electronics to ~10^{-3} sec), which are larger than the detector noise by a factor of 2 to 10^3. Typical spiking frequencies are 0.1 to 10/sec. The occurrence of spikes depends on the detector temperature, the background, and the previous illumination of the detector. The spiking characteristics vary from one manufacturer to another and even from one batch of detectors to another from a single vendor. Electronic supression techniques and/or the equivalent software processing that have been developed to remove the fast pulses resulting from interactions with nuclear particles will also remove these spontaneous spikes from the detector signal. Except in the cases of very frequent spiking (>10 sec) or long spikes ($>5 \times 10^{-3}$ sec), the signal can be recovered with only a minor penalty in noise.

The lowest NEP values reported for extrinsic detectors are about $3-10 \times 10^{-17}$ W/Hz$^{1/2}$ and are amplifier-noise-limited. These values were achieved by operating the detectors at high bias. As previously noted, this increases both the importance of the memory effects and the spontaneous spiking problems. Neither of these would seriously affect low-level photometry, but both would trouble scanning instruments such as IRAS and rapid-scan spectroscopic systems. Figure 5 summarizes current results on the performance of detectors considered for IRAS and/or the Spacelab 2 IRT.

E. PHOTON DETECTORS BEYOND 120 μm

1. *Germanium*

The long-wavelength limit of photon detectors currently used in astronomy is set by the ionization energy E_i of the common shallow acceptors in Ge. This acceptor ionization energy, which corresponds to 120 μm in Ge:Ga, can be reduced by the application of a uniaxial stress along the (100) crystallographic direction. A photoconductive detector constructed using this technique (Haller *et al.*, 1979) has a long-wavelength cutoff at 210 μm, NEP = 6×10^{-17} W/Hz$^{1/2}$, and a $G\eta$ product (in a cavity) of 0.3. Although the use of a mechanically stressed semiconductor as a detector is an unconventional technique, it appears to be sufficiently well controlled for applications in space astronomy. The benefits obtained by extending photoconductive technology to longer wavelengths are indicated by the fact that the NEP of this detector is about two orders of magnitude lower than for a bolometer at the same operating temperature of 2 °K and about one order of magnitude lower than the best bolometer made at any temperature. Unfortunately, the stress technique cannot be extended to longer wavelengths in Ge.

2. *III–V Semiconductors*

In some compound semiconductors, such as GaAs and InSb, impurity ionization energies are smaller than in Ge, so extrinsic photoconductivity at longer wavelengths is possible. Photoconductivity in GaAs has been

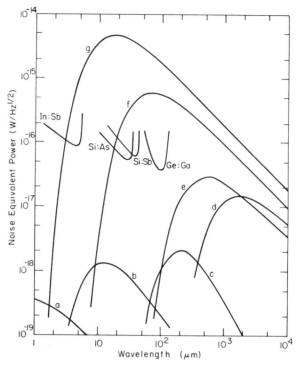

Fɪɢ. 5   The measured low-background NEPs of detectors that will be flown on the IRAS and Spacelab 2 IRT missions in 1982–83 compared with astrophysical background limits taken from Fig. 1. Because of the larger throughput and bandwidth used in the IRAS and Spacelab 2 IRT missions, the background limitations anticipated are closer to the detector performance than the limiting NEP curves here. The plot illustrates, however, the detector improvement required before the full benefit of diffraction-limited space missions can be obtained.

extensively investigated (Stillman *et al.*, 1970). A rather smeared-out photoconductive onset edge has been observed, which extends beyond 300 μm. The performance of this device in high backgrounds, such as those encountered in planetary astronomy, is relatively good, but its low-background performance is not competitive with bolometric detectors. The poor sensitivity in low background and that observed in other extrinsic III–V photoconductors is probably due to residual impurities. Preparation of these compounds with adequate purity is likely to require a major effort.

In the lowest carrier-concentration InSb available, impurity state overlap and strong impurity band effects are seen. Long-wavelength detectors and mixers have been developed from InSb and are often called "hot electron bolometers" (Putley, 1977). The mobility in n-type InSb increases with increasing electron temperature. The resistance of the sample thus reflects the heating of the electron plasma by absorbed photons. This type of detector operates efficiently at frequencies in the neighborhood of the electron plasma frequency, which is generally around $\lambda \simeq 1$ mm. The best detector performance obtained thus far is not competitive with modern composite bolometers. The performance as a mixer will be discussed in Section V.

F. ARRAYS

1. Advantages of Arrays

In many instruments, an array of detectors has potential advantages in integration time or signal-to-noise ratio over a single detector element. When the system is arranged so that parallel sampling of the scene is accomplished by the n array elements, an observation with a fixed signal-to-noise ratio can be made in n^{-1} the time required by a single detector. Alternatively, for a given integration time, an array with parallel sampling could provide a $n^{1/2}$ improvement in the signal-to-noise ratio. These improvement factors apply for the ideal case where the scientific mission requires at least n data elements, where the sensitivity of each array element is equal to that of the discrete detector, and where no additional noise sources are introduced by the array. Furthermore, parallel sampling discriminates against nonstationary types of noise, which can strongly affect sequentially sampled systems.

Most of the infrared detector development programs currently sponsored by the Department of Defense are aimed toward combining large numbers of detectors (in either one- or two-dimensional arrays) with suitable signal-processing electronics. Such a combination is sometimes referred to as an "integrated array." Besides the general sensitivity improvements with arrays previously listed, many types of integrated array promise additional practical advantages. They allow time averaging of infrared signals at the detector, rather than after amplification. They also have the ability to do on-chip multiplexing, which means that each detector element no longer requires an individual set of hand-wired electrical leads. Furthermore, on-chip power dissipation is often vanishingly small because a single output amplifier can be used. For large arrays, costs may be lower because automated photolithographic production techniques are used. These techniques also allow smaller detectors and higher focal-plane filling factors than are possible with discrete detectors.

Many different types of integrated detector arrays are being developed (Milton, 1977). A majority of these approaches use charge-coupled devices (CCDs) to collect the photo-generated charge from the individual detector elements and transport this charge to an output amplifier. A CCD is a metal–insulator–semiconductor (MIS) structure in which potential wells, capable of accumulating, storing, and transferring charge, can be created beneath the metal electrodes. In an integrated array using a CCD, the photocurrents generated in the individual detector elements are collected in the form of charge in such potential wells. All types of photodetectors (extrinsic PC, intrinsic PC, and intrinsic PV) can be used. The details of how the detector photocurrent is converted into CCD charge packets, however, differ for each case. At the end of an accumulation period, voltage is applied to an adjacent electrode, and the charge packet is transferred laterally into a newly created adjacent potential well. Charge packets from all array elements are shifted through the device in this manner to an output amplifier, where the stream of charge packets produces proportional output voltages. (For a two-dimensional array, a second transfer process in the perpendicular direction can be used.) This time-dependent voltage is then amplified and conveyed out of the cryogenic environment. Detailed aspects of CCD operation include the electrode clocking pattern, which typically includes multiple phases and sequential, overlapping clock voltage pulses for efficient charge transfer, and the various schemes for injection of charge into the transfer section.

Another approach to integrated detector arrays using an MIS structure is the charge-injection device (CID). In contrast to arrays with CCD readout, detector elements in a two-dimensional CID array are individually addressable (random access) by row and column. Readout is accomplished by injecting the charge contained in the selected potential well into the substrate and sensing the substrate current. Further multiplexing electronics may be integrated with the basic CID in a hybrid structure.

The basic sources of detector noise in array elements are the same as for discrete detectors. Performance of a given element may, however, be significantly worse because of the detailed requirements of array fabrication. Besides the noise of the detectors themselves, there is also the effective amplifier noise of the array electronics. For an array using a CCD, the CCD-related noise sources can be divided into three parts arising (1) from the process of injecting the detector photocurrent into the CCD, (2) from the process of charge transfer along the CCD, and (3) from the readout of charge by the output amplifier.

The most highly developed CCD imagers are intrinsic Si devices for visible light. In this case, the charge is generated directly in the wells of the

CCD. In these visible CCD arrays, charge-transfer and readout noise have been reduced to very low levels. In the analogous infrared-sensitive charge-coupled devices, the so-called IRCCDs, the performance achieved so far is much worse and is unlikely to approach that of visible Si devices in the near future. When infrared photon detectors of various types are coupled to Si CCDs, the dominant noise source is often the injection process and is much larger than any noise source in visible CCD imagers. The currently available integrated infrared arrays have fairly high readout noise and may also have short charge-storage times, yielding NEPs that are substantially worse than those of present discrete detectors. As readout noise and storage time improve, monolithic detector arrays may have a major impact upon low-background infrared observations. The sensitivity of charge-integrating detectors will be discussed in detail in Section VI.A.5.

The Department of Defense is providing substantial funding for integrated array development. Mosaic focal plane arrays composed of large numbers of multielement chips are planned. Because these arrays are potential components in future surveillance systems, much of the performance data, and in some cases the actual hardware, are currently classified.

It should be recognized that array requirements for astronomical applications often differ significantly from requirements for Department of Defense applications, which have guided infrared array developments to date. Detection of faint, infrared astronomical objects will require long integration times, perhaps in the seconds-to-minutes range, as opposed to the shorter integration times typically used in defense systems. The astronomical application does not require sophisticated on-chip processing functions. An astronomical array should be "calibratable" because scientific measurements, being photometric in nature, must not only detect an object but also indicate its absolute brightness. Versatile space telescopes will require liquid-helium cooling, so low detector operating temperatures will be available for detectors at any wavelength. Thus detectors could be used for astronomical purposes that might not be practical for long-duration military missions. Current monolithic arrays have pixel sizes of approximately 0.1×0.1 mm. For some astronomical applications, a larger pixel size may be desirable. This will probably be accompanied by some increase in NEP. Larger pixel sizes appear necessary at longer wavelengths if the use of several pixels to cover a diffraction-limited image is to be avoided.

Perhaps one area of difference in emphasis between defense and astronomical requirements is in the need for extremely low values of NEP for low-background astronomy. Because the NEPs of discrete detectors are improving, it is difficult to predict when (or whether) the performance of integrated arrays will surpass the performance of arrays of discrete detectors in such applications.

In general, although progress in some areas has been quite rapid, a large amount of work on materials, processing, and characterization remains to be done before the promise of integrated detector arrays will be fully realized. At present, for example, the estimated NEP at low background of the best integrated extrinsic Si arrays (using a 0.1-sec integration time) is at least a factor of ten larger than the NEP of a state-of-the-art extrinsic Si discrete detector. Therefore, an integrated array operated under the most favorable conditions would require $\gtrsim 10^2$ times more detectors to break even with a discrete array using the same integration time.

2. Status of Development

Three general types of infrared arrays utilizing CCDs are now under development. In the monolithic–extrinsic design, a doped Si substrate provides the infrared-sensitive medium, and charge transfer is carried out in a lightly doped Si CCD layer at or near the surface. In the monolithic–intrinsic array, an intrinsic semiconductor substrate, such as InSb, is used for detection, charge transfer, and readout. The hybrid array uses separate substrates for detection and charge transfer. Indium "bumps" are commonly used in hybrid arrays to electrically and mechanically interface the detector substrate to a Si CCD multiplexer. Within each general class, a large number of specific design approaches have been pursued (Ando, 1978).

The most highly developed type of infrared detector array is the monolithic–extrinsic array (Nelson, 1977). The operation of 32 × 32 test chips of Si:Ga and Si:In has been successfully demonstrated. Work on Si:As arrays for lower backgrounds is also in progress. A reasonable amount of high-background laboratory characterization of the Si:Ga and Si:In arrays has been carried out, although not much is known about monolithic–extrinsic array performance in low background and at liquid-helium temperatures. Data on Si:In arrays indicate that sensitivities are considerably poorer than those achieved by discrete detectors. Responsivities substantially below those of single detectors have been reported. Fast-interface-state noise, which arises in the charge-transfer process, dominates at low background, with typical values an order of magnitude or more above those achieved by intrinsic Si CCDs.

Developmental efforts toward monolithic–intrinsic IRCCDs have largely concentrated on InSb and HgCdTe. Prototype 1 × 20 InSb IRCCDs have recently been produced, but their electrical and optical characteristics have not been fully established. Intrinsic HgCdTe arrays are under development. Promising charge-transfer efficiencies have recently been demonstrated. Intrinsic arrays are also being developed in sandwiched, multilayer form, where related semiconductor alloy layers perform different functions, e.g., infrared detection in InAsSb and charge transfer in InGaSb. Although the promise of low NEP exists for IRCCDs, substantial problems remain. The technology for these intrinsic materials is far less advanced than that for silicon, and the high-surface-state densities are producing leakage currents as large as 10^5 times that achieved by extrinsic silicon CCD arrays. It appears that five, or perhaps ten years of development will be required before high-performance intrinsic CCDs become available.

A number of detector materials (InSb, InAsSb, PbSnTe, HgCdTe) are being developed in hybrid arrays where the well-developed silicon CCD technology is exploited. Each detector cell requires an individual microconnection to the adjacent multiplexer terminal, and the hybrid design must accommodate the differential thermal contraction of the detector and multiplexer materials. Reasonable progress has been made in indium-bonding techniques, and a 98% interconnect yield on a 32×32 InAsSb array has been reported. Interface circuitry problems are more acute in the hybrid approach, and difficulties with gate-threshold nonuniformities have been encountered. Hybrid arrays are at an intermediate state of development between monolithic–intrinsic and monolithic–extrinsic arrays.

As previously discussed, the CID represents a second general type of integrated array. Most CID development has been done on InSb, with some work on extrinsic Si CIDs. Excellent performance at moderate background has been demonstrated in an InSb CID at 77 °K. Preliminary tests suggest that substantial improvement should be possible with lower operating temperatures.

G. QUANTUM COUNTERS

The idea behind infrared quantum counting (and the parametric upconversion technique discussed in Section V.F) is that of converting the wavelength of the infrared signal into a spectral region where good detectors exist. Parametric upconversion is a coherent process that preserves the phase of the incident electromagnetic radiation, whereas quantum counting is incoherent and only detects the incident power. At present, neither technique is competitive with PV, PC, or bolometric detection. In principle, if formidable materials problems can be solved, upconversion and quantum counting could yield ideal receivers.

The concept behind a quantum counter is the use of a multilevel atomic, molecular, or solid-state system that has been pumped into a long-lived excited state as a medium to absorb infrared radiation. An infrared photon either ionizes the system so that the emitted electron can be detected or causes the emission of an energetic photon that can be detected with a photon counter. The detection process is incoherent and therefore has no limit on the number of modes that can be detected. The quantum-counting process, in principle, introduces no additional noise beyond the photon noise of the infrared signal and background (with the appropriate quantum efficiency) because there is no counter output without an infrared input photon. Initially, the approach was to convert infrared into visible radiation, thereby allowing the use of photomultipliers, image tubes, and film detectors. This concept can be broadened to allow for conversion from one infrared wavelength to another. This might prove advantageous because detector array technology is likely to become available for some infrared wavelengths but not others.

The early work on quantum counters used the level structure of trivalent rare earth ions in various host lattices. Difficulties were encountered in finding systems that had long-lived pumped levels with strong IR absorption and were also transparent to the output fluorescence. Values of NEP = 3×10^{-9} W/Hz$^{1/2}$ obtained for signals at $\lambda \approx 2$ μm were far from being competitive with direct detection. Recent work has utilized the Rydberg states of alkali atoms pumped by tunable dye lasers (Ducas *et al.*, 1979; Gelbwachs and Wessel, 1980).

IV. Bolometers

Bolometers are devices that respond to heating produced by absorbed radiation. Bolometric detectors have three essential parts: an absorbing surface, a thermometer, and a thermal link to a heat sink. The design of these elements depends on the wavelength range over which the bolometer is to be operated, the size of absorbing element required, the background infrared power, and the temperature of the heat sink. Bolometers have had extensive use because of their response throughout the infrared spectrum and their high absorptive efficiency (typically from 0.5 to 1). Because maximum sensitivity is achieved at very low operating temperatures, in this discussion we shall treat only bolometers operated at 2 °K and below. For applications with severe cooling constraints, such as deep-space probes, properly optimized higher temperature (50–100 °K) bolometric detectors might be significantly better than existing detectors.

Small liquid-helium-temperature bolometers with doped semiconductor thermometers have been important for high-background observations at infrared wavelengths $\lesssim 100$ μm. These generally make use of the absorption in the thermometric material. (A thin layer of black-absorbing paint is sometimes used to reduce the large reflectivity of the semiconductor.) The most important current applications of bolometers are at submillimeter wavelengths where no sensitive photon detectors exist. At submillimeter wavelengths, the most efficient bolometers are composite structures, using a metal-film-absorbing element on the reverse side of a thin transparent dielectric substrate in contact with a small thermometric element (Nishioka *et al.*, 1978). A commonly used example is a sapphire or diamond substrate with a bismuth or a nickel–chromium–alloy film absorber (thermometers will be discussed next). There is an optimum surface resistance of the absorbing film for a given dielectric substrate. For example, on a sapphire substrate with index of refraction equal to 3, a film with surface resistance 200 Ω/square gives a frequency-independent absorptivity of 50%.

In low-background environments, the fundamental limit to bolometer sensitivity is given by thermodynamic fluctuations in the heat exchanged with the low-temperature reservoir. This energy fluctuation leads to a lower limit on the noise equivalent power of a bolometer: NEP = $(4kT^2\mathscr{G})^{1/2}$. Here T is the operating temperature, $\mathscr{G}$ is the thermal conductance between the bolometer and the heat sink and k Boltzmann's constant. An important operational parameter, which is usually set by the experiment, is the thermal time constant $\tau = \mathscr{C}/\mathscr{G}$, where $\mathscr{C}$ is the bolometer heat capacity. If the heat capacity of the bolometer is dominated by dielectric substances, as will often be the case for temperatures 0.3 °K $\lesssim$ T $\lesssim$ 100 °K, the heat capacity will be proportional to T^3. It follows that the sensitivity of a low-temperature bolometer can be expressed as

$$\text{NEP} = FT^{5/2}\tau^{-1/2}.$$

Values for the parameter F in the liquid-^{4}He temperature range are approximately 1×10^{-16} W/Hz$^{1/2}$ · °K$^{5/2}$ · sec$^{1/2}$ for small, short-wavelength, Ge-doped bolometers and 3×10^{-16} W/Hz$^{1/2}$ · °K$^{5/2}$ · sec$^{1/2}$ for large, submillimeter-wavelength, composite bolometers. These values are within a factor of 2 of the theoretically expected value $(4k\mathscr{C}/T^3)^{1/2}$. A Si integrated circuit version of the composite bolometer is projected to have $F \simeq 1 \times 10^{-16}$ W/Hz$^{1/2}$ · °K$^{5/2}$ · sec$^{1/2}$. Once F is determined for a given bolometer structure, this NEP expression can be used to extrapolate the bolometer performance in low background to other values of T and τ. In practice, τ is adjusted by proper selection of the bolometer thermal conductance $\mathscr{G}$.

An ideal thermometric element for a bolometer should be able to measure the instantaneous temperature of the absorbing element without contributing any fluctuations of its own. The most commonly used (and only commercially available) thermometric element is a Ge chip heavily doped with Ga. At low infrared background, optimum operation is obtained with very small values of bias current. Under these conditions, it has been possible to construct thermometers that make a negligible contribution to the noise of bolometers operated at low audio frequencies in both the liquid-^{4}He and -^{3}He temperature ranges. Current-dependent $1/f$-noise in the thermometer places a limit on the lowest frequency at which bolometers can be operated efficiently. At present, this limit is 5–10 Hz for Ge:Ga thermometers suitable for low-background bolometers. Superconducting thermometers are a possible alternative to semiconducting thermometers and have demonstrated good laboratory performance to below 1 Hz. However, superconducting thermometers are more complicated to use than semiconducting thermometers because of the need to stabilize their operating temperatures and the need to provide low-noise amplifiers that match their low impedance. It therefore remains desirable to reduce low-frequency noise in semiconducting thermometers for applications benefiting from modulation frequencies below 10 Hz.

Figure 6 shows the performance achieved in negligible background with state-of-the-art bolometers at temperatures of 1.2 and 0.3 °K. Limits to detector NEP from several astrophysical background sources and from cold and warm telescopes (from Fig. 1) are also shown for comparison.

In applications where the radiant background is not negligible, limits to bolometer performance are set by the incident infrared background power. Depending on a number of system and bolometer parameters, this limitation can arise in either of two ways: the statistical fluctuations in the rate of energy deposition by background photons and the average temperature rise above the path temperature caused by the absorption of background power. The effects of photon fluctuations are fundamental and unavoidable. To avoid the latter effect, the thermal conductance $\mathscr{G}$ must be made large enough to keep the temperature rise small. This, in turn, sets a lower limit to the thermodynamic noise contribution $(4kT^2\mathscr{G})^{1/2}$ that can be achieved. For arbitrary spectral filtering and background source temperature, there is a bolometer operating temperature below which the photon fluctuation noise always dominates. For bolometers operated at 1.2 °K or below, the photon fluctuation limits from the astrophysical sources illustrated in Fig. 6 are more restrictive than the sensitivity limits set by background power loading.

The optimization of a bolometer for high-background applications is a complicated problem that cannot be treated in detail here. Because low heat capacity is not critical in high-background bolometers, it is generally easier to fabricate an optimized bolometer for a high-background than for a low-background application. However, optimum performance at high background requires operation at higher bias power levels, which can increase problems with current noise in the thermometric element.

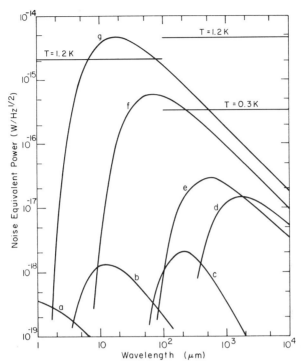

FIG. 6   Representative NEPs of state-of-the-art bolometers operated in negligible background at temperatures of 1.2 and 0.3 °K compared with the various background limits from Fig. 1. A time constant of 10 msec is used in all cases. For other time constants, the NEP scales as $\tau^{-1/2}$. The bolometers chosen for this illustration are a conventional Ge bolometer for wavelengths less than 100 μm and a composite bolometer (Ge thermometer on a sapphire substrate) for wavelengths greater than 100 μm. The break at 100 μm corresponds roughly to practice but does not represent a sharp cutoff for either device.

In the discussion of the fundamental limits to low-background bolometer performance, it was shown that the achievable bolometer sensitivity varies extremely rapidly with operating temperature. The time required for a low-background measurement varies as the fifth power of the operating temperature. For example, a bolometer operated at 0.2 °K could obtain data in 5 min that would require a year of integration with a bolometer operated at 2 °K. Though this extrapolation may be somewhat optimistic, in that the metallic elements of any bolometer will dominate the heat capacity at sufficiently low temperature and cause the NEP to vary only as $T^{3/2}$, it nevertheless is a matter of great importance for the development of low background satellite systems that refrigerators be developed that can operate bolometric detectors at temperatures below the liquid-^{4}He temperature range. A number of such refrigerators have been developed for other purposes, and with a suitable development effort, it should be possible to adapt such techniques to the space environment. It should also be noted that there are potential problems associated with bolometer operation at lower temperatures that need experimental investigation.

The best present bolometers are fabricated by hand, with the result that there are relatively limited sources and low yield of good-quality detectors. For applications benefiting from arrays of more than a few detectors, development of more automated production techniques of high-quality bolometers would be an important step.

The broad spectral response of bolometers places a strong demand on spectral filters for rejection of unwanted radiation. For submillimeter photometry, there is particular need for development of bandpass filters to operate at low temperatures and to provide sharp band definition, high in-band efficiency, and strong out-of-band rejection.

One additional problem, which bolometer users have not had to face in low-altitude applications, is the rejection of unwanted signals and increased noise arising from energy deposited by charged particles and gamma rays. Single events can produce pulses of substantial signal-to-noise ratio in bolometers operated at and below liquid-^{4}He temperatures. Though some work has been done, experience is needed with pulse discrimination and/or suppression techniques that are applicable to bolometers.

V.   Coherent Detectors

Two classes of detector are used for infrared signals. The incoherent (direct) detectors discussed thus far use infrared square-law devices in which the ouput signal is proportional to the input power. This class includes photon detectors in which the energy of a photon is converted into the current of an electron in the output, bolometric detectors, which are directly sensitive to the input power, and classical diode detectors or rectifiers, which produce a dc current proportional to the square of the infrared voltage.

The second major class of infrared detector uses a coherent mixing process to amplify the signal or to convert it to a frequency at which amplification and/or detection are more convenient. Heterodyne downconverters, parametric amplifiers, and upconverters belong in this category. In coherent detectors, the signal electric field is mixed with a pump or local oscillator (LO) field. Only that incoming signal mode coinciding with the pump or local oscillator mode contributes coherently to the output. In practice, therefore, these devices receive only a single mode, whatever nonlinear element or coupling scheme is used. In nearly all astronomical systems, the output of the linear coherent device is subsequently detected (rectified) to yield a dc receiver output.

In many infrared astronomical experiments, a choice must be made between the use of an optical spectrometer (such as a grating or a Fabry–Perot interferometer) followed by a square law detector or the use of a heterodyne downconverter followed by radio-frequency filters and detectors. Short wavelengths and wide bandwidths generally favor the optical spectrometers, whereas long wavelengths and narrow bandwidths favor the heterodyne technique. The tradeoff between these approaches will be discussed in detail in Appendix I.

A. FIGURES OF MERIT FOR COHERENT DETECTORS

Although the figures of merit for coherent detectors are different from those used for incoherent (direct) detectors, some analogies are possible. In an astronomical application in which the output is eventually rectified, the concept of an NEP can be used to describe the performance of a coherent receiver system. As is the case with background-limited direct detectors, the minimum detectable power depends on both the infrared bandwidth B and the postdetection bandwidth. When $kT_N \gg hc/\lambda$, the NEP is often expressed in terms of an equivalent noise temperature T_N, i.e.,

$$\text{NEP} = kT_N B^{1/2}.$$

We thus think of the receiver noise as if it were Johnson noise from a matched input resistor or blackbody at the physical temperature T_N. Although the NEP improves with decreasing B, measurement efficiency is lost if B is reduced to less than the band spread of the signal of astronomical interest.

Because coherent detectors respond to incoming signals with a specific phase relationship to the LO or pump, there is an uncertainty of ± 1 photon in the detector over a time B^{-1}. For a system with an infrared bandwidth B, this corresponds to an unavoidable quantum noise limit to the NEP of $(hc/\lambda)B^{1/2}$. Quantum-noise-limited operation can be expressed in terms of an equivalent noise temperature $T_N \simeq hc/\lambda k$, or $\sim 15\,°\text{K}$ for $\lambda = 1$ mm and $\sim 1500\,°\text{K}$ for $\lambda = 10\,\mu\text{m}$. At near-millimeter wavelengths, the sky temperature for ground-based telescopes is considerably larger than $hc/\lambda k$, so ideal (quantum-noise-limited) receivers are not required. Ambient temperature space telescopes, however, can easily have effective emission temperatures $\leq 15\,°\text{K}$, so that quantum-noise-limited receivers can be used to advantage at all infrared wavelengths. For infrared wavelengths $\lambda < 100\,\mu\text{m}$, where thermal noise makes a negligible contribution to receiver noise, the NEP of a coherent receiver can be expressed in terms of a system quantum efficiency η_s as

$$\text{NEP} = (hc/\lambda\eta_s)B^{1/2}.$$

A numerical factor of 2 is sometimes seen in this expression depending on the type of mixing element being used.

Receiver developers sometimes define a minimum detectable power or NEP in units of watts per hertz (rather than watts per square root of hertz), which is equal to the NEP in the preceding expression with $B = 1$ Hz. Although this parameter is relevant to certain heterodyne applications such as communications and radar, it does not directly represent the performance of an astronomical system. The more useful procedure is to specify an NEP in units of watts per square root of hertz by including the actual bandwidth required by the measurement or to specify the noise temperature or the system quantum efficiency.

B. HETERODYNE MIXERS

The heterodyne mixer, or downconverter, is the most generally useful coherent infrared detector. Any of the direct detectors described previously can be operated as mixers for heterodyne receivers. In practice, however, the response speed of the detector response must be adequate to provide a useful receiver bandwidth B. In the heterodyne application, the output of a relatively strong infrared LO is incident upon the square law device in addition to the smaller signal of interest. Thus, in addition to the rectified dc output, the mixer output has ac components at the difference frequency $|\nu_S - \nu_{LO}|$ and, in some cases, at harmonics and higher combination frequencies. In a heterodyne receiver the output is coupled to an intermediate frequency amplifier, which is sensitive only to a relatively narrow difference frequency bandwidth from $\nu_{IF} - (B_{IF}/2)$ to $\nu_{IF} + (B_{IF}/2)$. Because relative frequencies are preserved in the downconversion process, such receivers generally use intermediate frequency filters rather than infrared filters to define the wavelength band being observed. The output of a heterodyne mixer used to measure lines that are narrow compared with B_{IF} is subdivided by means of an array of radio-frequency filters followed by individual square-law detectors. The number of filter channels equals the number of infrared spectral elements observed simultaneously.

Infrared frequencies in the lower sideband, $\nu_{LO} - \nu_{IF} + (B_{IF}/2)$ to $\nu_{LO} - \nu_{IF} - (B_{IF}/2)$, and in the upper sideband, $\nu_{LO} + \nu_{IF} - (B_{IF}/2)$ to $\nu_{LO} + \nu_{IF} + (B_{IF}/2)$, appear superimposed in the output of usual types of infrared mixers. In broad-band radiometric experiments, both the upper and the lower sidebands provide useful information. For spectroscopy, the line of interest usually appears only in one sideband. Because antenna and mixer noise from both sidebands is downconverted, the useful figure of merit for spectroscopy is usually a single-sideband noise temperature, which is twice the double-sideband value used to characterize a receiver for broad-band radiometry.

The mixer parameter that is analogous to the responsivity of a direct detector is the conversion efficiency (conversion loss L^{-1}). Conversion efficiency is the ratio of the intermediate frequency power coupled out to the IF amplifier divided by the available signal power. Because the conversion efficiency of most heterodyne mixers is substantially less than unity, the noise in the IF amplifier becomes an important issue. The noise in a near-millimeter heterodyne receiver system is often expressed in the form of a receiver noise temperature T_N by referring all noise sources to the receiver input. Thus

$$T_N = T_M + LT_{IF},$$

where T_M and T_{IF} are the mixer and IF amplifier noise temperatures, respectively.

The development of monochromatic infrared laser oscillators has made possible the extension of heterodyne techniques to wavelengths $\lambda \ll 1$ mm. Most of the research in this area has been dedicated to developing receivers for coherent narrow-band sources such as are used in infrared radar and communication systems. In astrophysics, infrared heterodyne receivers have proved useful for high-resolution spectroscopy and spatial interferometry, primarily near $\lambda = 10$ μm. In principle, the technique promises quantum-limited coherent detection in narrow bandwidths throughout the infrared. Depending on the application, coherent receiver performance for $\lambda \lesssim 10$ μm can suffer, however, by comparison with direct detection receivers as will be discussed in Appendix I.

C. Antenna Coupling and Optical Coupling

It is convenient to make a distinction between optically coupled and antenna-coupled devices. Both types of devices can be used as either incoherent or coherent detectors. In optically coupled devices, the infrared signal is imaged directly on the device surface, which has dimensions comparable to or large compared with one wavelength. There are no inherent throughput limitations to this type of coupling. When an optically coupled device is used as a coherent detector, the requirement, discussed earlier, for overlap with the pump or LO field applies, and the throughput is restricted to the single mode value of λ^2. Antenna-coupled (or waveguide-coupled) devices, by contrast, are small compared with one wavelength and respond to the infrared current or voltage appearing at the output terminals of an antenna. In such devices, the antenna theorem limits the throughput to a single electromagnetic mode (diffraction-limited beam) with one polarization, even for incoherent detectors.

Optically coupled photon detectors, or bolometric detectors, can thus be either single- or multimode devices. Small diodes, or small antenna-coupled bolometric detectors, on the other hand, are limited to single-mode operation. The coupling efficiencies of antenna-coupled devices are usually discussed in terms of antenna and device impedances. These are closely analogous to the reflection and absorption coefficients used in the discussion of the coupling (or quantum) efficiencies of optically coupled devices.

All antenna-coupled devices for submillimeter wavelengths share a serious coupling problem. Waveguide techniques that provide efficient coupling to microwave diodes over a wide range of impedances become prohibitively difficult for $\lambda \ll 1$ mm. Matching to low-impedance diodes is especially troublesome. It is probable that some kind of quasi-optical coupling, possibly with components fabricated by optical lithography, will be important in the future. Many structures have been suggested, a few of which have been explored experimentally. Despite considerable recent progress, the efficiencies achieved are generally poor by microwave standards. Uncertainties in the performance of matching structures greatly complicate the evaluation of high-frequency diodes.

D. DIODE MIXERS FOR NEAR-MILLIMETER WAVES

Coherent detector technology at near-millimeter wavelengths comes primarily from extending microwave diode heterodyne mixer technology to shorter wavelengths (McCall, 1977). Molecular line astronomy has provided much of the motivation for the development of low-noise-mixer receivers for wavelengths near 1 mm. Microwave diodes are only now being optimized for submillimeter wavelengths. Because the quantum noise limit to mixer noise temperature in this wavelength range is low compared with the noise temperature of available IF amplifiers, it is important to maximize conversion efficiency. The ideal nonlinear element for this purpose is a switch that can be driven between high- and low-resistance states with the LO. In principle, such a device can have a double sideband conversion efficiency approaching $L^{-1} = 1$.

1. Schottky Barrier Diodes

The GaAs Schottky barrier diode is the most commonly used mixer from microwave to submillimeter wavelengths. In order to minimize both noise and series resistance, the barrier is usually produced on a thin epitaxial layer of lightly doped GaAs grown on a heavily doped (low-resistance) GaAs substrate. Thermally activated conduction produces an exponential I–V curve of the form $I = I_0[\exp(eV/kT) - 1]$. The curvature parameter e/kT, which equals 40 V^{-1} at 300 °K, can be increased and mixer noise reduced by cooling the device below 300 °K.

The equivalent circuit of a Schottky diode at near-millimeter wavelengths is quite complicated. Some of the major issues can be understood, however, in terms of a simplified model, which considers the junction to be a nonlinear resistor shunted by a capacitance C, both in series with a linear series (or spreading) resistance R_S. Because the junction capacitance must discharge through R_S twice each cycle, there is an important cutoff frequency $(2\pi R_S C)^{-1}$, which limits high-frequency operation. This frequency has been extended to ~1 THz ($\lambda \simeq 300$ μm) by using diodes with micrometer dimensions to reduce C and by using a very thin epitaxial layer and a heavily doped substrate to reduce R_S. A variety of parasitic loss effects involving skin effects and the plasma frequency in the semiconductor should become important when $\lambda \ll 1$ mm.

Cutoff wavelengths as short as 30 μm have been reported in submicrometer-sized Schottky diodes with uniform heavy doping (McColl, 1977). The heavy doping reduces the thickness of the Schottky barrier until tunneling currents are dominant and increases the plasma frequency.

The Schottky barrier diode is a very successful device, which has been optimized carefully for wavelengths longer than 1 mm. Careful work at near-millimeter wavelengths is now beginning (Clifton, 1977, Fetterman et al., 1978). The device has some inherent limitations, however, that should be mentioned. The amount of LO power (P_{LO}) required to drive a diode mixer depends on the curvature of the I–V curve or, more precisely, on the bias power required to drive the junction between the high- and low-impedance states. A Schottky diode mixer operated at 300 °K requires $P_{LO} \gtrsim 10^{-3}$ W. Such high power levels are difficult to produce at arbitrary submillimeter wavelengths. Another problem is that the noise in presently available Schottky diode mixers does not continue to decrease as the temperature is reduced to the liquid-He range. The thermally activated current decreases with temperature until the essentially linear tunneling current becomes important.

Considerable improvement in Schottky diode performance has been made recently by using the techniques of molecular-beam epitaxy to optimize the doping profile (Linke et al., 1978). When doped sufficiently, the substrate undergoes a Mott transition to a metallic state. A Schottky barrier device with a steep doping profile can thus combine a thick depletion layer (with low capacitance and small tunneling current) with a small spreading resistance. A DSB mixer-noise temperature of 104 °K was obtained at 81 GHz from such a Mott diode mixer operated at 18 °K. The LO power required was reduced to one-sixth that required by a 300 °K Schottky diode mixer. Although this development is very promising, it is not yet clear whether the conventional Schottky diode mixer will approach the quantum noise limit at near-millimeter wavelengths.

2. Superconducting Mixers

Mixing at millimeter wavelengths has been explored in a variety of superconducting tunneling devices. The effective nonlinearities of all of the superconducting mixers are very large compared with those of the conventional Schottky diode. The required LO power is therefore reduced by a factor $>10^5$. The problem of how to supply LO power is greatly alleviated but with a corresponding reduction in the saturation level of the device. In most cases, however, the dynamic range is large enough for astronomical purposes.

In one class of device, the nonlinearity used arises from the peaks in the density of states of the superconductor at the edge of the superconducting energy gap. If tunneling is between a superconductor and a normal conductor, there is a sharp onset of (quasi-particle) tunnel current when the bias voltage exceeds the half-energy-gap voltage. When the tunneling is between two superconductors, the onset of quasi-particle tunneling is at the full-gap voltage, which, for example, is 2.6 mV for Pb. Such devices can be given the general name of quasi-particle diodes.

The first of the quasi-particle diodes to be developed was a Schottky barrier device made from superconducting Pb on GaAs (McColl, 1977). This "super Schottky" has an essentially exponential I–V curve down to 1 °K where $e/kT = 11,600$ V^{-1}. At a wavelength of 3 cm, it gave the lowest mixer noise temperature (DSB $T_M = 3$ °K) ever reported. The DSB conversion efficiency was 0.40. Because of its inherently low cutoff frequency, however, the GaAs super Schottky diode appears to be limited to operation at wavelengths considerably longer than 1 mm. Super Schottky diodes have also been made on very thin Si membranes. Shorter cutoff wavelengths are expected but have not yet been demonstrated.

Quasi-particle mixers have also been made from superconductor–insulator–superconductor (SIS) tunnel junctions of the type used for Josephson effect devices (Shen *et al.*, 1980). The nonlinearity in the I–V curve of this device is even sharper than for the super Schottky diode. Quantum effects in the junction response make conversion gain ($L^{-1} > 1$) possible. Laboratory results for Pb alloy junctions at 36 and 115 GHz gave DSB $T_M \leq 3$ °K and 40 °K, respectively. The former result is approximately equal to the quantum noise. Receivers are now being constructed for use on radio telescopes at wavelengths from 2.5 to 3.8 mm (80–120

GHz). This type of junction has several promising features. It is an evaporated film device and is thus compatible with integrated circuit technology. The series spreading resistance is zero so there is no cutoff frequency in the usual sense. Intensive development of the type of junction required is being independently funded because of the interest in Josephson effect digital computers. The very low mixer noise temperatures of present quasi-particle mixers and the possibility of conversion gain suggest that these devices could provide the basis for quantum-noise-limited receivers at near-millimeter wavelengths.

If the problem of coupling to very broad bandwidths can be solved adequately, there may be important applications of quasi-particle diodes as direct detectors. The NEP = 3×10^{-16} W/Hz$^{1/2}$ with quantum efficiency $\eta = 0.5$ reported for a 1.5 °K SIS detector at 36 GHz (Richards *et al.*, 1980) suggests that the SIS quasi-particle diodes may prove to be competitive detectors of single-mode, broad-band, near-millimeter radiation.

A second class of superconducting devices is based on the Josephson effect (Richards, 1977). It makes use of the nonlinear properties of the electron pair current in SIS tunnel junctions and related devices such as Nb point contacts. Josephson devices can be operated in a reactive mode as any of several kinds of parametric amplifier or in a resistive mode as either of two kinds of mixer. The parametric amplifiers, however, show noise-amplification phenomena, which have thus far prevented useful low-noise operation. The ac Josephson oscillations can, in principle, be used as an internal LO for a heterodyne mixer. Their linewidth, however, is too wide for effective use in astronomical line receivers. The heterodyne mixer with external LO is the most promising Josephson effect device for submillimeter astronomy. To operate in this mode, a Josephson junction must have negligible capacitance at the gap wavelength (~300 μm in Pb or Nb). Consequently, operation has been limited thus far to relatively unstable point–contact junctions. This device has relatively large DSB noise (ten to twenty times the ambient temperature) and high DSB conversion efficiency from 0.6 to 2.8. The IF amplifier problem is therefore much less serious than with conventional mixers. Very good laboratory results have been obtained from the Josephson effect mixer with external LO at a variety of millimeter and submillimeter wavelengths. A receiver is now being constructed for use on a radio telescope at $\lambda = 2.5$ mm.

3. Status of Infrared Diodes

Although the diodes just discussed have shown considerable promise for the near-millimeter region, it is likely that new diodes will have to be invented in order to approach quantum-noise-limited operation in the 100-μm region. Problems with heating in submicrometer-sized structures and the lack of intense LO sources seem to favor highly nonlinear mixers. Present high-frequency devices such as metal–oxide–metal (MOM) diodes have relatively weak nonlinearities. Systematic investigations of mixing in tunnel structures suggest that it is desirable to increase the device nonlinearity until photon-assisted tunneling effects become important at the operating wavelength. The superconducting devices, for example, are optimum in this sense at wavelengths close to the energy-gap wavelength $\lambda \simeq 300$ μm, room-temperature Schottky devices at $\lambda \simeq 40$ μm, and MOM diodes at $\lambda \simeq 1$ μm.

4. Hot Electron Mixer

In the absence of low-noise, broad-bandwidth diode mixers at submillimeter wavelengths, very useful performance has been obtained for $\lambda > 300$ μm by operating the liquid-helium-temperature InSb hot-electron bolometer discussed in Section III.D as a waveguide-coupled heterodyne mixer (Phillips and Jefferts, 1973). The relative simplicity of this device has allowed it to be rather well optimized at a variety of wavelengths. The available bandwidth, however, is severely limited by electron relaxation times to $B \lesssim 2$ MHz. At such low frequencies, relatively low IF amplifier-noise temperatures $T_{IF} \simeq 4$ °K are available, so receiver-noise temperatures are very good. The limited bandwidth has not proved too serious for molecular line spectroscopy in cases where line frequencies are precisely known in advance. When wide frequency bands must be searched, however, the receiver figure of merit varies as $T_N^{-1}B^{1/2}$, so wider band, higher noise Schottky diode receivers give comparable performance.

E. PV and PC Mixers

Most of the developments in sensitive, coherent receivers for $\lambda \lesssim 100$ μm have focused on optically coupled PV and PC mixers (Keyes and Quist, 1970; Teich, 1970; Arams *et al.*, 1970). The optimization of infrared devices for heterodyne mixer applications is quite different than for the same device used as an incoherent detector. A heterodyne mixer operates in the strong field of a local oscillator, that is, in a high-radiation background. The emphasis in mixer design is on high quantum efficiency, high speed so as to obtain a large bandwidth B, and good performance at high power so as to permit the application of enough LO power to obtain good conversion efficiency.

The performance of a mixer with a photon detector as the nonlinear element can be described in a general way, which is similar to that for an incoherent detector. The oscillating IF current at the output of the mixer caused by a monochromatic signal is

$$I_{IF} = (G\eta e\lambda/hc)(2P_{LO}P_{SIG})^{1/2},$$

where P_{LO} and P_{SIG} are, respectively, the local oscillator power and signal power incident on the mixer and G the photoconductive gain ($G = 1$ for PV detectors). This current must be compared with the effective noise current of the IF amplifier and with the current noise produced in the detector by the photon flux of the local oscillator. The resulting system NEP can be written as

$$NEP = [(\alpha hc/\lambda\eta) + (2kT_{IF}/R_{IF}P_{LO})(hc/G\eta e\lambda)^2]B^{1/2},$$

where the first term in the square bracket arises from the irreducible shot noise of the local oscillator-induced photocurrent. The parameter α in the shot–noise term is equal to 1 for photodiodes, and ranges between 1 and 2 in photoconductors because of the additional noise from the lack of correlation between charge generation and recombination in PC devices. The second term arises from noise in the IF amplifier, which is characterized by a noise temperature T_{IF} and an equivalent input resistance R_{IF}.

This expression shows directly the requirement for quantum-limited performance, namely that the second term in the bracket be smaller than the first. The necessary conditions are a combination of large LO power, low-noise IF amplifiers, and high-gain mixers. The overriding factor affecting ultimate performance is the quantum efficiency. Good measuring efficiency also requires the system bandwidth be as wide as the frequency range of interest. Currently available PV and PC mixers have difficulty meeting this condition for many experiments.

Both PV and PC devices have been used as mixers in heterodyne receivers. Extrinsic photoconductors used for incoherent detectors generally have long charge-carrier relaxation times $\sim 10^{-7}$ sec. In order to increase their bandwidth for use in heterodyne receivers, they can be heavily doped with compensating impurities, which, however, reduces the photoconductive gain. Among the extrinsic materials that have seen use in heterodyne mixers are Ge:Au, Ge:Hg, Ge:Cu, Ge:Ga, and Si:As. Besides the normal photoconductivity extending in wavelength to $\lambda_c = hc/E_i$, negative-donor-ion-state (D^-) photoconductivity in extrinsic silicon may prove useful in the 100–500-μm range.

Nearly quantum-noise-limited performance and bandwidths of 2×10^9 Hz are currently obtainable for wavelengths <12 μm at 77 °K using HgCdTe photodiodes. These have been used to wavelengths as long as 18 μm by operating at liquid-helium temperatures with reduced electrical bandwidths.

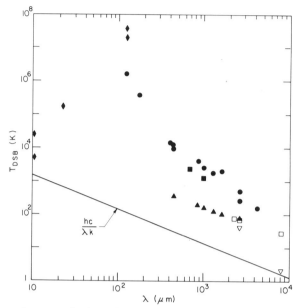

FIG. 7 Recently reported values of double-sideband heterodyne receiver noise temperature (solid symbols) and mixer noise temperature (open symbols). The broad-band diode receivers are about two orders of magnitude above the quantum noise limit in the near-
millimeter range. The fractional bandwidth of these receivers can be as large as $B\lambda/c \simeq 0.1$.
The lower noise temperatures of the hot-electron mixers are partially compensated by their
small fractional bandwidths ($B\lambda/c \simeq 10^{-5}$). The absence of good mixers for $\lambda \simeq 100$ μm is
apparent. Quantum-noise-limited PV and PC mixers can be made for $\lambda \lesssim 10$ μm wherever
sufficient LO power is available. The fractional bandwidths of these mixers are $<10^{-4}$. ◆,
PV and PC; ●, Schottky; ■, Josephson; ▲, InSb-thermal; ▼, quasiparticle.

A summary of present heterodyne receiver performance is given in Fig.
7. No good heterodyne receivers exist at present in the wavelength band
from 24 to 300 μm.

F. PARAMETRIC UPCONVERTERS

An infrared upconverter is a coherent device that is analogous in some
ways to an optically coupled, heterodyne downconverter (Boyd, 1977).
The infrared signal frequency ν_S is mixed with a pump frequency ν_P from
an intense laser source in a medium with a nonlinear index of refraction.
The pump and signal waves generate a polarization wave at the sum (output) frequency $\nu_0 = \nu_S + \nu_P$. The attractiveness of upconversion (discussed previously in the context of quantum counters) is the possibility of
the availability of better detectors for ν_0 than for ν_S. If the pump-, signal-,
and sum-frequency waves can be made to satisfy the phase-matching con-

dition that arises from momentum conservation in the dielectric, the amplitude of the sum-frequency wave will increase with the distance through the dielectric until all photons at ν_S have been converted to photons at ν_0. The conversion efficiency, defined as the ratio of the number of photons at ν_0 to that at ν_S, depends, in practice, on the pump power, the nonlinear polarizability of the dielectric, and the length of dielectric over which a phase match can be maintained.

In principle, the upconversion process is noise free because there can be no sum-frequency photon unless there is a signal (or background) photon present. Upconversion can be a broad-band process but, because of dispersion in the dielectric, the phase-matching condition can only be satisfied in practice over a narrow wavelength band. The major difficulty in upconversion is in finding materials that have small absorption at all three wavelengths, large nonlinear polarizability, appropriate dispersion and/or birefringence, and sufficient material homogeneity to permit phase matching over long interaction lengths. Present upconverters (Boyd, 1977) have conversion efficiencies of $\sim 10^{-6}$ for signal $\lambda = 10.6$ μm and $\sim 10^{-3}$ for $\lambda \sim 3$ μm, and most have suffered from an excess noise source that is not fully understood. The resulting NEPs are not competitive with those of incoherent PV and PC receivers.

Imaging with upconversion is possible because the phase-matching conditions impose a one-to-one correspondence between the direction of propagation of the signal-frequency and sum-frequency waves. Thus, upconversion is not strictly a single-mode process but does preserve the uniqueness of individual modes.

VI. Important Supporting Technologies

A. Amplifier Noise Limitations of PV and PC Detectors

Most high-performance PV and PC detectors have resistances on the order of 10^{10} to 10^{13} Ω when operated under low-background conditions. These high impedances require the use of high load resistances and cold preamplifiers. The better types of detectors are limited in low infrared background by the equivalent current noise of the amplifier referred to its input $i_{n,a}$, which has dimensions of amperes per square root of hertz. Under these conditions, the NEP of a PV or PC detector can be written as

$$\text{NEP} = i_{n,a}/S = (hc/e\lambda)(i_{n,a}/G\eta),$$

where $G = 1$ for a PV detector. Gains in sensitivity can come only from decreasing $i_{n,a}$ or increasing the $G\eta$ product.

The equivalent current noise of a preamplifier (voltage follower or transimpedance) employing a load resistor R_L at temperatture T_L and a field-effect transistor (JFET or MOSFET) is

$$i_{n,a} = [(4kT_L/R_L) + i_{n,F}^2 + (v_{n,F}/Z)^2]^{1/2}.$$

In this equation, $i_{n,F}$ is the gate current noise of the FET, $v_{n,F}$ the series voltage noise of the FET (at the low frequencies of usual interest, $v_{n,F}^2$ varies approximately as $1/f$), and Z the parallel impedance of the detector, FET, load resistor, and stray capacitance. In addition, there may be an excess current noise contribution from the load resistor. With the very high values of detector and load resistance that are commonly used, the impedance Z is capacitive above a characteristic frequency, which is typically ≤ 1 Hz. The minimum capacitance possible is that of the FET alone. If the capacitance of the detector is larger than that of the FET, then the contribution to $i_{n,a}$ due to $v_{n,F}$ becomes relatively larger and the limiting NEP is thereby degraded. This effect is of particular importance for PV detectors because their capacitance can be many picofarads. Thus, it is often important to use an FET having low voltage noise with PV detectors.

1. JFETs

Because the voltage noise of a good JFET is typically 10^{-2} that of a MOSFET, JFETs are generally preferred over MOSFETs for use with PV detectors and most bolometers. The voltage noise and, especially, the current noise of a JFET decrease as it is cooled below room temperature. However, currently available Si JFETs do not operate below about ~ 80 °K. Amplifiers have been built with FETs at higher temperatures than the detectors and load resistors. However, detectors for wavelengths $\lambda \gtrsim 20$ μm must be shielded from the thermal radiation of the preamplifier while maintaining electrical performance. With large numbers of detectors, this approach of warm (~ 80 °K) amplifiers with cold (≤ 10 °K) detectors becomes cumbersome. The JFETs with the lowest voltage noise tend to have large input capacitance ($\gtrsim 10$ pF) and nonnegligible current noise (typically 10^{-17}–10^{-16} A/Hz$^{1/2}$). Development of better JFETs, particularly ones with lower operating temperature, current noise, and input capacitance, would substantially improve the sensitivities achievable at low background of both PC and PV detectors.

2. MOSFETS

Long-wavelength, extrinsic PC detectors usually operate below 10 °K and have capacitances that are typically less than 1 pF. It has become customary to use MOSFETs with these detectors because they typically have low input capacitance (~ 3 pF) and low current noise and operate at low temperatures. Their voltage noise, which has a $1/f$ power spectrum, is much larger than for a JFET. Typical $v_{n,F}$ for a good-quality MOSFET is 300 nV/Hz$^{1/2}$ at 5 Hz.

Commercially available MOSFETs display a variety of properties below ~ 15 °K that are loosely grouped together and sometimes called the "open substrate phenomena." These include a marked drift in their dc operating point (typical drifts amount to ~ 500 μV over a period of several hours), increased $1/f$ noise at low frequencies (0.1–1 Hz), and very long recovery times (5–20 min) for turn-on transients and saturating noise spikes. Current investigations suggest that these effects are related to carrier freeze-out in the silicon substrate.

3. Load Resistors

Some problems encountered with many existing load resistors are that their voltage and temperature coefficients of resistance are large enough that these nonidealities can be sources of excess noise and calibration errors. These resistors often change in value by a factor 2 or more over normal variations in low-temperature operating conditions. Selected components do not have such large coefficients, but the yield is often small. More sensitive detectors in the future will require resistors with higher values of resistance and, in some cases, lower shunt capacitance than are currently available.

4. Frequency Response of Preamplifiers

With the exception of rapid-scan Fourier transform spectroscopy, most astronomical signals are in the frequency range from 0 to 10 Hz. However, the availability of a wider bandwidth allows a clean separation of infrared signals from both spontaneous detector spikes and spikes induced by energetic nuclear particles. The separation and the subsequent blanking of these spikes is important if one is to achieve the maximum sensitivity from a detector operated on a space platform. Use of a trans-impedance amplifier (TIA) does not improve the ratio of signal to noise at a given frequency, but does extend the frequency response over that obtainable with a simple voltage amplifier. The frequency response of a TIA is usually determined by the $R-C$ time constant of the load resistor. Large bandwidths thus require load resistors with small shunt capacitance.

5. Charge-Sensitive Amplifiers with Accumulation Mode Detectors

Rather than sensing the detector current continuously, it is possible to accumulate the photo-generated charge during an integration period τ_{int} and then read out the charge quickly with a charge-sensitive amplifier.

This method is in wide use with advanced detector arrays because of the possibility of multiplexing many detectors sequentially with a single amplifier. The NEP of such a combination of a detector with a charge-sensitive amplifier is

$$\text{NEP} = (hc/G\eta e\lambda)(2/\tau_{\text{int}})^{1/2}Q_{\text{rms}},$$

where Q_{rms} is the uncertainty in the amount of charge that is read out. We have assumed 100% readout efficiency. For example, at 10 μm with $Q_{\text{rms}} = 10^3$ electrons, $\tau_{\text{int}} = 1$ sec, and $G\eta = 1$, the NEP is 2.8×10^{-17} W/Hz$^{1/2}$.

At high backgrounds, where Q_{rms} is (it is hoped) determined by the statistics of the photocurrent, the NEP obtained is the same as with a continuous-current-sensing amplifier. We are concerned here with the low-illumination limit, where Q_{rms} is the result only of noise in the readout electronics. In this case, charge integration can be more or less sensitive than continuous current sensing using the same components and a modulation frequency of τ_{int}^{-1}. In principle, if $R_{\text{L}}/T_{\text{L}}$ is sufficiently large that Johnson noise is not important, the accumulation mode is of advantage only if the unavoidable, low-frequency, FET noise "power" spectral density $v_{n,F}^2$ varies more rapidly than $1/f$. Even if this criterion is not met, there are likely to be practical advantages to accumulation mode detectors if integration times longer than several seconds can be used. These potential advantages arise from limitations to available resistor values and cooling constraints.

Values of Q_{rms} thus far obtained with infrared detectors are typically 1000 electrons. However, values as low as ~25 electrons have been achieved in visible Si CCD devices. Attainment of such low values of Q_{rms} and values of $\tau_{\text{int}} > 1$ sec would yield substantial improvements in infrared sensitivity. Large values of τ_{int} require not only low photon background and low leakage but also sufficient charge-storage capacity in order to have the desired dynamic range.

B. CHARGED-PARTICLE-INDUCED NOISE IN
    INFRARED DETECTORS

A charged particle passing through an infrared detector loses an amount of energy E that depends on the path length within the detector, the chemical composition of the detector, and the particle's energy and type. A fraction of the energy deposited in a semiconductor produces electron–hole pairs. Experimentally, for a large number of semiconductors, the energy loss per pair created is about 3.5 times the band gap energy. Thus, a particle that loses energy E in a PV or PC detector creates ~$E/3.5E_g$ pairs. These pairs produce a current pulse, which is roughly equivalent to an incident photon pulse of $E/3.5\eta E_g$ photons. For intrinsic detectors, $E_g = hc/\lambda_c$, and for extrinsic detectors, it is E_g/E_i times larger. Thus, the equivalent photon input is $\sim E\lambda_c/3.5\eta$ hc for intrinsic detectors and $\sim (E\lambda_c/3.5\eta hc) \times (E_i/E_g)$ for extrinsics. However, the energy loss E in extrinsic detectors is much larger than in intrinsic detectors because the path length required for good quantum efficiency is many times larger. As a result, intrinsic detectors should be less vulnerable to charged-particle-induced noise than extrinsic detectors. Considerations similar to these for charged particles apply to gamma rays that interact with infrared detectors.

All the energy lost in a bolometer raises its temperature, and thus, the equivalent photon input (for an operating wavelength λ) is $E\lambda/hc$. Bolometers are usually intermediate in thickness between intrinsic and extrinsic detectors. Because they are used at long wavelengths, have large cross-sectional areas, and have slow response times, the sensitivity of bolometers may be substantially limited in space environments by particle and/or gamma-ray events.

Although the basic considerations discussed here may be useful as guidelines, it is very important that the vulnerability of all types of infrared detectors proposed for low-background space experiments be studied experimentally and that shielding and circumvention techniques be developed more fully.

1. *Shielding Techniques*

In near-earth orbit, energetic charged particles arise from several sources: the trapped-particle (Van Allen) belts, the primary cosmic rays, and, occasionally, particles from solar flare events. Moderate amounts of shielding (a few grams per centimeter squared) will stop most of the trapped electrons (and their bremsstrahlung) and the lower energy (≤ 50 MeV) protons. In dense parts of the radiation belts (e.g., in the region of the South Atlantic anomaly), however, this amount of shielding will permit substantial pulse rates and more elaborate shielding may be required. Virtually none of the high-energy particles in primary cosmic rays or high-energy trapped protons will be stopped by practical shielding.

Space missions with durations of one year are expected to be too short for charged particles to produce significant permanent degradation in present infrared detectors if they are used with appropriate shielding.

2. *Circumvention Techniques*

Many circumvention schemes have been developed to reduce the effect of charged-particle-induced noise. These schemes can be implemented through hardware or software but always use the rapid rise time of a charged-particle event to trigger a blanking mechanism. This blanking can range from simply gating off the data stream for a preset time to linearly fitting a straight line to the pre- and postevent outputs. In the case of IRAS, the detector output is clamped at its prepulse value for a time that depends on the magnitude of the pulse.

C. SPECTRAL FILTERS

Improved filters for the 20–1000-μm band are badly needed both for photometry and to serve as order-sorting filters for spectroscopic work. For wavelengths shorter than 20 μm, high-efficiency multilayer dielectric filters are available commercially. These filters can be cooled many times without serious degradation. Dielectric filters for the 20–30-μm band exist but generally have poor efficiencies. In addition, these filters suffer from delamination and changes in their optical properties upon cooling.

Crystal resonance bands are used for filtering in the 20–200-μm region. Band selection can be accomplished by using the crystals in reflection, in transmission, or by scattering. Although there are a large number of candidate crystals for this purpose, many of them are difficult to use because they are hygroscopic, fragile, and/or soft or because they have weak resonances leading to only slow variations of optical properties with wavelength.

Interference filters made with metal meshes at room temperatures have been used extensively at submillimeter wavelengths. Techniques are being developed to permit cooling of such filters without degradation of their properties. It also seems possible to build high-efficiency mesh filters for wavelengths shorter than those where mesh filters are in current use.

D. DETECTOR COOLING

Most high-performance infrared detectors must be cooled to temperatures $\lesssim 10$ °K. Some, but not all, of the short wavelength PV detectors are exceptions to this rule. Even though InSb PV detectors operate well up to ~70 °K, the very best performance is achieved at liquid-helium temperatures. Most infrared systems developed for defense purposes employ closed-cycle refrigerators to achieve the required temperatures. On the other hand, planned NASA programs use stored cryogenic liquids for cooling. The IRAS telescope, for example, is to be cooled by superfluid helium for a mission duration of about a year.

More advanced programs may use bolometers or other detectors requiring cooling below the region ($T \gtrsim 1$ °K) accessible with liquid ^{4}He. Several schemes have been identified for achieving these low temperatures in the space environment. Temperatures $\gtrsim 0.3$ °K can be reached with pumped liquid-^{3}He refrigerators similar to those that have been used at balloon and aircraft altitudes. Other low-temperature laboratory techniques such as adiabatic demagnetization and dilution refrigeration could be adapted to produce temperatures ≤ 0.01 °K. These approaches should be studied with the goal of developing simple, reliable refrigerators.

The best heterodyne receivers require cooling to temperatures ≤ 10 °K. Schottky diode mixers and some types of IF amplifiers (masers and most paramps) place a large load (≥ 10 mW) on the cooling system. Superconducting mixers and FET amplifiers, by comparison, have much less power dissipation.

E. SPECIAL REQUIREMENTS FOR HETERODYNE RECEIVERS

1. IF Amplifiers

Intermediate frequency amplifier noise contributes significantly to the noise in all near-millimeter receiver systems and in receivers at shorter wavelengths that are starved for LO power. Improvements in IF amplifier technology will clearly have to accompany mixer improvements if lower noise systems are to be available in the future.

Typical conversion efficiencies for near-millimeter mixers are $0.1 < L^{-1} < 0.2$ (the larger values obtained with the Josephson and SIS mixers are exceptional). Because LT_{IF} enters the expression for receiver noise temperature, it follows that values of $T_{IF} \leq 2$ °K are required for a receiver to approach quantum-noise-limited operation at $\lambda = 1$ mm and 20 °K at $l = 100$ μm. Such values are difficult to obtain with the broad-band ($B_{IF} > 10^8$ Hz) amplifier technology available today.

Most IF amplifiers in current use operate in the low gigahertz frequency range. This choice allows $B_{IF} \geq 10^8$ Hz, even with resonant impedance matching. Significant separation between the signal and LO frequencies is desirable for strongly pumped mixers, such as conventional Schottky diodes and PC mixers, so as to avoid noise from the pump source. This is not an important issue for the superconducting mixers with their small P_{LO} requirement.

Cooled GaAs FET amplifiers are very attractive for space astronomy applications because of their simplicity and low power dissipation. Noise temperatures as low as 15 °K have been achieved. Parametric amplifiers with $T_{IF} \simeq 12$ °K and masers with $T_{IF} \simeq 6$ °K are more complicated devices with larger power dissipation.

Besides the need for broad-band, low-noise amplifiers, there is a related need for compact, multichannel, IF filter banks. Efficient heterodyne spectroscopy with space telescopes requires broad-band filter banks with lower power consumption, bulk, and weight than those currently used on

the ground. Acoustooptic techniques are one promising method of achieving these goals.

2. Local Oscillators

The available sources of LO power limit the application of heterodyne
techniques to infrared astronomy at many wavelengths. At near-
millimeter wavelengths, the severity of the problem depends on the type
of mixer being used. Room-temperature Schottky barrier diodes require
$P_{LO} \simeq 10^{-3}$ W, which is available at most submillimeter wavelengths only
from backward wave oscillators such as the O-type Carcinotron.
Although these oscillators make reasonably satisfactory LO sources for
Schottky diode mixers for $\lambda \gtrsim 450$ μm (670 GHz), they are bulky, expensive, and require large, well-regulated power supplies. There has been
little development of backward-wave oscillators during the past 25 years.
The LO power required by the superconducting mixers is sufficiently
small (10^{-8}–10^{-9} W) that it can be obtained by harmonic generation from
klystron or other millimeter-wave oscillators over the expected operating
wavelength range of the devices.

Mixers for $\lambda < 300$ μm generally require LO power in the milliwatt
range. Molecular-discharge lasers and CO_2-pumped molecular lasers are
the only practical sources of continuous-wave LO power for a wide
range of infrared wavelengths. Although several effects can be used to
tune these laser LOs over narrow-wavelength ranges, they are essentially
fixed-wavelength sources. Although many lines are available, most heterodyne receiver developments have taken place at the wavelengths of a
few well-known lasers: ~10 μm (CO_2), 118 μm (H_2O), 337 μm (CN^-), etc.
Heterodyne spectroscopy is now limited to frequency ranges of a few
gigahertz on either side of the laser frequency. Continuously tunable
infrared diode lasers are becoming available for $\lambda \lesssim 20$ μm. The shorter
the wavelength, the better the state of development. The amounts of
infrared power available from such LO sources is generally so small that
receivers are IF amplifier-noise-limited at NEPs which are significantly
above the quantum noise limit.

Appendix I.   Comparison of Heterodyne Detection with
Direct Detection

A. SPECTROSCOPY

The elements of an astronomical heterodyne spectrometer are shown in
Fig. 8a. Under ideal circumstances, when the noise of the heterodyne receiver is dominated by shot noise from the local oscillator, the

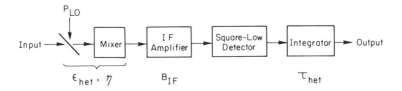

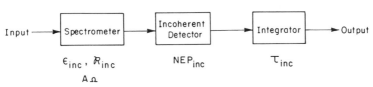

FIG. 8 Functional diagrams of a heterodyne receiver (a) and an incoherent receiver (b). In the latter case, the box labeled spectrometer could, for example, be infrared filters, a prism, a grating spectrometer, a Fabry–Perot interferometer, or a Fourier transform spectrometer. Important properties affecting system performance are shown below the relevant elements. The effect of IF amplifier noise in the heterodyne receiver has been absorbed into η. The quantum efficiency of the incoherent detector has been absorbed into NEP_{inc}.

quantum-noise-limited minimum detectable power incident on the optical mixer per unit postdetection bandwidth is

$$\text{NEP} = (hc/\eta\lambda)B^{1/2},$$

where η is the quantum efficiency of the mixer and $B = 2B_{\text{IF}}$. The signal power from a source of spectral radiance L_λ (in watts per meter squared steradian micrometer) incident on the mixer is

$$P_{\text{SIG}} = \epsilon_{\text{het}}L_\lambda\lambda^2\,\Delta\lambda,$$

where $\Delta\lambda = \lambda^2 B/c$. Here ϵ_{het} is the optical efficiency of the system, which includes reflection losses, wave front distortion, chopping losses, etc., and also accounts for the sensitivity of a heterodyne receiver to only one polarization. Unless an array of mixers is used, the throughput of the receiver is limited to the single mode value of λ^2 by the coherent mixing process. The integration time required to obtain a given signal-to-noise (S/N) ratio is, in general,

$$\tfrac{1}{2}(S/N)^2(\text{NEP}/P_{\text{SIG}})^2,$$

which, for this heterodyne receiver example, becomes

$$\tau_{\text{het}} = \tfrac{1}{2}[(S/N)h/\epsilon_{\text{het}}\eta L_\lambda]^2(c^3\mathcal{R}/\lambda^9),$$

where the resolving power $\mathscr{R} = \lambda/\Delta\lambda = c/\lambda B$. The very rapid dependence of the integration time τ_{het} on wavelength in this expression is an important feature of heterodyne spectrometers.

The integration time τ_{inc}, required to achieve a given signal-to-noise ratio with the generalized incoherent spectrometer shown in Fig. 8b is

$$\tau_{inc} = \tfrac{1}{2}((S/N)NEP_{inc}/\epsilon_{inc}L_\lambda A\Omega)^2(\mathscr{R}^2/\lambda).$$

Here NEP_{inc} is the NEP of the detector employed in the incoherent spectrometer, ϵ_{inc} the spectrometer efficiency, $A\Omega$ the spectrometer throughput, and $\mathscr{R}$ the resolving power.

A useful comparison of the two techniques is to calculate the ratio of the integration times to achieve the same S/N for equal resolving power. This yields

$$\frac{\tau_{het}}{\tau_{inc}} = \left(\frac{\epsilon_{inc}}{\epsilon_{het}}\frac{hc/\lambda}{\eta NEP_{inc}}\frac{A\Omega}{\lambda^2}\right)^2\frac{c}{\lambda\mathscr{R}},$$

where $c/\lambda\mathscr{R}$ is equal to the bandwidth B. As long as there is no limit imposed by the solid angle subtended by the source, this ratio is only limited by practical constraints on the throughput of the incoherent spectrometer. An interesting case (which favors the heterodyne system somewhat) is to assume that we are viewing a point source for which the effective $A\Omega = \lambda^2$. The ratio of the integration times then becomes

$$\tau_{het}/\tau_{inc} = (\epsilon_{inc}hc/\epsilon_{het}\eta NEP_{inc}\lambda)^2(c/\lambda\mathscr{R}).$$

Typical heterodyne receivers have $\epsilon_{het} \simeq 0.1$ with mixer quantum efficiencies $\eta \simeq 0.5$. An astrophysically interesting resolving power might be $\mathscr{R} \simeq 10^5$ to resolve a typical Doppler-broadened line. A Fabry–Perot or Michelson interferometer, or an echelle spectrometer, with comparable resolution could reasonably have $\varepsilon_{inc} \simeq \varepsilon_{het}\eta$. For an incoherent detector with an NEP of 10^{-16} W/Hz$^{1/2}$, the crossing point for the two techniques occurs at $\lambda \simeq 23$ μm (and at $\lambda \simeq 11$ μm for a resolving power of 10^6). If the detector NEP is 10^{-17} W/Hz$^{1/2}$ the crossing point is at $\lambda \simeq 110$ μm for $\mathscr{R} = 10^5$ and $\lambda \simeq 50$ μm for $\mathscr{R} = 10^6$. High resolution and long wavelengths favor the heterodyne receiver. Furthermore, resolving powers much greater than 10^5 are difficult to achieve efficiently with incoherent techniques. Thus, it is likely that for those scientific problems where such high resolving powers are necessary, heterodyne spectrometers will be attractive over a wide wavelength range.

A technique that is used with both heterodyne and direct-detection spectrometers is to employ multiple outputs so as to scan many resolution elements simultaneously. The time needed to search for a line (in frequency) or to measure its profile is then reduced by the number of channels. The relative case of making multiple-filter banks following the IF amplifiers is an asset for heterodyne spectrometry and highlights the need for broad-band mixers and IF amplifiers.

B. RADIOMETRY

Both coherent and incoherent techniques are used to measure the brightness of a source with a broad spectrum. In this case the crossover between the two techniques is at wavelengths near 1 mm, where the noise in available heterodyne receivers is considerably above the quantum limit. The infrared bandwidth B of the heterodyne receiver is generally limited by the IF amplifier to a value that is significantly less than that of an incoherent radiometer. The equations just derived to compare the times required for the two types of instrument to perform a spectroscopy experiment can be used for radiometry, if we distinguish between the resolving powers used and include a factor f, which is the ratio of the noise in the heterodyne receiver to the idealized quantum noise, i.e.,

$$\tau_{het}/\tau_{inc} = (\epsilon_{inc} f h c A \Omega / \epsilon_{het} NEP_{inc} \lambda^2)^2 (c \mathcal{R}_{het} / \lambda^3 \mathcal{R}_{inc}^2).$$

If we assume a point source so that $A\Omega = \lambda^2$ for both receivers, then

$$\tau_{het}/\tau_{inc} = (\epsilon_{inc} f h / \epsilon_{het} NEP_{inc})^2 (c^3 \mathcal{R}_{het} / \lambda^3 \mathcal{R}_{inc}^2).$$

Parameters for a presently available GaAs Schottky diode mixer receiver at $\lambda \simeq 1$ mm are $f/\epsilon_{het} = 70$ and $\mathcal{R}_{het} = 150$ ($B = 2$ GHz). A good composite bolometer at $T = 0.3$ °K will have $NEP_{inc} \simeq 10^{-15}$ W/Hz$^{1/2}$ with $\epsilon_{inc} \simeq 0.3$ and $\mathcal{R}_{inc} = 3$. The latter system is close to being background-limited on a 300 °K telescope. These numerical values give $\tau_{het} \simeq 90\tau_{inc}$.

The limiting sensitivity for both systems at $\lambda = 1$ mm is determined by fluctuations in the 3 °K cosmic background. If ideal receivers of both kinds were available with the resolving powers just given, then $\tau_{het} = \mathcal{R}_{het}\tau_{inc}/\mathcal{R}_{inc} \simeq 50\tau_{inc}$. Thus, ideal coherent detectors are not competitive with ideal incoherent detectors as radiometers for $\lambda \simeq 1$ mm, unless their IF bandwidth can be increased.

Appendix II. List of Symbols and Acronyms

A	Detector area
$A\Omega$	Throughput
B	Predetection (infrared) bandwidth of a heterodyne receiver
B_{IF}	IF bandwidth of heterodyne receiver
$\mathcal{C}$	Bolometer heat capacity
C	Capacitance of detector or amplifier
c	Speed of light
CB	Conduction band of a semiconductor

CCD	Charge-coupled device
CID	Charge-injection device
D^*	D-star; $D^* = A^{1/2}/\text{NEP}$ (cm $\cdot$ Hz$^{1/2}$/W)
DSB	Double sideband
E_g	Semiconducting band gap; also superconducting band gap
E_i	Impurity ionization energy
F	Quality factor for bolometers (W/Hz$^{1/2}$ $\cdot$ K$^{5/2}$ $\cdot$ sec$^{1/2}$)
f	Ratio of coherent receiver noise to quantum noise
FET	Field-effect transistor; see JFET, MOSFET
$\mathcal{G}$	Thermal conductance between bolometer and heat sink
G	Photoconductive gain
Ge:XX	Extrinsic germanium; XX can be chemical symbol for specific dopant, e.g., Ge:Ga
h	Planck's constant
I–V	Current–voltage
I_{IF}	IF current of heterodyne mixer
IF	Intermediate frequency
IR	Infrared
IRCCD	Infrared-sensitive charge-coupled device
i_n	Current noise; total current noise (A/Hz$^{1/2}$)
$i_{n,a}$	Amplifier contribution to i_n
$i_{n,d}$	Detector contribution to i_n
$i_{n,e}$	Excess noise contribution to i_n
$i_{n,F}$	Gate current noise of an FET
JFET	Junction field-effect transistor; cf. MOSFET
k	Boltzmann's constant
L^{-1}	Conversion efficiency of heterodyne receiver
L_λ	Spectral radiance of a source of infrared radiation (W/m^2 $\cdot$ sr $\cdot$ μm)
LO	Local oscillator
MIS	Metal–insulator–semiconductor structure
MOM	Metal–oxide–metal structure
MOS	Metal–oxide–semiconductor structure
MOSFET	Metal–oxide–semiconductor field-effect transistor; cf. JFET
$\dot{N}$	Rate of photons incident on detector (photons/sec)
NE$\dot{N}$	Rate of photons incident on a detector that is required to produce a signal-to-rms-noise ratio of unity in the detector's output; the noise is measured in a 1-Hz bandwidth (photons/sec $\cdot$ Hz$^{1/2}$)
NEP	Noise equivalent power, which is the value of monochromatic power incident on a detector that is required to produce a signal-to-rms-noise ratio of unity in the detector's output; the noise is measured in a 1-Hz bandwidth (W/Hz$^{1/2}$). In other documents, monochromatic NEPs often appear with a subscript λ. A NEP may also be given for radiation with a particular spectrum (usually a blackbody with a specified temperature). An alternative convention is to specify the noise bandwidth whereby the units of NEP become watts. Depending on usage, the value for the incident power may be either the steady-state value or the rms value of a modulated signal
NEP$_{het}$	NEP of a heterodyne receiver or spectrometer
NEP$_{inc}$	NEP of an incoherent receiver or spectrometer
NEP$_{PC}$	NEP of a photoconductive detector

NEP_{PV}	NEP of a photovoltaic detector
n	Number of detectors in an array
P_{LO}	Local oscillator power of a coherent receiver
P_{SIG}	Signal power
PC	Photoconductive
PV	Photovoltaic
Q_{rms}	Root-mean-square readout noise of a charge-sensitive amplifier
$\mathscr{R}$	Resolving power of a spectrometer; may have subscript indicating incoherent or heterodyne
R	Resistance
R_{IF}	Equivalent noise resistance of an IF amplifier
R_L	Load resistance of current amplifier for PV or PC detector
R_S	Spreading resistance of a diode mixer
R_0	Zero-bias impedance of a PV detector
S	Responsivitity of PV or PC detector (A/W)
SIS	Superconductor–insulator–superconductor structure
Si:XX	Extrinsic silicon; XX can be chemical symbol for specific dopant, e.g., Si:As
S/N	Signal-to-noise ratio
T	Temperature
T_A	Ambient temperature
T_{IF}	Noise temperature of IF amplifier
T_M	Mixer-noise temperature
T_N	Noise tempertature; also, specifically, receiver-noise temperature
TIA	Transimpedance amplifier
V	Voltage
VB	Valence band of a semiconductor
$v_{n,F}$	Series voltage noise of an FET
Z	Amplifier input impedance
α	Generation–recombination noise parameter
ϵ	Optical efficiency
ϵ_{het}	Efficiency of heterodyne receiver or spectrometer
ϵ_{inc}	Efficiency of incoherent receiver or spectrometer
η	Quantum efficiency. Two different definitions of quantum efficiency exist. The responsive quantum efficiency (RQE) compares the observed responsivity of a detector to that which would occur if every incident photon contributed equally, i.e., $S = G\eta e\lambda/hc$. The detective quantum efficiency (DQE) compares the detector noise with that which would occur if there were only ideal photon noise, i.e., $NEP_{PC} = (4\dot{N}/\eta)^{1/2}$. Both RQE and DQE can be defined for coherent detectors as well as for incoherent detectors
λ	Wavelength
$\Delta\lambda$	Wavelength bandwidth
λ_c	Cutoff wavelength; longest wavelength of response
ν	Infrared frequency
$\Delta\nu$	Infrared bandwidth; $\Delta\nu = B$
ν_{IF}	IF frequency of heterodyne receiver
ν_{LO}	LO frequency of heterodyne receiver
ν_O	Output frequency of upconverter
ν_P	Pump frequency of upconverter

ν_S	Signal frequency of upconverter
τ_{het}	Integration time of heterodyne receiver or spectrometer
τ_{inc}	Integration time of incoherent receiver or spectrometer
τ_{int}	Integration time of accumulation mode device

ACKNOWLEDGMENTS

This chapter is an outgrowth of a report to the NASA Office of Space Sciences by the Infrared Detector Subpanel of the Management Operations Working Group for Airborn Astronomy. The members of this subpanel were: N. W. Boggess, NASA Headquarters; L. T. Greenberg, The Aerospace Corporation; M. G. Hauser, NASA Goddard Space Flight Center; J. R. Houck, Cornell University; F. J. Low, University of Arizona, C. R. McCreight, NASA Ames Research Center; D. M. Rank, University of California, Santa Cruz; P. L. Richards, University of California, Berkeley; and R. Weiss, Massachusetts Institute of Technology.

Although the authors of this chapter took primary responsibility for the written text of the NASA report, much that is of value has come from the individual and collective knowledge of the panel members. Responsibility for any errors in this chapter, however, must rest with the authors.

REFERENCES

Ando, K. J. (1978). *NASA Tech. Rep.* **NASA-CR-152169.**

Arams, F. R., ed. (1973). "Infrared-to-Millimeter Wavelength Detectors." Artech House, Dedham, Massachusetts.

Arams, F. R., Sard, E. W. Peyton, B. J., and Pace, F. P. (1970). *In* "Semiconductors and Semimetals" (R. K. Willardson and A. C. Beer, eds.), Vol. 5, pp. 409–434. Academic Press, New York.

Blaney, T. G. (1978). *J. Phys. E* **11,** 856–881.

Boyd, R. W. (1977). *Opt. Eng.* **16,** 563–568.

Bratt, P. R. (1977). *In* "Semiconductors and Semimetals" (R. K. Willardson and A. C. Beer, eds.), Vol. 12, pp. 39–142. Academic Press, New York.

Clifton, B. J. (1977). *IEEE Trans. Microwave Theory Tech.* **MTT-25,** 457–463.

Ducas, T. W., Spencer, W. P., Vaidyanathan, A. G., Hamilton, W. H., and Klepner, D. (1979). *Appl. Phys. Lett.* **35,** 382–384.

Emmons, R. B., Hawkins, S. R., and Cuff, K. F. (1975). *Opt. Eng.* **14,** 21–30.

Fetterman, H. R., Tannenwald, P. E., Clifton, B. J., Parker, C. D., Fitzgerald, W. D., and Erickson, N. R. (1978). *Appl. Phys. Lett.* **33,** 151–154.

Gelbwachs, J. A., and Wessel, J. E. (1980). *IEEE Trans. Electron Devices* **ED-27,** 99–108.

Gillett, F. C., Dereniak, E. L., and Joyce, R. R. (1977). *Opt. Eng.* **16,** 544–550.

Haller, E. E., Hueschen, M. R., and Richards, P. L. (1979). *Appl. Phys. Lett.* **34,** 494–496.

Hudson, R. D., Jr., and Hudson, J. W., eds. (1975). "Infrared Detectors." Halsted Press, Stroudsburg, Pennsylvania.

Keyes, R. J., ed. (1977). "Optical and Infrared Detectors." Springer-Verlag, Berlin and New York.

Keyes, R. J., and Quist, T. M. (1970). *In* "Semiconductors and Semimetals" (R. K. Willardson and A. C. Beer, eds.), Vol. 5, pp. 321–359. Academic Press, New York.

Kingston, R. H. (1978). "Detection of Optical and Infrared Radiation." Springer-Verlag, Berlin and New York.

Kruse, P. W. (1977). *In* "Optical and Infrared Detectors" (R. J. Keyes, ed.), pp. 5–69. Springer-Verlag, Berlin and New York.

Linke, R. A., Schneider, M. V., and Cho, A. Y. (1978). *IEEE Trans. Microwave Theory Tech.* **MTT-26**, 935–938.

Longo, J. T., Cheung, D. T., Andrews, A. M., Wang, C. C., and Tracy, J. M. (1978). *IEEE Trans. Electron Devices* **ED-25**, 213–232.

McColl, M. (1977). *Proc. Soc. Photo-Opt. Instrum. Eng.* **105**, 24–34.

Milton, A. F. (1977). *In* "Optical and Infrared Detectors" (R. J. Keyes, ed.), pp. 197–228. Springer-Verlag, Berlin and New York.

Nelson, R. D. (1977). *Proc. Soc. Photo-Opt. Instrum. Eng.* **124**, 91–101.

Nishioka, N. S., Richards, P. L., and Woody, D. P. (1978). *Appl. Opt.* **17**, 1562–1567.

Phillips, T. G., and Jefferts, K. B. (1973). *Rev. Sci. Instrum.* **44**, 1009–1014.

Putley, E. H. (1977). *In* "Semiconductors and Semimetals" (R.l K. Willardson and A. C. Beer, eds.), Vol. 12, pp. 143–168. Academic Press, New York.

Richards, P. L. (1977). *In* "Semiconductors and Semimetals" (R. K. Willardson and A. C. Beer, eds.), Vol. 12, pp. 395–440. Academic Press, New York.

Richards, P. L., Shen, T. M., Harris, R. E., and Lloyd, F. L. (1980). *Appl. Phys. Lett.* **36**, 480–482.

Shen, T-M., Richards, P. L., Harris, R. E., and Lloyd, F. L. (1980). *Appl. Phys. Lett.* **36**, 777–779.

Spicer, W. E. (1977). *Appl. Phys.* **12**, 115–130.

Stillman, G. E., Wolfe, C. M., and Dimmock, J. O. (1970). *In* "Semiconductors and Semimetals" (R. K. Willardson and A. C. Beer, eds.), Vol. 5, pp. 169–290. Academic Press, New York.

Teich, M. C. (1970). *In* "Semiconductors and Semimetals" (R. K. Willardson and A. C. Beer, eds.), Vol. 5, pp. 361–407. Academic Press, New York.

Willardson, R. K., and Beer, A. C., eds. (1970). "Semiconductors and Semimetals," Vol. 5. Academic Press, New York.

Willardson, R. K., and Beer, A. C., eds. (1977). "Semiconductors and Semimetals," Vol. 12. Academic Press, New York.

Wolfe, W. L., and Zissis, G. J., eds. (1978). "The Infrared Handbook." US Govt. Printing Office, Washington, D.C.

CHAPTER 4

Metal–Semiconductor Junctions as Frequency Converters

M. V. Schneider

Bell Laboratories
Crawford Hill Laboratory
Holmdel, New Jersey

List of Symbols

A	Junction area
A^{**}	Modified Richardson constant
B	Thermionic saturation current
C	Junction capacitance
C_0	Zero-bias junction capacitance
C_{PL}	Capacitance of planar junction
ΔC	Parasitic junction capacitance
C_P	Parasitic device capacitance
D	Tunneling saturation current
d	Thickness of epitaxial layer
E_{max}	Maximum electric field strength
$E(x)$	Electric field strength in depletion layer
f	Radio frequency
f_C	Cutoff frequency of diode
f_S	Signal frequency
Δf	Receiver bandwidth in signal band
G_a	Antenna gain
g	Junction conductance
g_n	Normalized junction conductance
h	Planck's constant = 6.63×10^{-34} J sec
I_k	Modified Bessel function of the first kind of order k
I_0	Direct bias current
I_S	Saturation current
i	Diode current
i_N	Normalized diode current
i_S	Video source current
i_{THERM}	Thermionic emission current
i_{TUNNEL}	Tunneling current
K_S	Sensitivity constant
k	Boltzmann's constant = 1.38×10^{-23} J/°K
L	Mixer conversion loss
L_I	Image conversion loss
L_{IF}	Intermediate frequency conversion loss
L_{rf}	Radio-frequency conversion loss
L_S	Signal conversion loss
m	Diode factor
m^*	Effective electron mass
N_D	Donor doping concentration
n	Ideality factor
n_0	Normalized doping concentration
P_{IF}	Intermediate frequency power
P_{min}	Minimum detectable power
$P_{mixer,min}$	Minimum detectable signal power (heterodyne detection)
P_{rf}	Incident rf power
$P_{rf,min}$	Minimum detectable signal power (direct detection)
P_S	Signal power
P_{shot}	Available shot noise power
P_T	Transmitted power

q	Electron charge = 1.6×10^{-19} C
R	Junction resistance
R_d	Average junction resistance
R_B	Dynamic resistance
R_{epi}	Series resistance in epitaxial layer
R_L	Resistance per unit length in linear junction
R_S	Diode series resistance
r	Radius
T	Physical temperature of junction
T_a	Antenna temperature
T_{eq}	Equivalent noise temperature
T_{IF}	Noise temperature of IF postamplifier
$T_m(\text{SSB})$	Single-sideband mixer noise temperature
T_r	Receiver noise temperature
T_{shot}	Shot noise temperature of junction
T_{sys}	System noise temperature
δT_{min}	Minimum detectable temperature
t_0	Observation time
U	Circumference
u	Normalized voltage
V	Externally applied voltage
V'	Linear voltage transformation
V_B	Breakdown voltage of junction
V_D	Diffusion voltage
V_P	Punchthrough voltage
w	Depletion layer width
α	Parameter between 0 and 1
α_n	Ionization rate for electrons
α_p	Ionization rate for holes
δ	skin depth
ε_0	Free-space permittivity = 8.85×10^{-12} F/m
ε_r	Relative dielectric constant
μ_0	Free-space permeability = 1.257×10^{-6} H/m
μ_r	Relative permeability
ψ	Electrostatic potential
ρ	Resistivity
ρ'	Charge density
ρ_{epi}	Resistivity of expitaxial layer
Φ_B	Barrier height
$\Delta\Phi$	Barrier height lowering by image force
ω_D	Dielectric relaxation frequency
ω_{IF}	Intermediate frequency
ω_p	Plasma frequency
ω_S	Scattering frequency
ω_s	Signal frequency
ω_p	Pump frequency

I. Introduction

The detection and amplification of electromagnetic radiation in the microwave and millimeter-wave frequency range are of central importance for terrestrial and satellite communication links, for remote sensing of stationary and moving objects, and for a large number of scientific investigations. A widely used detection scheme is to collect the incoming radiation by a parabolically shaped reflector or a lens, to couple the wave into a suitable metallic or dielectric transmission line, and to concentrate the field into a detecting element. One of the most versatile detectors in use today is a rectifying metal–semiconductor junction commonly known as a Schottky diode or hot-carrier diode. The nonlinear current–voltage characteristic of such a diode is used to convert the millimeter-wave signal to a much lower frequency where it can be readily processed, detected, and demodulated. The conversion is achieved by applying a relatively large microwave or millimeter-wave voltage (local oscillator or pump frequency) to the terminals of the Schottky diode. The combination of the signal and the local oscillator produces, among other frequencies, a difference frequency (intermediate frequency, or IF), which can be separated by filtering. The electrical network that combines the various frequencies and includes the rectifying metal–semiconductor junction is called a Schottky mixer.

For most practical applications, it is desirable that the conversion to a lower frequency, commonly referred to as downconversion, be achieved with the smallest possible generation of electrical noise in the diode, with a small loss of signal power (conversion loss), and minimum distortion. Such an optimum design requires an understanding of the current transport mechanisms through the junction, the electrical properties of the diode and the surrounding diode package, and the electrical properties of the embedding network at all the sum and difference frequencies of the signal, the local oscillator, and the harmonics of the local oscillator. Recent progress in understanding these various device and network parameters and new developments in the fields of semiconductor crystal growth and thin-film lithography have resulted in frequency converters with very low noise temperatures and excellent conversion characteristics. Such circuits are in active use at a large number of radio observatories and are incorporated as front ends in terrestrial radio links and satellite communication systems. They are also used in radars, in imaging radiometers, and for detecting trace molecules in the atmosphere.

Note that metal–semiconductor junctions used in frequency converters have a number of properties that make them attractive for use in many other scientific investigations: The junctions are sensitive to light, nuclear radiation, temperature, and pressure. They can display a negative electrical resistance, which may be used for amplification or oscillation. They also provide solid-state scientists with a powerful tool for the study of the bulk and surface properties of semiconductor materials. The improved understanding of these material properties has led to better devices, and the improved device performance has, in turn, motivated further efforts in solid-state research. This feedback process may explain the rapid progress achieved in the performance of frequency converters. Many improvements of the junctions and the associated circuits are also related to advances in material growth, material modification, and pattern generation. These techniques have been essential in achieving new results in the areas of molecular-beam epitaxy (MBE) and very large scale integration (VLSI). Advances in these high-technology areas will be of direct benefit in attaining optimal performance in microwave and millimeter-wave frequency converters.

II. Origin of the Metal–Semiconductor Junction Mixer

Nonlinear current–voltage characteristics and unilateral conduction through contacts between metal wires and various metal sulfides were already observed more than 100 years ago by Braun (1874). For his pioneering work, Braun was awarded, along with Marconi, the Nobel Prize in Physics in 1903 (Braun, 1967). Although the rectification process was not clearly understood, substantial progress was nevertheless achieved during the next few decades in the development of plate rectifiers used for large-scale power rectification and of point-contact rectifiers, which were applied in the early days of radio telegraphy as detectors of high-frequency signals. A description of these early developments can be found in the monograph by Henisch (1957). About 50 years after Braun's discovery, Schottky (1923) proposed that the rectification process depended on the existence of a potential barrier at the interface of the contact between a metal and a semiconductor. This concept led Schottky and Spenke to the postulation of a stable space charge in the semiconductor that allows mobile carriers with sufficient energy to pass over the potential barrier (Schottky and Spenke, 1939; Schottky, 1941). They proposed that the current through this space-charge, or depletion, region was governed by drift and the diffusion of carriers originating in the bulk of the semiconductor. This model was improved by Bethe (1942), who showed that the carriers do not only require sufficient energy to climb over the barrier but also a sufficiently large momentum perpendicular to the metal–semiconductor interface to be accepted by the metal. Bethe's model is with minor modifications still accepted today, and because of its analogy with the emission of a carrier from a conductor into vacuum, it is referred to as the "thermionic emission model."

The development of nonlinear, two-terminal devices made it possible to devise circuits for detecting weak electromagnetic signals. Early in this century, it was discovered that nonlinear devices could be used to down-convert radio frequencies to much lower frequencies. This could be achieved by letting the nonlinear device generate the beat frequency between the radio signal and a fixed pump frequency. Armstrong (1924) and Schottky (1925) independently discovered the basic mixer circuits that are now an essential part of almost every receiver in the frequency range from a few hundred kilohertz up to several hundred gigahertz. Substantial efforts were made during the Second World War to improve mixers in the microwave range for use in radar receivers, as described by Torrey and Whitmer (1948) and Pound (1948). These receivers still had very large noise temperatures because the detecting elements were point contacts, which are not true metal–semiconductor junctions but hybrids between Schottky and p–n junctions. The electrical properties of these point-contact devices were difficult to reproduce and their response time was relatively slow because the holes in a semiconductor have a much lower mobility than electrons. Low-noise mixers only became available after a number of advances were made in the areas of material growth and planar processing of semiconductor surfaces. Major achievements were the fabrication of stable evaporated contacts by Archer and Atalla (1963), the discovery of silicide formation by Kahng and Lepselter (1965), the fabrication of honeycomb diodes (Irvin and Young, 1965; Burrus, 1966), and beam-leaded diodes (Lepselter, 1966).

More recently, research has continued in three major directions. New types of junctions have been grown using molecular-beam epitaxy (Cho, 1979; Panish, 1980; Panish and Cho, 1980) to deposit semiconducting layers with accurately controlled doping profiles. An analysis of the conversion loss and the noise of microwave and millimeter-wave mixers has shown that the performance of the circuit can be predicted from measurable device and circuit parameters (Held and Kerr, 1978). Finally, the use of strip transmission-line circuits combined with subharmonic pumping has resulted in new types of mixers, which have shown good performance (Carlson *et al.*, 1978). These efforts and other projects in progress at many laboratories may result in optimized receivers suitable for many scientific and communications applications.

III. The Metal–Semiconductor Junction

A. Current Transport Mechanisms and Current–Voltage Characteristics

The various transport mechanisms by which electrons or holes may pass from a metal to a semiconductor or vice versa determine the current–voltage characteristic of the junction. Both linear and nonlinear characteristics are required to achieve frequency conversion, i.e., linear contacts to establish an ohmic contact to a semiconductor chip and nonlinear, or blocking, contacts to generate the sum or difference frequencies of two periodic voltages applied to the junction.

The analysis of scattering and transport mechanisms of the carriers in a metal–semiconductor junction is relatively complicated, as shown in various reviews and treatises by Spenke (1958), Padovani (1971), Rideout (1978), Rhoderick (1978), Salardi *et al.* (1979), and Sze (1981). There are, however, some simplifications possible, which describe the electron transport with sufficient accuracy for most practical applications such as switching and frequency conversion. A common feature of these simplified models is that the externally measured current is the sum of a thermionic emission current and a tunneling current, both currents being exponential functions of the externally applied voltage V. The various currents flowing across a forward-biased barrier are shown in the schematic band diagram of Fig. 1. The parabolically shaped energy barrier in the n-type semiconductor material determines the electron flux flowing into the metal at a forward bias V. At sufficiently high temperatures, the carriers reach the metal by thermionic emission over the top of the barrier, but at low temperatures, tunneling becomes predominant. Tunneling effects in thin barriers have been reviewed by Wolf (1975), Duke (1969), and Tsui (1981).

An equation for the current–voltage characteristic in terms of a sum of two exponentials was published in 1932 by Wilson. The external current i was given by

$$i = \text{const}\{\exp(\alpha qV/kT) - \exp[(\alpha - 1)qV/kT]\}, \tag{1}$$

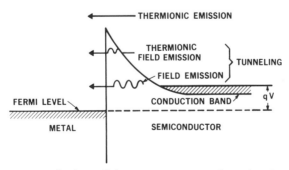

FIG. 1 Transport mechanisms of electrons across a metal–semiconductor barrier. The junction is at forward bias and the doping concentration of donors in the semiconductor is sufficiently high to cause degeneracy. (From Schneider, 1980. Reprinted from *IEEE Transactions on Microwave Theory and Techniques* **28**, 1169–1173. © 1980 IEEE.)

where q is the electron charge, k the Boltzmann constant, T the physical temperature of the junction, and α a parameter that is independent of voltage and lies between 0 and 1.

A similar equation based on a much more advanced understanding of current-transport mechanisms was later derived by Rideout and Crowell (1970) and by Rideout (1975). The current was also a sum of two exponential terms and was given by

$$i = I_S\{\exp(qV/nkT) - \exp[(n^{-1} - 1)qV/kT]\} \tag{2}$$

where I_S is a saturation current and n an ideality factor ≥ 1. The major part of the first term in Eq. (2) is caused by thermionic emission, which is the spilling of thermally excited electrons over the potential barrier. The second term, which is a reverse flux, is mainly due to tunneling, which is composed of field emission and thermionic field emission. For the special case $n = 1$, which represents a diode with negligible tunneling, one obtains the diode equation described in the treatise by Sze (1981):

$$i = I_S[\exp(qV/kT) - 1]. \tag{3}$$

The saturation current I_S is a function of the diode area A, the barrier height Φ_B, and the modified Richardson constant A^{**} and is given by

$$I_S = A^{**}AT^2 \exp(-q\Phi_B/kT), \tag{4}$$

where $\Phi_B = (kT/q) + V_D + \Phi_F - \Delta\Phi$ and V_D is the diffusion voltage, Φ_F the position of the Fermi level relative to the bottom of the conduction band, and $\Delta\Phi$ the image force lowering of the barrier (Sze, 1981).

Both Eqs. (1) and (2) can be simplified if we introduce a diode factor m defined by

$$m = \frac{2}{n} - 1 = 2\alpha - 1 \tag{5}$$

and a normalized voltage u defined by

$$u = qV/2kT. \tag{6}$$

Equations (1) and (2) now become

$$i = I_S\{\exp[(m + 1)u] - \exp[(m - 1)u]\}, \tag{7}$$

which can be written in the product form as

$$i = 2I_S \exp(mu) \sinh u. \tag{8}$$

For the special case $m = 0$, which represents a balance between tunneling and thermionic emission, one obtains

$$i = 2I_S \sinh u. \tag{9}$$

A plot of the natural logarithm of the normalized current $\ln(i/I_S)$ as a function of the positive normalized voltage u is shown in Fig. 2. For $u \gg 1$, all traces are straight lines and for $m = 0$, the current becomes an odd function of the applied voltage as stated in Eq. (9). The same current–voltage characteristic is shown in Fig. 3 for positive and negative voltages.

Note that Eqs. (1) and (2) are not fulfilled for all metal–semiconductor junctions. For the general case, the thermionic emission current i_{therm} can be written in the form

$$i_{\text{therm}} = B[\exp(\beta V) - 1], \tag{10}$$

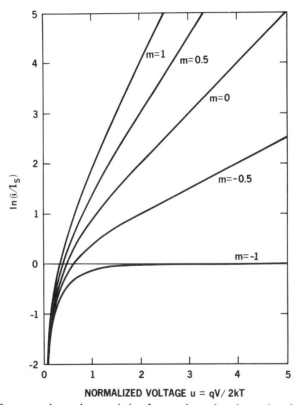

FIG. 2   Current–voltage characteristic of a metal–semiconductor junction at forward bias. The natural logarithm of the normalized current i/I_S is plotted as a function of the normalized forward voltage $u = qV/2kT$, with the diode factor m as a parameter. (From Schneider, 1980. Reprinted from *IEEE Transactions on Microwave Theory and Techniques* **28**, 1169–1173. © 1980 IEEE.)

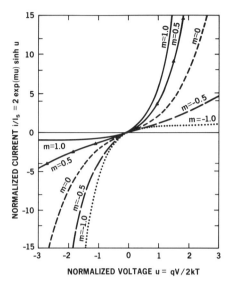

FIG. 3   Current–voltage characteristic at forward and at back bias. The curve labeled $m = 1$ applies to pure thermionic emission and $m = -1$ to pure tunneling. A symmetrical hyperbolic sine function is obtained for a diode factor $m = 0$. (From Schneider, 1980. Reprinted from *IEEE Transactions on Microwave Theory and Techniques* **28**, 1169–1173. © 1980 IEEE.)

where B is a saturation current and $\beta = q/kT$ as given in Eq. (3). The total current i is the sum of the thermionic emission current and the tunneling current, $i = i_{\text{therm}} + i_{\text{tunnel}}$. The tunneling current is large if the barrier is very thin, i.e., if the semiconductor is highly doped or if the semiconductor consists of a highly doped substrate covered by a very thin epitaxial layer with a lower doping concentration. The tunneling current i_{tunnel} is given by

$$i_{\text{tunnel}} = D[1 - \exp(-\gamma V)]. \qquad (11)$$

The saturation current D depends on the height and shape of the barrier. It can be made equal to B by making a linear voltage transformation $V' = V - V_0$, i.e., by applying a dc bias V_0 to the junction. The remaining coefficients β and γ can be made equal by choosing an appropriate junction temperature or barrier profile. It is therefore possible to approach the current–voltage characteristics of Eqs. (3) or (9) by proper choice of the materials, the doping profile of the semiconductor, and the junction temperature.

Junctions displaying the characteristics of Eqs. (8) or (9) can also be fabricated using heterostructures such as a $Ga_{1-x}Al_xAs$ layer sandwiched between two GaAs layers (Allyn *et al.*, 1980). Being a special case of the heterojunction superlattice devised by Dingle (1975), the device has been termed the "single-dingle diode." The ratio of thermionic to tunneling current in such a diode can be accurately controlled by a proper choice of the thickness of the $Ga_{1-x}Al_xAs$ layer and the profile of the index x, with $x = 0$ at both boundaries of the layer. Another type of junction which can have a symmetrical current–voltage characteristic is the planar doped barrier diode, or Malik junction (Malik *et al.*, 1980).

The various types of junctions and their basic properties are listed in Table I. The most widely used junctions for frequency converters are the Schottky barrier (Schottky, 1939) and the Mott barrier (Mott, 1939). Schottky diodes are used in room-temperature mixers, but their widest industrial application is their use as clamping diodes in transistor–transistor logic (TTL; Tada and Laraya, 1967) and integrated injection logic circuits (I^2L; Hewlett, 1975). They are also incorporated into metal–semiconductor field-effect transistors (MESFET, Liechti, 1975) and high-density IC circuits, which became feasible with the development of reliable ion-implantation methods and related techniques as described in the treatise by Hannay (1976).

Mott diodes are of interest in cooled mixers with very low noise temperatures (Linke *et al.*, 1978). These devices are characterized by a very thin intrinsic or low-doped epitaxial layer at the metal–semiconductor interface. The doping concentration in the epitaxial layer is so small that the

TABLE I

JUNCTIONS FOR FREQUENCY CONVERTERS

Diode	Characteristics	Reference
Braun	Point contact, hybrid between metal–semiconductor and $p-n$ junction	Braun (1874); Burrus (1966)
Schottky	Metal–semiconductor junction with thermionic emission over parabolic energy barrier	Schottky (1939, 1941)
Mott	Metal–semiconductor junction with thermionic emission over rectangular energy barrier	Mott (1938, 1939)
Wilson	Metal–semiconductor junction with balance between tunneling and thermionic emission	Wilson (1932)
Single-dingle	Heterojunction, material with higher band gap sandwiched between lower gap materials	Allyn *et al.* (1980)
Back diode	Abrupt $p-n$ junction with large reverse current (tunnel current)	Eng (1961); Shurmer (1964)
Esaki	Abrupt $p-n$ junction with negative resistance	Esaki (1958); Gysel (1971)
Malik	Planar doped barrier	Malik *et al.* (1980)

energy barrier is too thick at reverse bias to observe a reverse tunneling current. The energy barrier has the shape of a rectangle, and for back bias, the device resembles a parallel plate capacitor with a fixed, voltage-independent capacitance.

The Wilson and the single-dingle diode are useful in subharmonically pumped mixers requiring a local oscillator frequency that is approximately one-half that needed in a conventional mixer. These devices have electrical properties similar to those of a pair of Schottky or Mott diodes in an antiparallel connection.

B. JUNCTION CONDUCTANCE

The conductance is an important electrical property that determines the suitability of a junction for electrical engineering applications. The conductance g is

$$g = \partial i / \partial V = (q/2kT)(\partial i / \partial u). \tag{12}$$

From Eq. (7) we obtain

$$g = (qI_S/kT) \exp(mu)[m \sinh u + \cosh u]. \tag{13}$$

For the special case $m = 0$, this reduces to

$$g = (qI_S/kT) \cosh u. \tag{14}$$

From Eqs. (7) and (10) one obtains

$$g = (g/2kT)(m + 1)i + 2I_S \exp[(m - 1)u] \tag{15}$$

which, for $i \gg I_S$ and $m = 1$, gives the familiar result

$$g = qi/kT. \tag{16}$$

Table II lists the ideality factor, the current and the conductance for $m = +1, -1$, and 0, which are the three cases of major interest. A normalized current $i_n = i/2I_S$ and a normalized conductance

$$g_n = (kT/qI_S)g \tag{17}$$

are used to simplify the table.

A plot of the natural logarithm of g_n (ln g_n) as a function of the normalized voltage is shown in Fig. 4. Straight lines with a slope of $\pm 45°$ are obtained for the limiting cases $m = \pm 1$. The conductance for any intermediate case lies within the bounds of these straight lines.

C. CAPACITANCE AND BREAKDOWN VOLTAGE

Metal–semiconductor junctions that are back-biased have electrical properties similar to those of a parallel-plate capacitor. The capacitance

TABLE II

ELECTRICAL CHARACTERISTICS OF JUNCTIONS

Diode factor m	Ideality factor n	Normalized current $i_n = i/2I_s$	Normalized conductance $g_n = kTg/qI_s$
+1	1	$\exp u \cdot \sinh u$	$\exp(2u)$
0	2	$\sinh u$	$\cosh u$
−1	∞	$\exp(-u) \cdot \sinh u$	$\exp(-2u)$

of the junction is directly proportional to the area and indirectly proportional to the thickness of the depletion layer. The breakdown voltage is determined by the ionization rates for carriers in the junction region. The major significant difference between a parallel-plate capacitor and a metal–semiconductor junction is that the electrical field in the junction reaches a maximum at the interface of the metal and the semiconductor and that the capacitance is a function of the applied reverse bias. The high field intensity at the interface affects the breakdown voltage, and it causes a soft breakdown characteristic for planar diodes, which have very high

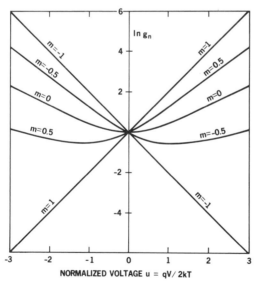

FIG. 4   Plot of the natural logarithm of the normalized device conductance $g_n = kTg/qI_s$ as a function of the normalized voltage $u = qV/2kT$. Straight lines are obtained for the limiting cases $m = \pm 1$. (From Schneider, 1980. Reprinted from *IEEE Transactions on Microwave Theory and Techniques* **28**, 1169–1173. © 1980 IEEE.)

field intensities at the periphery of the junction. The variable capacitance is used in reactive frequency converters and in frequency multipliers to generate harmonic multiples of a pump frequency.

The capacitance of the junction can be calculated by solving the one-dimensional Poisson equation subject to the appropriate boundary conditions

$$d^2\psi/dx^2 = -\rho'/\varepsilon_0\varepsilon_r \qquad (18)$$

where ψ is the electrostatic potential, ρ' the charge density, ε_0 the free-space permittivity (8.85×10^{-12} F/m), and ε_r the relative dielectric constant of the dielectric material ($\varepsilon_r = 12.9$ for GaAs). For a semiconductor with a donor doping concentration N_D, the charge density is given by

$$\rho' = q(N_D + p - n), \qquad (19)$$

where p and n are the hole and electron concentrations, respectively. For a constant donor doping concentration N_D, the electric field $E(x)$ in the junction is

$$E(x) = E_{max}[1 - (x/w)], \qquad (20)$$

where the maximum electric field E_{max} and the depletion layer width w are given, respectively, by (Rhoderick, 1978)

$$E^2_{max} = (2qN_D/\varepsilon_0\varepsilon_r)(V_D - V - (kT/q)), \qquad (21)$$

$$w = \{2\varepsilon_0\varepsilon_r(V_D - V - [kT/q])/qN_D\}^{1/2}. \qquad (22)$$

The quantity V_D is the diffusion voltage and V the externally applied voltage as shown in Fig. 5.

For practical applications, it is convenient to define a normalized doping concentration n_0, given by $n_0 = N_D/10^{17}$, where N_D is in units of reciprocal cubic centimeters. From Eq. (22) and for GaAs with $\varepsilon_r = 12.9$, one obtains, for the depletion layer thickness w in micrometers if the term kT/q is neglected,

$$w = 0.120[(V_D - V)/n_0]^{1/2} \quad \mu m. \qquad (23)$$

The junction capacitance C is given by

$$C = \varepsilon_0\varepsilon_r A/w, \qquad (24)$$

where A is the junction area. From Eq. (22) one obtains

$$C = A\{q\varepsilon_0\varepsilon_r N_D/2[V_D - V - (kT/q)]\}^{1/2}. \qquad (25)$$

A plot of $1/C^2$ versus the applied voltage V thus results in a straight line with an intercept $V_D - (kT/q)$ on the V axis. It is to be noted that Eqs. (21), (22), and (25) are only valid if the donor concentration N_D is constant

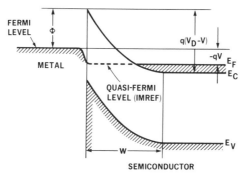

FIG. 5 Band diagram of a metal–semiconductor junction at back bias showing the electron quasi-Fermi level (IMREF), Fermi levels in the metal and the bulk semiconductor, depletion layer width w, and diffusion voltage V_D.

throughout the depletion layer. Most practical junctions are fabricated from epitaxial semiconductor material with a lightly doped layer of thickness d on a heavily doped substrate. This layered structure has the advantage that resistive losses in the semiconductor are minimized. The voltage for which the depletion layer thickness w becomes equal to the epitaxial layer thickness d is called the punchthrough voltage V_P. The junction capacitance remains almost constant above this voltage with a relatively small decrease with increasing voltage because the electric field penetrates a small distance into the heavily doped substrate. At sufficiently high back bias, the electric field in the junction becomes large enough to generate avalanche multiplication because of secondary and higher order ionization of holes and electrons from injected or thermally generated carriers. This process may result in a negative conductance (impact avalanche transit time operation), which can be used for amplification or for generation of rf oscillations (De Loach, 1969). Carrier multiplication is also used in sensitive photodiodes to produce gain (Melchior, 1977).

The junction capacitance given by Eqs. (24) and (25) is valid for mesa junctions, which do not show edge effects. Planar junctions have a parasitic capacitance ΔC, which is due to edge effects. The total capacitance C_{PL} of a planar junction is

$$C_{PL} = C + \Delta C, \qquad (26)$$

where C is given by Eqs. (24) and (25). For a circular junction with a radius r, one obtains, to a first approximation (Copeland, 1970),

$$\Delta C = (3w/2r)C = (3\pi/2)\varepsilon_0\varepsilon_r r. \qquad (27)$$

If the junction is not circular and has an area A and a circumference U, one obtains from Eqs. (26) and (27)

$$C_{PL} = C[1 + (3wU/4A)].$$ (28)

Junctions that are used in conventional frequency converters at ambient room temperature require a breakdown voltage between 5 to 7 V to prevent a breakdown of the diode pumped by the local oscillator. The breakdown voltage is determined by the condition (Sze and Gibbons, 1966)

$$\int_0^w \alpha_n \exp\left[- \int_0^x (\alpha_n - \alpha_p) \, dx'\right] dx = 1,$$ (29)

where α_n and α_p are the ionization rates for electrons and holes, respectively, and x is the axis parallel to the electric field in the junction. For GaAs with $\alpha_n = \alpha_p$, one obtains the much simpler expression

$$\int_0^w \alpha_n \, dx = 1$$ (30)

with an experimentally determined ionization rate

$$\alpha_n = \alpha_p = A_1 \exp\{-[A_2/E(x)]^2\},$$ (31)

where $E(x)$ is the electric field in the depletion layer, $A_1 = 35 \ \mu\text{m}^{-1}$, and $A_2 = 68.5 \ \text{V}/\mu\text{m}$. The upper limit of integration in both Eqs. (29) and (30) is given by the width of the depletion layer w, which can be calculated from Eq. (22). For epitaxial material, the upper limit of integration is equal to the epitaxial layer thickness d if the diode is punched through before reaching breakdown. For very accurate calculations, one also has to take into account that the ionization rates are dependent on the junction temperature and the orientation of the crystal (Pearsall et al., 1978).

Graphs for the depletion layer width w and the breakdown voltage V_B for GaAs as a function of the donor concentration N_D are shown in Fig. 6. The breakdown voltage under the condition of punchthrough is nearly independent of the donor concentration because the field intensity $E(x)$ is mainly determined by the thickness of the epitaxial layer and the applied voltage. One also concludes that an epitaxial layer thickness of 0.1 μm is sufficient to obtain a breakdown voltage of approximately 6 V. A more detailed analysis, which includes the effects of the rapidly decaying electric field in the heavily doped substrate, shows that the change in breakdown voltage shown in Fig. 6 is small for an epilayer with a thickness of 0.1 μm or higher. The breakdown voltage for thinner epilayers cannot be calculated from the junction geometry and N_D alone because of the onset of ballistic electron transport mechanisms and because the donor concen-

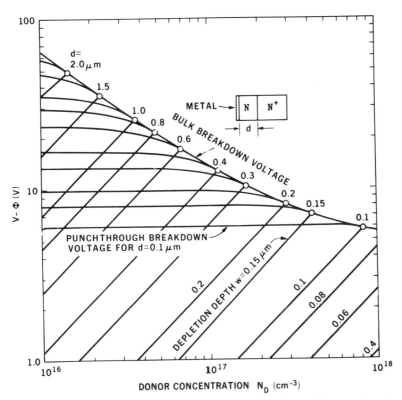

FIG. 6 Breakdown voltage in bulk GaAs and punchthrough breakdown voltage in epitaxial GaAs as a function of the donor concentration. The set of 45° lines gives the depletion layer thickness for a given applied voltage. The set of horizontal lines gives the breakdown voltage of punchthrough for a fixed thickness of the epitaxial layer.

tration cannot be assumed to remain constant over a distance of a few hundred angstroms.

The breakdown voltage of practical devices is often lower than the value computed from Eq. (29) because of high field intensities near the periphery of the device or surface leakage. Both surface leakage and edge breakdown can be suppressed by a guard ring, which surrounds the junction periphery (Haitz and Smits, 1966). For small junction areas, a moat-etched structure devised by Schneider et al. (1977) is preferable because the alignment of the guard ring would be too critical. This specific moat geometry is described in detail in Section III.E and a cross-sectional view of the device is shown in Fig. 11. The current–voltage characteristics of two diodes fabricated with a planar junction and a moat-etched junction are shown in Figs. 7a and b, respectively. The breakdown of the moat-

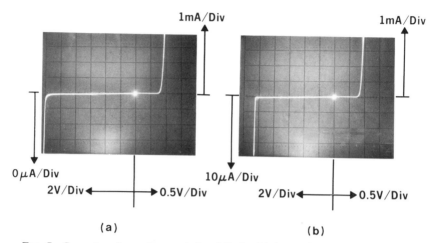

1mA/Div   **1mA/Div**

0 μA/Div   **10 μA/Div**
2V/Div ◄————►**0.5V/Div**   **2V/Div**◄————►**0.5V/Div**

(a)   (b)

FIG. 7 Current–voltage characteristic of diodes fabricated from molecular-beam epi-taxial GaAs. (a) Characteristic for planar diode showing soft breakdown characteristic. (b) Characteristic of moat-etched diode made by anodization of a planar structure (slice N277-87) to a depth of 400 Å and removal of the anodization (slice N277-86) by selective etching followed by metallization of the GaAs surface.

etched structure is much sharper than for the planar device shown in Fig. 7a. The breakdown voltage for the moat-etched structure is lower than that for the planar device and it can be reproduced more accurately be-cause high field intensities do not arise near the junction periphery of the diode.

D.   SERIES RESISTANCE

The equivalent electrical circuit of a metal–semiconductor diode con-sists of an intrinsic diode circuit, a series resistance, and external circuit elements such as a series inductance and parasitic capacitances. A sche-matic drawing of the equivalent circuit is shown in Fig. 8.

The series resistance R_S is voltage- and frequency-dependent and its major part is caused by the series resistance of the semiconducting sub-strate and the undepleted epitaxial layer. Secondary contributions are due to the resistance of the ohmic contact to the semiconductor and ohmic re-sistances in the external leads. The series resistance in the epitaxial layer and the semiconducting substrate depends on the junction geometry, the frequency, and, to a lesser extent, the applied voltage. Four basic junc-tion geometries in current use are shown in Fig. 9. Circular, hemispher-ical, elliptical, and linear junctions are shown in the figure. The spreading resistance of each of these junctions can be calculated by solving the La-place equation subject to the appropriate boundary conditions (Holm, 1967; Smythe, 1950). The result is a dc resistance which has to be modi-

$$i(V) = I_S \left[\exp\left(\frac{qV}{nkT}\right) - 1 \right] \qquad g = \frac{\partial I}{\partial V} = \frac{q}{nkT}(i + I_S),$$

$$C(V) = A \sqrt{\frac{\varepsilon_r \varepsilon_0 q N}{2(\phi - V)}}$$

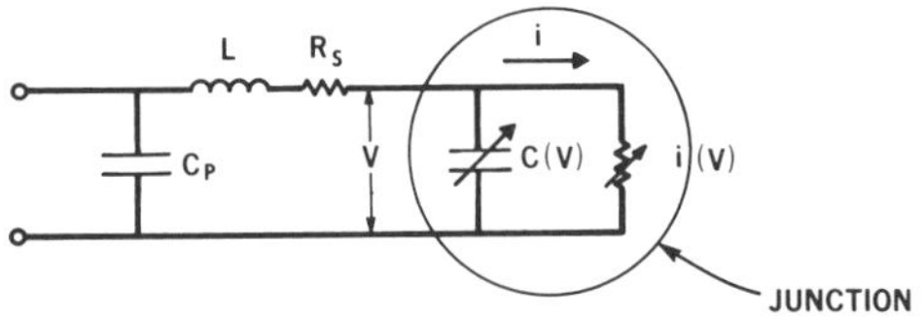

FIG. 8 Equivalent circuit of metal–semiconductor junction showing nonlinear junction conductance $g(V)$, nonlinear junction capacitance $C(V)$, and external elements series resistance R_s, series inductance L, and parasitic device capacitance C_p.

fied at rf frequencies because of the presence of skin effects. The dc values of these four junction types are listed in Table III, where $K(k)$ is the complete elliptic integral with modulus $k = (a - b)/(a + b)$. It is to be noted that the dc and rf resistance are equal if the distance between the junction and the ohmic contact is equal or less than a skin depth, which is given by Eq. (37). The series resistance at rf frequencies can be measured using the De Loach method (1964) or related techniques described by Schilder (1976). The series resistance of the linear junction is of particular interest because this geometry is used for transmission lines in monolithic integrated circuits and as a gate electrode in field-effect transistors. This

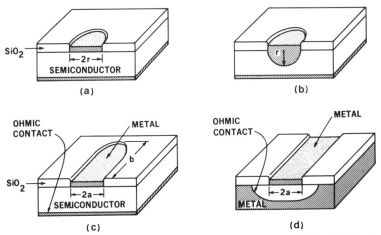

FIG. 9 Basic junction geometries used for frequency-converter diodes: (a) circular; (b) hemispherical; (c) elliptical; (d) linear. The circular, hemispherical, and elliptical junctions can be made by conventional planar photolithographic processing. The linear junction with an internal ohmic metal–semiconductor interface requires advanced three-dimensional processing.

TABLE III

JUNCTION DC SPREADING RESISTANCE

Junction	Resistance	Reference
Circular	$\rho/4r$	Henisch (1957)
Hemispherical	$\rho/2\pi r$	Holm (1967)
Elliptical	$[\rho/\pi(a + b)]\,K(k)$	Holm (1967)
Linear	$(\rho/\pi)\ln(4\beta)$	Eq. (36)

resistance can be determined by conformal mapping of the upper half of the z_1 plane shown in Fig. 10a into a semi-infinite vertical strip in the z plane shown in Fig. 10b. The mapping is achieved by the function

$$z_1 = a \sin z, \tag{32}$$

which transforms the line of length $2a$ centered at the origin of the x_1 axis into a line of length π centered at the origin at the x axis. Separating Eq. (32) into real and imaginary parts gives

$$x_1 = a \sin x \cosh y, \tag{33}$$

$$y_1 = a \cos x \sinh y. \tag{34}$$

If the rectangle of Fig. 10b with a length π and a height $2\gamma a$ is filled with a material of resistivity ρ, the resistance between the upper electrode ABC and the lower electrode DEF is given by $R_L = 2\rho\gamma a/\pi$, where R_L is

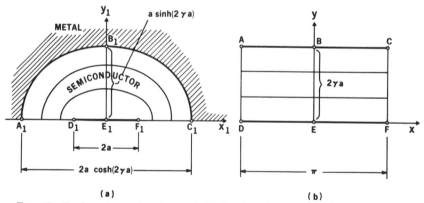

(a)   (b)

FIG. 10   Conformal mapping of upper half of z_1 plane into semi-infinite vertical strip in a z plane, using $z_1 = a \sin z$. (a) Upper half of z_1 plane with interface of metal–semiconductor junction $D_1E_1F_1$ and metal–semiconductor ohmic contact $A_1B_1C_1$. (b) Rectangle in z plane. The inside of the rectangle ABCFED is the conformally mapped image of the semiconductor area shown in Fig. 10a.

the resistance per unit length between two parallel and infinite long strip conductors with the cross section of Fig. 10b. This resistance is not changed by the conformal mapping process, and thus the resistance per unit length between the electrode $A_1B_1C_1$ and the linear junction $D_1E_1F_1$ remains the same. The transformed electrode $A_1B_1C_1$ is the arc of an ellipse with focal points D_1 and F_1 and with major and minor axis $a \cosh(2\gamma a)$ and $a \sinh(2\gamma a)$, respectively, as shown in Fig. 10b. It is assumed that the space between the electrodes $D_1E_1F_1$ and $A_1B_1C_1$ is filled with a semiconductor with resistivity ρ and that the remaining part of the upper half-plane is filled with metal. We now define the ratio between the minor axis of the ellipse and the junction length $2a$ by $\beta = \sinh(2\gamma a)/2$. The resistance R_L between the electrodes $D_1E_1F_1$ and $A_1B_1C_1$ in the semiconductor thus becomes

$$R_L = 2\rho\gamma a/\pi = (\rho/\pi) \text{ arc } \sinh(2\beta). \tag{35}$$

For $\beta \gg 1$, one obtains the simple equation

$$R_L = (\rho/\pi) \ln(4\beta). \tag{36}$$

For GaAs, with $\rho = 1 \times 10^{-3}$ Ω cm for a doping concentration of 2×10^{18} cm^{-3} and a ratio $\beta = 10$, one obtains, from Eq. (36), $R_L = 1.2 \times 10^{-3}$ Ω cm. If the minor axis of the ellipse is equal or smaller than the skin depth in the semiconductor, the dc and the rf resistance R_L will be the same. The skin depth in the semiconductor is given by

$$\delta = (\rho/\pi\mu_0\mu_r f)^{1/2} \tag{37}$$

where f is the radio frequency in hertz and $\mu_0 = 4\pi \times 10^{-7}$ H/m. For a frequency $f = 100$ GHz and $\rho = 1 \times 10^{-3}$ Ω cm, one obtains, with $\mu_r = 1$, $\delta = 5.0$ μm. Recently developed processing techniques make it possible to fabricate junction geometries with the cross section shown in Fig. 10a and with a length of the minor axis equal or less than a skin depth in the 100-GHz frequency range. The total series resistance of such a junction will be the sum of the contact resistance to the metal, the spreading resistance in the heavily doped semiconductor material, and the series resistance in the epitaxial layer. The spreading resistance in the heavily doped semiconductor is given by Eq. (35), and the series resistance in the epitaxial–semiconductor layer adjacent to the junction R_{EPI} is

$$R_{epi} = \rho_{epi}d/A, \tag{38}$$

where ρ_{EPI} is the resistivity of the epitaxial layer, d the thickness of the epitaxial layer, and A the area of the junction. For a circular junction on GaAs with a junction diameter of 2 μm, a doping concentration of 2×10^{17} cm^{-3} with a corresponding mobility of 4×10^3 cm^2/V sec and a resulting epitaxial layer resistivity of 8×10^{-3} Ω cm, and an epitaxial layer thickness of 0.1 μm, one obtains $R_{EPI} = 2.5$ Ω. The zero-bias junction capacitance C_0 calculated from Eq. (25) is $C_0 = 4.7$ fF (1 fF $= 10^{-15}$ F). The additional contributions depend on the junction geometry, the frequency, and the quality of the the ohmic contact. Calculations of these various contributions have been performed by Dickens (1967), Carlson et al. (1978), Calviello (1979), and Wrixon et al. (1979). Typical values obtained for the total series resistance of a circular junction on GaAs with a junction diameter of 2 μm are 5–10 Ω at a frequency of 100 GHz. Using a linear junction as just outlined with a low-resistivity substrate (Keen, 1980) and improved ohmic contacts recently developed by Stall et al. (1979), the total series resistance of a device with a zero-bias capacitance of 5 fF should range from 3 to 5 Ω at 100 GHz. The corresponding zero-bias cutoff frequency for such a diode is given by $f_c = 1/(2\pi RC_0)$ and is equal to 8 THz (8 $\times 10^{12}$ Hz), which means that the junction can be used for direct detection or frequency conversion into the submillimeter-wave frequency range.

Two undesired effects that limit the performance of the diode in the submillimeter-wave range are carrier inertia and dielectric relaxation time (Champlin and Eisenstein, 1978; Kelly and Wrixon, 1979). Both carrier inertia and dielectric relaxation lead to plasma oscillations that result in an increased series resistance near the plasma frequency ω_P given by

$$\omega_P = (\omega_S\omega_D)^{1/2} = (q^2N_D/m^*\varepsilon_r\varepsilon_0)^{1/2} \tag{39}$$

where ω_S and ω_D are the scattering frequency and the dielectric relaxation frequency, respectively, and m^* the effective mass of the carrier. The scattering frequency is given by $\omega_S = q/(m^*\varepsilon_r\varepsilon_0)$ and the dielectric relaxation frequency by $\omega_D = (\varepsilon_r\varepsilon_0\rho)^{-1}$. These effects are difficult to confirm by experimental techniques since the predicted plasma frequencies are in the range of 1 THz. However, several authors have claimed that the assumption of a constant ρ is inappropriate for a variable charge density and that the predicted relaxation times require substantial corrections (Mott et al., 1972; Gruber, 1973).

The series resistance of a junction can be further reduced by cooling the junction or by choosing a semiconductor material with a higher mobility. Attractive materials are InSb, InAs (McColl and Millea, 1976; Millea et al., 1976; Korwin-Pawloski and Heasell, 1975), and ternary and quaternary compounds such as $Ga_xIn_{1-x}As_yP_{1-y}$ (Oliver, 1980). Oliver reports a 50% higher mobility than GaAs at room temperature for the ternary system $Ga_xIn_{1-x}As$ grown by liquid-phase epitaxy. The material can be grown lattice-matched on an InP substrate and looks attractive for use as an improved mixer diode and also as an FET with increased switching speed and cutoff frequency.

E. DEVICE GEOMETRIES

The two basic device geometries in current use in low-noise mixers are the honeycomb structure devised by Young and Irvin (1965) and the beam-leaded diode invented by Lepselter (1966). Honeycomb diodes are mainly used in waveguide transmission lines and beam-leaded devices are suited for strip transmission line circuits. A cross-sectional view of a honeycomb diode chip with a whisker contact is shown in Fig. 11. The n on n^+ GaAs chip is typically 200 μm square by 100 μm thick. The epitaxial layer with a thickness of 0.1 μm and a Sn doping concentration of 1×10^{16} cm^{-3} covered with a protective coating of silicon dioxide with a thickness of 0.4 μm is desired for cryogenically cooled diodes to reduce tunneling and to enhance thermionic emission at low temperatures. For diodes operating at room temperature, the doping concentration of the epilayer is increased to about 2×10^{17} cm^{-3} to achieve a relatively small tunnel current and a small series resistance in the epilayer. Ohmic contacts are formed on the side or the bottom of the chip and an array of moat-shaped junctions, which are surrounded by the silicon dioxide layer, are formed on the epilayer by depositing a metal layer by electroplating or

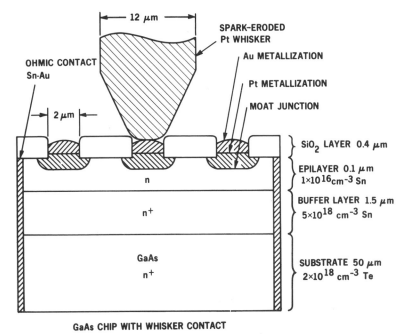

GaAs CHIP WITH WHISKER CONTACT

FIG. 11 Cross-sectional view of honeycomb diode with whisker showing GaAs substrate with buffer layer, epitaxial layer, and protective SiO₂ coating. The ohmic contact is made on both sides of the chip to reduce the rf resistance of the diode.

evaporation. An individual junction is contacted randomly by a spark-eroded or electrolytically etched metal whisker. An electron micrograph of a junction is shown in Fig. 12. The junction has the shape of a dumbbell to achieve a lower series resistance compared to a standard circular-shaped junction with the same area. Other junction geometries that result in a reduced series resistance are the cross and stripe geometries devised by Wrixon (1974) and described in more detail by Kelley and Wrixon (1980), the interconnected array diodes described by McColl *et al.* (1977), and the bathtub diodes by Schneider *et al.* (1977). These junctions can be readily contacted with an electrolytically etched whisker (Dozier and Rodgers, 1964) or a spark-eroded microelectrode (Schneider and Berger, 1979). The semiconductor chip and the whisker are mounted on separate metal pins or studs, which are then inserted into a reduced-height wave-guide through holes in the top and bottom wall of the waveguide. Other successful mounting schemes are the triple-quartz substrate mount developed by Kerr (1975) and the notch-front diode mount used by Schneider and Carlson (1977) with a diode chip and a whisker soldered to the gold conductor stripes of a strip transmission line.

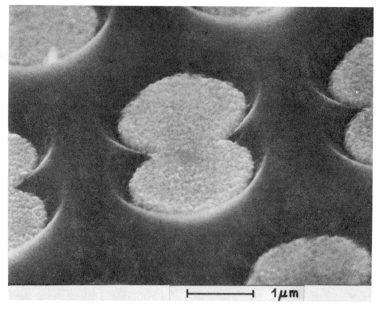

FIG. 12 Electron micrograph of dumbbell junction with a diameter of 1.5 μm on the bottom of the junction. The structure can be considered as two diodes in parallel with a resulting reduced spreading resistance of a factor $\sqrt{2}$ compared to a single circular diode with the same area. The photograph shows the metallization on the bottom of the dumbbell with the sloped sides of the surrounding SiO_2 layer.

For planar circuits such as microstrip and stripline mixers, diodes can be fabricated directly on a high-resistivity semiconducting substrate or they can be provided with beam leads, which are subsequently bonded to the metallization of a strip transmission line circuit. A cross-sectional view of a beam-leaded device is shown in Fig. 13 (Calviello, 1979). The junction metallization is an evaporated Ta–Au film and the surrounding protective coating is a SiO_2–Si_3N_4 layer. An alloyed GeAu–Au film forms an ohmic contact, which is very close to the junction to reduce the series resistance caused by the skin effect. The air gap between the beam lead attached to the junction and the Si_3N_4 layer reduces the parasitic shunt capacitance to the heavily doped n^+ substrate. Another device geometry with small parasitics is the MBE beam-lead diode devised by Ballamy and Cho (1976). This device, shown in Fig. 13b, is fabricated on a high-resistivity GaAs substrate on which a silicon dioxide layer is grown using standard vapor-phase deposition techniques. A window is subsequently defined in the SiO_2 layer and a GaAs layer is deposited by molecular-beam epitaxy on the slice. The resulting thin film is a single-crystal GaAs layer of device quality inside the window and polycrystalline insulating GaAs outside of the window. This unique growth characteristic is used to fabricate a junction and an ohmic contact in the window area and to deposit the beam leads using the polycrystalline insulating GaAs as a mechanical support.

A further development in using conducting pockets of GaAs on an insulating substrate is the surface-oriented diode developed by Clifton (1977), Murphy et al. (1977), and Immorlica and Wood (1978). These authors have used proton bombardment to convert a conducting GaAs layer into high-resistivity material in the bombarded regions. Using suitable masks, pockets of conducting GaAs containing the junction and the ohmic contact were defined, and the structures were subsequently processed into devices. The surface-oriented diodes are useful in integrated monolithic receivers in which antennas, transmission lines, and other circuit components are integrated on the same GaAs wafer (Murphy and Clifton, 1978). The diodes have small parasitics and they have been incorporated into harmonic mixers with signal frequencies up to 670 GHz (Clifton, 1979).

IV. Junction, Device, and Circuit Fabrication

A. MICROSCIENCE AND MICROELECTRONIC STRUCTURES

The essential building blocks of a low-noise mixer are characterized by their small size. For example, epitaxial layers are grown with a thickness of a few hundred angstroms, passivating layers have a depth measured in nanometers, junction sizes are in the order of micrometers, and strip trans-

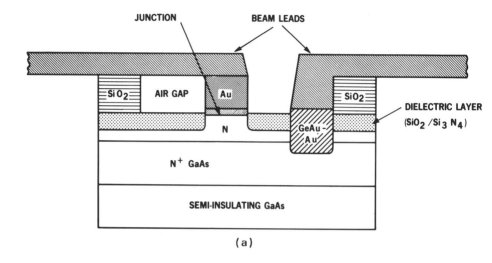

(a)

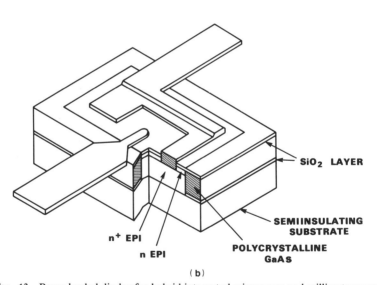

(b)

FIG. 13   Beam-leaded diodes for hybrid integrated microwave and millimeter-wave circuits. (a) Beam-lead structure after Calviello *et al.* (1979). The cross-sectional view shows a
layered structure with an air gap under the junction beam lead to reduce parasitic capacitance. The ohmic contact (GeAu–Au) is very close to the junction to reduce the skin resistance in the device. (b) MBE beam-leaded diode after Ballamy and Cho (1976). The junction
and the ohmic contact are fabricated on the active epitaxial area on the center of the chip.
This active area is surrounded by polycrystalline insulating GaAs, which gives mechanical
support to both beam leads.

mission line circuits require tolerances of several micrometers. The importance of developing structures with small physical sizes was already recognized by Feynman in his fundamental paper "There's Plenty of Room at the Bottom" (1961). More recently, the need for constructing micron and submicron structures with accurately controlled dimensions and material compositions has resulted in new types of research facilities (Wolf, 1979) that are exclusively devoted to research in microscience and to exploratory work on microelectronic structures. The basic processes used in such a facility are directly applicable to advancing the state of the art in low-noise junction mixers and other low-noise electronic components, such as Josephson switching devices, quasi-particle mixers, field-effect transistors, and monolithic integrated circuits.

The basic processing and fabrication steps used to generate small devices and circuits are listed in Table IV. They can be classed into three categories comprising material growth, material modification, and pattern generation. Most of these processes can be performed using different techniques; for example, alloying can be achieved by laser irradiation (Bell, 1979; Salathe *et al.*, 1979; Badertscher *et al.*, 1980), spark alloying (D'Angelo and Verlangieri, 1980), or indirect electric heating (Kelly and Wrixon, 1978). Another example where several alternative choices are available is shown in Fig. 14, where small features in a suitable resist

TABLE IV

TECHNIQUES FOR FABRICATING DEVICES AND CIRCUITS

Category	Technique
Material growth	Epitaxial growth (MBE, VPE, LPE)
	Evaporation
	Sputtering
	Electroplating
	Plasma deposition
Material modification	Diffusion
	Annealing
	Alloying
	Ion implantation
	Proton bombardment
	Plasma oxidation
	Anodization
Pattern generation	Lithography (UV, x-ray, electron-beam)
	Etching (chemical, ion, plasma)
	Machining
	Laser cutting
	Spark erosion

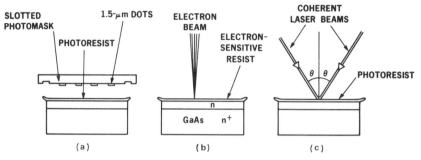

FIG. 14 Techniques for fabricating planar junction geometries on a semiconductor slice: (a) photolithography; (b) electron-beam lithography; (c) laser interference. The surface of the slice is coated with an appropriate resist material, which is then exposed by UV light through a mask, by a scanned electron beam, or by interfering laser beams. The periodicity of the pattern generated by the laser beams is $\lambda/(2 \sin \theta)$.

material are fabricated by photolithography, electron-beam lithography, or interference of coherent laser beams. Figure 15 shows in more detail the electron-beam lithographic process together with another technique known as x-ray lithography. These microelectronic fabrication techniques will not be treated in detail in this chapter, however, a specific example

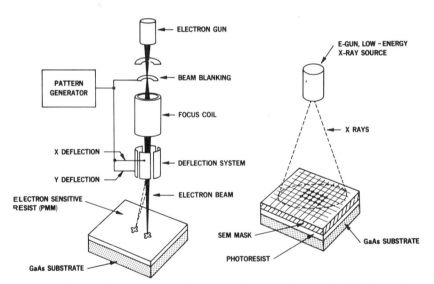

FIG. 15 Schematic diagram of pattern generation on semiconductor slice by electron-beam and x-ray lithography. Electron lithographic structures are made in a modified scanning electron microscope, which is equipped with a computer-controlled pattern generator. For x-ray lithography, a mask is required, which is first produced by electron lithography.

describing the required processing steps to fabricate junction devices and thin-film circuits will be given in the following sections.

B. MATERIAL GROWTH AND JUNCTION FABRICATION

The semiconductor material required for high-quality microwave and optoelectronic devices can be grown by vapor-phase epitaxy, liquid-phase epitaxy, or molecular-beam epitaxy (MBE). Molecular-beam epitaxy offers several advantages because it allows the growth of ultrathin layers with precisely controlled doping profiles. The process can be described as an epitaxial growth of compound semiconductor films by the reaction of one or more thermal molecular beams with a crystalline surface under ultrahigh vacuum conditions (Cho and Arthur, 1975; Joyce 1979; Cho, 1979; Panish, 1980). Although MBE is related to vacuum evaporation, it offers a substantially improved control over the incident atomic or molecular fluxes. Both n-type and p-type doped layers can be grown using effusion cells with suitable doping materials and calibrated flux rates. The doping concentrations that have been obtained are, in some cases, substantially higher than concentrations achieved by liquid-phase epitaxy (Ilegems, 1977). The MBE growth process can also be used to deposit a single-crystal metal film on an epitaxial semiconducting film in one continuous and uninterrupted process without breaking the vacuum (Cho and Dernier, 1978). Therefore MBE is ideally suited for the growth of metal–semiconductor junctions with a nearly ideal interface, which does not have a native oxide layer or any absorbed contaminants.

An alternative successful crystal growth method is the organometallic vapor-phase-deposition technique under reduced pressure (Lacombe *et al.*, 1977; Duchemin *et al.*, 1977, 1978). Impurities are removed in the gas phase using a stream of hydrogen at reduced pressure, which results in a relatively small autodoping and abrupt transitions from the epitaxial to the bulk material. A schematic cross-sectional view of an MBE growth system is shown in Fig. 16. The single-crystal substrate on which the epitaxial layer is to be deposited is inserted into the ultrahigh vacuum chamber through an air lock. The substrate is surrounded by a metal shroud cooled with liquid nitrogen. A number of effusion cells with individual shutters are used to generate the molecular beams of the constituents of the semiconductor materials and for the various dopants. A typical doping profile of an epitaxial layer that has resulted in low-noise mixer diodes is shown in Fig. 17. The growth of MBE layers with similar profiles is described by Harris and Woodcock (1980). The surface doping concentration of the Sn dopant is relatively low (3×10^{16} cm^{-3}) in order to minimize conduction by electron tunneling and thereby to minimize shot noise (Viola and Mattauch, 1973a,b). The doping concentration increases rap-

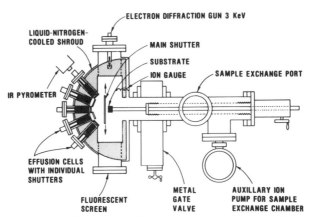

FIG. 16 Cross-sectional view of MBE system with substrate, shutters, and effusion cells for evaporation of the elements of the compound semiconductor and the dopants. The main chamber is surrounded by a liquid-nitrogen-filled shroud to achieve pressures in the range of 10^{-10} Torr (Cho, 1979).

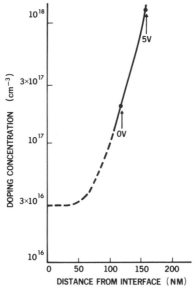

FIG. 17 Doping profile of Sn-doped epitaxial GaAs layer (slice N280) deposited by MBE on a heavily doped GaAs substrate (Schneider et al., 1977). The doping concentration near the surface increases by about an order of magnitude over a distance of 500 Å. The 0-V point marks the depth to which the layer is depleted at a bias of 0 V. The dashed portion of the curve is inferred from measurements on a thicker sample. (From Linke, Schneider, and Cho, 1978. Reprinted from *IEEE Transactions on Microwave Theory and Techniques* **26,** 935–938. © 1978 IEEE.)

idly after the first 60 nm to 3×10^{18} cm^{-3}. The concentration is measured by means of a Miller profiler (Miller, 1972; Miller *et al.*, 1977), which uses capacitance versus voltage data to deduce the doping profile in the sample. The resolution of the profiler is determined by the Debye length in the sample, which is given by $(\varepsilon_r\varepsilon_0 kT/q^2N_D)^{1/2}$. The Debye length at room temperature in GaAs is 400 Å for a doping concentration of 10^{16} cm^{-3} and 150 Å for 10^{17} cm^{-3}. The resolution of the Miller profiler is approximately two to three times the Debye length in the sample.

The processing steps used to fabricate an MBE diode from single-crystal Al (001) grown in situ on GaAs (001) are shown in Fig. 18 (Cho *et al.*, 1982). The sequence starts after the removal of the slice from the MBE system with an ion-etching step to remove the native Al oxide from the top surface. The remaining Al layer has a thickness of 1000 to 2000 Å. A thin film of Cr/Au is subsequently evaporated in a vacuum system over the total area of the slice to protect the Al layer from oxidation. Another

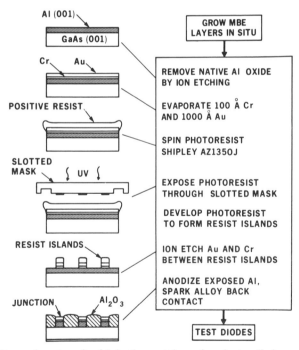

FIG. 18 Processing steps for fabricating metal–semiconductor diodes using a single-crystal Al(001) film grown in situ on (001) GaAs by MBE (Cho *et al.*, 1982). The most important fabrication step is the selective anodization of the Al film to form islands of metal surrounded by an Al$_2$O$_3$ insulating layer.

material combination that can be used is a Ti/W layer, which separates the Au from the Al film and inhibits chemical reactions between the Au and the Al (Cunningham *et al.*, 1970; Harris *et al.*, 1976; Ghate *et al.*, 1978; Nowicki *et al.*, 1978; Hill, 1980). Positive photoresist is spun on the slice, exposed with a UV source through a slotted mask, and developed to form an array of islands which are covered with photoresist. The Cr/Au film between the islands is etched off by reactive ion etching with a mixture of 90% Ar and 10% O_2. The relatively small amount of oxygen in the ion beam acts as an etch stop after the Au film is removed, which means only a negligible amount of the underlying Al will be removed after the Au removal because the etch rate of an oxidized Al surface is very small. After removing the slice from the ion-etch apparatus, the photoresist is dissolved in acetone and the exposed Al layer, which surrounds the gold islands, is oxidized by boiling the slice in deionized water. This results in a swelling of the Al film as the oxygen is incorporated into the layer to form Al_2O_3. Alternate techniques to form Al_2O_3 are anodization (Bayraktaroglu and Hartnagel, 1978) or plasma oxidation. The anodization can be performed in an electrolyte that is a mixture of tartaric acid and glycolethylene. The Al is converted into oxide at a rate of 9.9 Å/V and the Al_2O_3 film growth is 14.8 Å/V, which results in a swelling factor of 1.5. The final step of the processing sequence is to form an ohmic back contact by spark alloying as described by D'Angelo and Verlangieri (1981).

The current–voltage characteristic of a completed diode from a slice processed as shown in Fig. 18 is given in Fig. 19, which is a plot of the logarithm of the diode current as a function of the applied forward voltage. The device is made from a slice with a substrate doping concentration of 2×10^{18} cm^{-3}, an epilayer thickness of 400 Å with a doping concentration of 1×10^{16} cm^{-3}, and a single-crystal Al layer with a thickness of 2000 Å. The measurements are performed at room temperature, where tunneling contributions over the barrier can be neglected. The ideality factor of the diode is derived from the slope of the current–voltage characteristic by assuming that the second term of Eq. (2), which represents the tunnel reverse current, can be neglected. This assumption is valid because the doping concentration in the epitaxial layer is relatively small and the epitaxial layer thickness sufficiently thick to suppress tunneling currents from the metal into the semiconductor. The ideality factor $n(T)$ is thus given by

$$n(T) = q/kT \, \partial V/\partial(\ln i), \tag{40}$$

where $q/kT = 40$ V^{-1} for $T = 290$ °K. The characteristic of a Western Electric WE-20N silicon $n-p-n$ transistor is plotted for comparison purposes. The dc series resistance of the MBE Schottky diode with a junc-

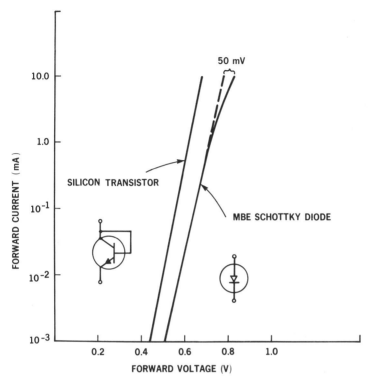

FIG. 19 Current–voltage characteristics of MBE Schottky diode made from single-crystal Al(001) grown in situ on GaAs (001). The junction diameter of the diode is 2.5 μm. The zero-bias cutoff frequency calculated from $R_S = 5 \Omega$ and $C_0 = 12$ fF is $f_c = 1/(2\pi R_S C_0) = 2.6$ THz, where $n = 1.15$ and $V_B = 8.2$ V. The characteristic of a Western Electric 20N silicon n–p–n transistor is shown for comparison.

tion diameter of 2.5 μm is derived from the deviation of the characteristic straight line as shown in Fig. 19.

An alternative method for processing a slice is to start with a GaAs substrate, to grow an epitaxial GaAs layer on the substrate, and to deposit an SiO$_2$ layer on top of the epilayer. An array of holes is subsequently etched into the SiO$_2$ layer using lithographic techniques. The holes are etched down to the GaAs epilayer, the epilayer is cleaned using suitable etching and rinsing techniques, and a metal layer is deposited on the bottom of the hole by electroplating or by evaporation. This method has the disadvantage that the GaAs surface on the bottom of the holes has to be etched and cleaned before the deposition. This process, which is difficult to reproduce often shows hysteresis in the voltage–current characteristic because of trapped carriers at the metal–semiconductor interface.

C. CIRCUIT FABRICATION

There are four basic circuit types that are used to combine the signal and local oscillator frequencies, to apply these frequencies to the diode terminals, and to extract the resulting IF. The circuits can be built with

  (i)   strip transmission lines,
  (ii) coaxial lines or waveguides,
  (iii) dielectric image lines, and
  (iv) quasi-optical structures.

A large variety of fabrication techniques are used to build various components of the circuits, such as filters, couplers, resonators, fixed or adjustable shorts, and dc bias circuits. Strip transmission line circuits are generally made using thin-film deposition techniques of metals on a low-loss dielectric substrate. The specific circuit pattern is generated with a photomask and the diodes are bonded or soldered to the conductor pattern on the substrate. This approach is advantageous if a large number of circuits with highly reproducible performance are required because the same photomask can be used over and over again. The other three remaining circuit types are made by micromachining or by assembling layered metal dielectric structures (Itoh, 1981). These structures are generally preferred for producing a small number of identical mixer circuits. Because of the large variety of fabrication methods, only a representative example of a thin-film fabrication process will be described.

The electrical properties of strip transmission line circuits as a function of the conductor geometry have been treated in a number of fundamental papers and textbooks (Wheeler, 1965; Pucel *et al.*, 1968; Schneider, 1969a,b; Sobol and Caulton, 1974; Mittra and Itoh, 1974; Howe, 1974; Gupta *et al.*, 1979; Denlinger, 1980). The four important line properties needed to design a circuit are the impedance, the guide wavelength, the attenuation, and the quality factor of the transmission line. It was originally believed that the lines would be too lossy at frequencies above 10 GHz and that waveguides and quasi-optical structures would be the only alternatives at millimeter-wave frequencies. It was shown, however, that microstrip and suspended stripline circuits show a relatively small attenuation and a good quality factor in the millimeter-wave frequency range if the line is built on a low-loss dielectric substrate with a relatively small dielectric constant and if the transmission line is properly shielded to reduce the losses from excitation of higher order modes (Schneider *et al.*, 1969). Such lines have been proven to be useful in radio relay systems (Penzias and Schneider, 1976), in frequency converters at 100 GHz (McMaster *et al.*, 1976), and in circuits up to 230 GHz (Carlson and

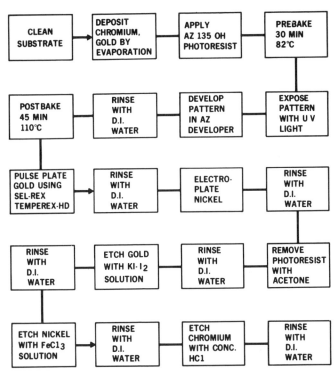

FIG. 20   Processing steps for fabricating stripline conductor patterns by evaporation of a chromium–gold film on quartz substrate, photolithographic pattern generation, and selective electroplating.

Schneider, 1980). Figure 20 shows the processing steps that are used to fabricate a stripline conductor pattern on a quartz substrate. The processing sequence starts with a planar fused quartz substrate and ends with a microstrip conductor pattern on the substrate. A detailed flowchart of the process listing all the cleaning, deposition, and etching steps is shown in Fig. 21. Figure 22 is a photograph showing a top view of the microstrip conductor pattern on the quartz substrate together with the conductor profile measured with a moving stylus across the surface of the substrate. The linewidths are reproducible to ± 2 μm for a conductor thickness of 3.2 μm. The minimum conductor thickness that is required for a low-loss circuit is approximately two skin depths. It is to be noted that the attenuation in a microstrip transmission line reaches a minimum for a conductor thickness of about three skin depths. This effect is caused by the interference of the electric field from the top and the bottom of the conductor,

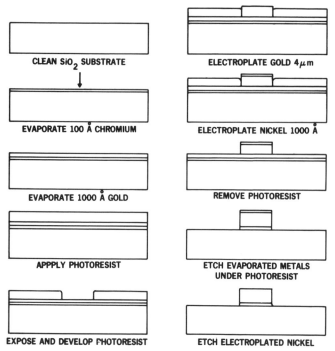

FIG. 21   Flowchart of microstrip fabrication process showing deposition process, selective plating steps, and etching of areas surrounding the strip conductor. A similar technique using low-cost conductor materials (Kanthal/Cu) is described by Kerns *et al.*, 1979).

which results in a relatively more uniform current distribution through the metal compared to the highly nonuniform exponential decay of current in a thick bulk sample (Horton *et al.*, 1971).

V.   Mixer Circuits

A.   CLASSIFICATION OF MIXERS

Diode mixers can be classified according to the electrical properties of the network seen by the diode, the nonlinear characteristics of the mixer diode, or the types of transmission lines used to combine the various input and output frequencies. A logical classification system based on the electrical network seen by an idealized junction was developed by Saleh (1971). The classification depends on the different terminations seen by the diode at the undesired out-of-band frequencies, i.e., all the possible combinations of the sum and difference of integer multiples of the local

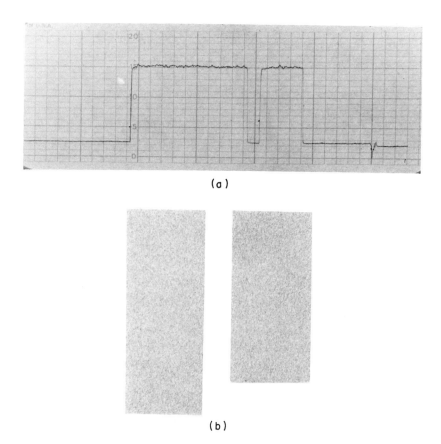

(a)

(b)

FIG. 22   Microstrip conductor pattern on quartz substrate fabricated as shown in Figs. 20 and 21 showing the conductor pattern (b) with 52-μm-wide capactive coupling gap and the associated conductor profile (a) measured by the moving stylus method, where the horizontal is 50 μm/div and the vertical is 2580 Å/div.

oscillator and the signal frequency with the exception of the IF. Saleh defined four types of mixers, which are designated as Z, Y, H, and G, based on the types of electrical network matrices that are used to analyze the conversion characteristics of the circuit. The Z-mixer is defined as a circuit in which all the out-of-band frequencies are open-circuited, and its performance can be analyzed using the impedance matrix (Z matrix) of the circuit. The Y-mixer can be evaluated using the admittance matrix (Y matrix), and it is defined as a circuit in which all the out-of-band frequencies are short-circuited. It was correctly pointed out by the author that only the Y-mixer can be readily realized at high signal frequencies be-

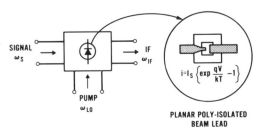

FIG. 23   Schematic drawing of conventional single-diode mixer with planar polyisolated beam lead diode (Ballamy and Cho, 1976). The IF, $\omega_{IF} = \omega_S - \omega_{LO}$, is substantially lower than the signal frequency. The image frequency, $\omega_i = \omega_{LO} - \omega_{IF}$, is terminated reactively or resistively inside the mixer circuit.

cause the junction capacitance of the diode tends to short out the harmonics of the pump and signal frequency, i.e., the undesired out-of-band frequencies.

The two remaining classification schemes are based on the nonlinear properties of the mixing element or the types of transmission lines that couple the fields to the diode. The two fundamentally different nonlinear device characteristics and the associated circuits are schematically shown in Figs. 23 and 24. Figure 23 shows a conventional single-diode mixer using one diode with an exponential characteristic as the mixing element, whereas Fig. 24 shows two diodes in an antiparallel connection with a hyperbolic sine current–voltage characteristic. This two-diode mixer can also be realized with a single diode if the tunneling current is sufficiently large, as was discussed previously in Section III.A. The nonlinear device with a hyperbolic sine characteristic and its associated circuit can be pumped by approximately half the conventional local oscillator frequency, as will be described in the following section. This property is advantageous at high millimeter-wave or at submillimeter-wave frequencies where solid-state pump oscillators are not readily available.

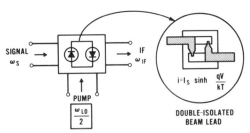

FIG. 24   Two-diode mixer built with two diodes in an antiparallel connection. The diode pair has a hyperbolic sine characteristic and it is pumped by a local oscillator frequency, which is about half the conventional pump frequency.

B. FREQUENCY CONVERSION

The nonlinear characteristic of the junction is used for frequency conversion by applying a pump frequency ω_p and a signal frequency ω_s to its terminals. Let us assume that the normalized pump voltage u defined by Eq. (6) across the terminals of the junction is given by

$$u = U_p \cos \omega_p t. \tag{41}$$

The voltage is thus constrained to be sinusoidal, whereas the current i can have all the possible harmonics. The current and the conductance can be computed from the following series expansions:

$$\exp(z \cos \theta) = I_0(z) + 2 \sum_{k=1}^{\infty} I_k(z) \cos(k\theta), \tag{42}$$

$$\sinh(z \cos \theta) = 2 \sum_{k=0}^{\infty} I_{2k+1}(z) \cos[(2k + 1)\theta], \tag{43}$$

$$\cosh(z \cos \theta) = I_0(z) + 2 \sum_{k=1}^{\infty} I(2k)(z) \cos(2k\theta), \tag{44}$$

where I_k is the modified Bessel function of the first kind of order k. From Eqs. (7), (41), and (42), one obtains for the junction current

$$i = I_S\left[A_0 + 2 \sum_{k=1}^{\infty} A_k \cos(k\omega_p t)\right] \tag{45}$$

and for the conductance

$$g = (qI_S/2k)\left[B_0 + 2 \sum_{k=1}^{\infty} B_k \cos(k\omega_p t)\right], \tag{46}$$

where

$$A_k = I_k[(m + 1)U_p] - I_k[(m - 1)U_p], \tag{47}$$

$$B_k = (m + 1)I_k[(m + 1)U_p] - (m - 1)I_k[(m - 1)U_p]. \tag{48}$$

For a conventional frequency converter with $m = 1$, the second term in both Eqs. (47) and (48) vanishes. For all values of the diode factor m except $m = 0$, all positive integer multiples of the pump frequency appear in the diode current and the diode conductance.

The special case $m = 0$ is of interest because $A_k = 0$ for $k = 0, 2, 4, \ldots$ and $B_k = 0$ for $k = 1, 3, 5, \ldots$. The resulting current and the conductance are

$$i = 4I_S \sum_{k=0}^{\infty} I_{2k+1}(U_p) \cos[(2k + 1)\omega_p t] \tag{49}$$

$$g = (qI_S/kT)\left[I_0(U_p) + 2 \sum_{k=1}^{\infty} I_{2k}(U_p) \cos(2k\omega_p t)\right].$$ (50)

From Eqs. (49) and (50), one concludes that

(i) there is no direct current flowing through the junction,

(ii) the device current contains only odd-order harmonics of the pump frequency, and

(iii) the conductance contains only even-order harmonics.

These electrical properties, which can be obtained either with a single diode with $m = 0$ or with a diode pair in antiparallel connection, make the device suitable for use in subharmonic mixers (Carlson *et al.*, 1978) or subharmonic upconverters.

If a small signal $U_s \cos \omega_s t$ is applied to the diode terminals simultaneously with the pump voltage given by Eq. (41), the resulting diode current and the conversion characteristics of the diode can be calculated from the harmonic series expansions of g given by Eqs. (46) and (50). A detailed analytic treatment giving the power that is converted to the intermediate frequency, $\omega_{IF} = \omega_p \pm \omega_s$, is described extensively in the standard work by Saleh (1971).

C. Definitions and Characterizations

The most important parameters that describe a diode mixer are the single-sideband noise temperature $T_m(\text{SSB})$, the conversion loss L_S at the signal frequency, and the conversion loss L_I at the image frequency. The quantity $T_m(\text{SSB})$ is a measure of the noise that is internally created in the mixing process and that cannot be observed directly. It is the physical temperature of a resistor that would have to be placed at the signal input port of a fictitious, completely noiseless mixer to generate the observed output noise power of the real mixer. The conversion losses L_S and L_I are power conversion losses between the signal and the IF port and between the image and the IF port, respectively. The conversion loss L_S in decibels is given by

$$L_S(dB) = 10 \log(P_s/P_{IF})$$ (51)

where P_S and P_{IF} are the signal and IF powers measured at the respective terminals of the mixer under matched conditions. The conversion loss L of the mixer is the power ratio between the input ports (signal and image) and the IF port if equal power is fed into both input ports. It is given by

$$1/L = (1/L_S) + (1/L_I).$$ (52)

A circuit for which $L_S \ll L_I$ is called a *single-response mixer* and a circuit with $L_S \approx L_I$ is called a *dual-response mixer*.

The single-sideband noise temperature of a receiver, i.e., a mixer followed by an IF amplifier is given by

$$T_r(SSB) = T_m(SSB) + L_S T_{IF}, \tag{53}$$

where T_{IF} is the noise temperature of the IF postamplifier. For some applications, a double-sideband noise temperature is used, which is defined by

$$T_r(DSB)/L = T_r(SSB)/L_S. \tag{54}$$

From Eq. (52), one concludes $L \leq L_S$, which results in $T_r(DSB) \leq T_r(SSB)$. For a dual-response receiver with $L_S = L_I$, one obtains, using Eq. (52),

$$T_r(DSB) = \tfrac{1}{2}T_r(SSB). \tag{55}$$

Manufacturers of microwave components and systems often prefer to characterize a receiver by a noise figure instead of a noise temperature. The calculation of a noise figure requires the specification of a reference temperature, which is 290 K. The single-sideband noise figure is defined as

$$F_{SSB}/L_S = (1/L) + (T_{SSB}/290L_S) \tag{56}$$

and the double-sideband noise figure by

$$F_{DSB} = 1 + (T_{DSB}/290). \tag{57}$$

Both Eqs. (56) and (57) are valid for the mixer and receiver noise figure if a mixer and receiver noise temperature is used in each case, respectively. A schematic diagram showing the relationship between noise temperature and noise figure is shown in Fig. 25. A more detailed treatment of the various noise parameters and their relationships is given by Mumford and Scheibe (1968) and by Carlson *et al.* (1978).

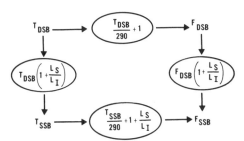

FIG. 25 Noise relationships in dual-response receivers for single-sideband (SSB) and double-sideband (DSB) noise temperature T, where L_S and L_I are the signal and image conversion losses, respectively. The relationships apply to both receiver and mixer noise temperature or noise figure. (From Carlson, Schneider, and McMaster, 1978. Reprinted from *IEEE Transactions on Microwave Theory and Techniques* **26**, 706–715. © 1978 IEEE.)

The minimum detectable temperature of a receiver is of substantial interest to any designer of a millimeter-wave system. This minimum detectable temperature is (Kraus, 1966)

$$\delta T_{\min} = K_S T_r / (\Delta f\, t_0)^{1/2}, \tag{58}$$

where K_S is a sensitivity constant which is of the order of 1, Δf the bandwidth of the receiver in the signal band, and t_0 the detection or integration time. From the fundamental relationship $P = kT\, \Delta f$, where P is the noise power available from a resistive network and $k = 1.38 \times 10^{-23}$ J/K, one obtains for the minimum detectable power

$$P_{\min} = kT_r (\Delta f / t_0)^{1/2}. \tag{59}$$

From this, for the minimum observation time t_0, one obtains

$$t_0 = \Delta f (kT_r / P_{\min})^2. \tag{60}$$

It can be seen that the observation time is reduced in direct proportion to the squared receiver noise temperature. This explains why a reduction of T_r is highly desirable.

For practical applications, it is necessary to determine the operational noise temperature or the system noise temperature of the receiver. The system noise temperature T_{sys} is given by

$$T_{sys} = T_a + T_r, \tag{61}$$

where T_a is the antenna temperature. If the antenna is directed toward free space, one would expect T_a to be equal to the 3 °K cosmic background radiation (Penzias and Wilson, 1965). In practice, T_a is much higher because of the contributions due to atmospheric attenuation and spillover from the ground, which is at a temperature of about 300 °K.

The receivers that are used for different applications such as communications, radar, and radiometry have to be evaluated according to various performance criteria as shown in Table V. The development of receivers with a low system noise temperature, T_{sys}, is therefore essential to attain good performance for communications applications, for radar, and for remote sensing.

D. NOISE MECHANISMS

The two basic noise mechanisms that determine the noise temperature of a metal–semiconductor junction mixer are shot noise caused by the carrier transport over the potential barrier of the junction and thermal noise due to any resistive losses in the diode and the surrounding electrical network. Fundamental contributions on diode and mixer noise have been made by Uhlir (1958), Dragone (1968), Viola and Mattauch (1973a,b), Keen (1977), and, more recently, by Held and Kerr (1978), and

TABLE V

PERFORMANCE OF RECEIVERS

Receiver type	Performance criteria
Communications	carrier/noise $\approx G_{a}/T_{sys}$[a]
Radar	range $\approx (P_{T}/T_{sys})^{1/2}$ [b]
Radiometry	$\delta T \approx T_{sys}/(\Delta f\, t_{0})^{1/2}$

[a] G_{a} is antenna gain.
[b] P_{T} is transmitted power.

by Kerr (1979a,b). These various studies show that shot noise is the predominant noise mechanism and that the pumped diode can be represented by an equivalent lossy network with an equivalent network temperature of $n/2$ times the physical temperature of the diode, where n is the ideality factor defined in Eq. (2).

A simple derivation of the equivalent network temperature can be made by writing Eq. (7) in the form

$$i = I_1 - I_2 = I_S \exp[(m + 1)u] - I_S \exp[(m - 1)u], \qquad (62)$$

where I_1 is mainly thermionic and I_2 mainly tunneling current flowing through the junction. In order to calculate the shot noise, we have to add currents I_1 and I_2 because they generate noise independently:

$$\overline{\Delta i^2} = 2q(I_1 + I_2)\, \Delta f, \qquad (63)$$

where Δf is the bandwidth determined by the electrical transmission properties of the external embedding network of the metal–semiconductor junction. The available shot noise power is

$$P_{shot} = \overline{\Delta i^2}/4g = (q/2g)(I_1 + I_2)\, \Delta f \qquad (64)$$

and, from Eq. (62),

$$P_{shot} = (qI_S/g) \exp(mu) \cosh u\, \Delta f. \qquad (65)$$

The final result, using Eq. (13), is

$$P_{shot} = [kT/(1 + m \tanh u)]\, \Delta f. \qquad (66)$$

The equivalent shot noise temperature of the junction T_{shot} is defined by

$$T_{shot} = P_{shot}/k\, \Delta f. \qquad (67)$$

For $u \gg 1$ and the special case $m = 0$, we obtain, from Eqs. (66) and (67),

$$T_{shot} = T. \qquad (68)$$

For $u \gg 1$ and pure thermionic emission with $m = 1$, we obtain the well-known equation derived by Kerr (1979a,b)

$$T_{shot} = \tfrac{1}{2}T. \tag{69}$$

The thermal noise generated in the series resistance of the junction is statistically independent of the shot noise, and the contributions of both shot noise and resistive noise can be linearly added to obtain the total equivalent noise temperature T_{eq} of the junction. If one assumes, in accordance with Viola and Mattauch (1973a,b), that the current–voltage characteristic of the diode can be represented by the single exponential term

$$i = I_S \exp(V/V_0), \tag{70}$$

one obtains, using Eqs. (64) and (67),

$$T_{shot} = qV_0/2k. \tag{71}$$

The assumption that the diode current i can be represented by a single exponential term is sufficiently accurate if the transport mechanism of the carriers across the junction is mainly thermionic or if it is only due to field emission. For thermionic transport, one obtains, by assuming $i \gg I_S$ from Eq. (3),

$$V_0 = kT/q. \tag{72}$$

For pure field emission with all electrons originating at the Fermi level or below the Fermi level in the semiconductor, one obtains, for V_0 (Padovani, 1971),

$$V_0 = (h/2\pi)(N_D/\varepsilon_r\varepsilon_0 m^*)^{1/2}, \tag{73}$$

where h is Planck's constant (6.63×10^{-34} J sec) and m^* the effective mass of the electrons in the semiconductor. The ratio of effective mass to real mass for electrons in GaAs is $m^*/m = 0.068$. The ratio qV_0/k determines whether thermionic or field emission is predominant. For $T \gg qV_0/k$, the electron transport across the barrier is caused mainly by thermionic emission and for $T \ll qV_0/k$, it is caused by field emission. It is convenient to define a normalized doping concentration n_0 given by $n_0 = N_D/10^{17}$, where N_D is in units of one per cubic centimeter, to determine which transport mechanism is predominant. From Eq. (73), with $\varepsilon_r = 12.9$ for GaAs, one obtains

$$qV_0/k = 75.5(n_0)^{1/2} \quad \text{K}, \tag{74}$$

which gives, for the equivalent shot noise temperature, using Eq. (71) for the case of pure field emission,

$$T_{shot} = 36.2(n_0)^{1/2} \quad \text{K}. \tag{75}$$

One concludes that the minimum equivalent shot noise temperature that can be achieved by cooling a diode with a normalized doping concentration $n_0 = 1$ ($N_D = 10^{17}$ cm^{-3}) is 36.2 K. To achieve a lower equivalent shot noise temperature, one may reduce the normalized doping concentration in the epitaxial layer or choose a semiconductor with a higher effective electron mass m^* and a higher relative dielectric constant ε_r.

The total equivalent noise temperature T_{eq} of the diode, which includes the contributions of shot noise, is (Viola and Mattauch, 1973a,b)

$$T_{eq} = \frac{qV_0}{2k} \frac{R}{R + R_S} + T \frac{R_S}{R + R_S}, \tag{76}$$

where R_S is the series resistance of the diode, R the junction resistance, and T the physical temperature of the diode. For most practical applications, the junction resistance is approximately 50 Ω. The second term in Eq. (76) is relatively small for cryogenically cooled diodes with physical junction temperatures of 20 K or below.

Note that there are additional noise mechanisms that increase the equivalent noise temperature given by Eq. (76). These additional contributions, commonly referred to as hot-electron excess noise, are due to carriers scattered into states with a different momentum but equal energy as they traverse the potential barrier of the metal–semiconductor junction. Electrons in GaAs that occupy a state in the central conduction valley and have sufficient energy may be scattered by a lattice phonon into one of the upper conduction valleys. This intervalley scattering process results in electrical noise because some of the carriers may not reach the metal conductor even if their initial energy and momentum would allow a crossing of the barrier with certainty. The process also results in a change of the saturation current and a knee in the current–voltage characteristic, as discussed by Pellegrini and Di Leo (1977) and Salardi et al. (1979) and as independently observed by Kollberg and Staehlberg (1981). Intervalley scattering is reduced if the electron transport across the junction is ballistic, i.e., if the barrier is thin compared to the electron mean free path in the device material (Shur and Eastman, 1979, 1980; Maloney, 1980; Eastman et al., 1980).

E. Mixer Theory

The fundamental principles of frequency conversion in diode mixers have been studied in depth by Dragone (1968), Barber (1967), Saleh (1971), Held and Kerr (1978), and Kerr (1979a,b). These authors show that an accurate prediction of the mixer performance is possible if the electrical properties of the diode and the impedance across the diode terminals is known at the signal, the image, the IF, and all harmonics of the local oscillator frequency. The network parameters cannot be readily established at millimeter-wave frequencies. However, it is possible to perform accurate measurements on a scaled model at lower frequencies or to measure the embedding impedance using a high-directivity directional coupler as described by Hagstrom and Kollberg (1980).

The three basic conclusions that can be drawn from the mixer theories are that

(a) the cutoff frequency of the diode $[f_C = 1/(2\pi R_S C_0)]$ should be as high as possible,

(b) the diode terminals should be short-circuited at the image and all harmonics of the pump frequency, and

(c) the average junction resistance should be approximately equal to the reactance of the diode at zero bias.

An accurate calculation of the conversion loss of a specific mixer can be obtained using the computer program created by Siegel and Kerr (1979), which performs a nonlinear analysis to determine the diode conductance and capacitance waveforms produced by the local oscillator. A small-signal linear analysis is then used to find the conversion loss, port impedances, and input noise temperature of the mixer. These authors show that the parametric effects of the voltage-dependent capacitance of a conventional Schottky diode may be either detrimental or beneficial on the mixer performance, depending on the diode and circuit parameters.

For some applications, it is sufficient to estimate the conversion loss L from the product

$$L = L_{rf}L_{IF}, \tag{77}$$

where L_{rf} is the conversion loss caused by the power dissipated in the diode with a series resistance R_S at the signal frequency and L_{IF} the conversion loss due to R_S at the intermediate frequency. The rf conversion loss L_{rf} is given by

$$L_{rf} = 1 + \frac{R_S}{R_d} + \frac{R_d f_S^2}{R_S f_C^2} \tag{78}$$

where $f_C = 1/(2\pi R_S C_0)$ is the zero-bias cutoff frequency of the diode. The IF conversion loss L_{IF} is

$$L_{IF} = 1 + (R_S/R_0), \tag{79}$$

where R_0 is the IF output impedance, R_d the average junction resistance (approximately 50 Ω), f_S the signal frequency, and f_C the cutoff frequency of the diode $[f_C = 1/(2\pi R_S C_0)]$. The equations can be derived from the simple voltage divider circuit shown in Fig. 26. It is to be noted, however, that the diode conversion loss has to be multipled by the conversion loss of the electrical embedding circuit to obtain the conversion loss of the

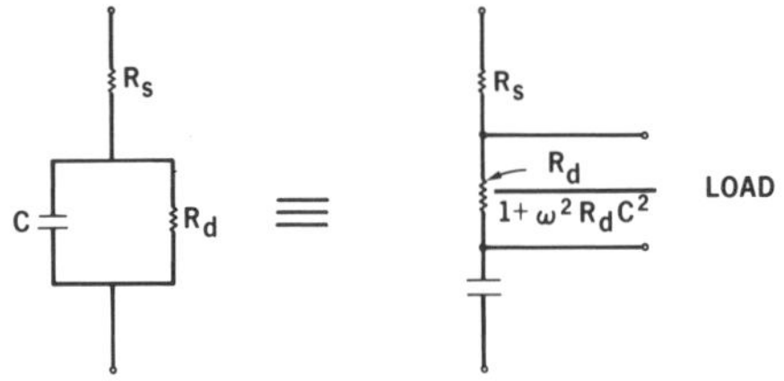

$$L = \frac{\text{TOTAL RESISTANCE}}{\text{LOAD RESISTANCE}} = \frac{R_s + \dfrac{R_d}{1 + \omega^2 R_d^2 C^2}}{\dfrac{R_d}{1 + \omega^2 R_d^2 C^2}} = 1 + \frac{R_s}{R_d} + R_s R_d (\omega C)^2.$$

FIG. 26 Voltage divider circuit for calculating the signal conversion loss of a mixer diode. The same circuit is applicable for calculating the loss associated with the IF network.

complete circuit. This problem has been treated in detail by Wrixon and Kelly (1978).

If the conversion loss of the circuit has been determined by analytic techniques or by performing loss measurements on the actual circuit, the single-sideband mixer noise temperature is given by

$$T_m(\text{SSB}) = [L_S - (L_S/L_I) - 1]T_{eq}. \tag{80}$$

If shot noise is the predominant noise mechanism, T_{eq} is given by Eq. (68) or (69). For the special case $L_S = L_I$ and $m = 1$, one obtains

$$T_m(\text{SSB}) = [(L_S/2) - 1]T, \tag{81}$$

where T is the physical temperature of the diode junction. One concludes from Eq. (81) that the cooling of the diode results in a linear decrease of the mixer temperature with decreasing junction temperature if thermionic emission is the predominant transport mechanism at all temperatures. This is possible by using junctions with a low doping concentration in the epitaxial layer or with devices with a small barrier height.

F. CIRCUIT REALIZATION

Three mixer circuits in common use today are shown in Figs. 27–30. The most common structure is the waveguide mixer block in which the diode is mounted in a reduced-height waveguide section as shown in the cross-sectional view of Fig. 27. A conical taper is used to match the receiver horn followed by a conventional rectangular waveguide to the reduced height section. The low waveguide height is chosen to reduce the length and, therefore, the inductance of the contact whisker/post combination. The diode junction on the diode chip is contacted by the whisker and a co-axial low-pass filter is used to couple the IF frequency to any

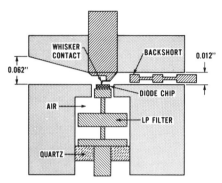

FIG. 27 Cross-sectional view of mixer block with Schottky barrier mounted in a reduced-height waveguide section. The low-pass (LP) filter is a co-axial structure, which blocks the signal frequency and transmits the intermediate frequency to the terminal, which is supported by a quartz washer. (From Linke, Schneider, and Cho, 1978. Reprinted from *IEEE Transactions on Microwave Theory and Techniques* **26**, 935–938. © 1978 IEEE.)

FIG. 28 An open 100-GHz mixer block with associated components. The waveguide channel with a linear taper is milled into the block shown in the lower part of the figure. (From Linke, Schneider, and Cho, 1978. Reprinted from *IEEE Transactions on Microwave Theory and Techniques* **26**, 935–915. © 1978 IEEE.)

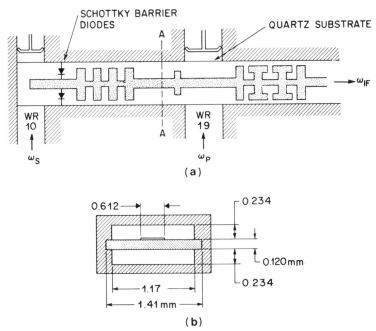

FIG. 29 (a) Top view and (b) cross-sectional view of a subharmonically pumped stripline circuit with signal input waveguide, pump waveguide, and IF output port. (From Carlson, Schneider, and McMaster, 1978. Reprinted from *IEEE Transactions on Microwave Theory and Techniques* **26**, 706–715. © 1978 IEEE.)

subsequent post amplifier. Tuning is accomplished by means of a noncontacting waveguide backshort, which consists of alternating low- and high-impedance sections. An open 100-GHz mixer block with the associated components built by Linke *et al.* (1978) is shown in Fig. 28. The separation of signal and pump frequency is made externally by a quasi-optical filter, which is not shown in this figure.

A substantially different circuit in which the local oscillator can be injected from the IF side is shown in Fig. 29. This circuit developed by Schneider and Snell (1975) and McMaster *et al.* (1976) consists of two beam-leaded or two notch-front diodes (Schneider and Carlson, 1977), which are mounted in an antiparallel connection on a strip transmission line. A similar circuit has been also developed by Cohn *et al.* (1975). The advantages of these circuits compared to the conventional waveguide structure with a single diode are as follows:

(a) The local oscillator frequency is half that of the conventional LO frequency (subharmonic pump).

├──────────────┤
50 mm

Fig. 30 Subharmonically pumped millimeter-wave mixers developed at the Institute of Applied Physics, University of Bern (Kunzi and Berger, 1978). The mixers are described in detail in the text.

(b) The filter design is less difficult (planar photoetched stripline filters).

(c) The AM pump noise is suppressed (Henry *et al.*, 1976).

(d) There is no dc return needed for the diodes.

(e) There is an inherent self-protection of the diodes against large peak voltage burnout.

(f) Fourth subharmonic pumping is possible with only minor modifications required.

(g) The output impedance at the IF terminal is relatively small and of the order of 50 Ω.

Therefore, it is possible to build a 100-GHz receiver with a local oscillator frequency of only about 50 or 25 GHz, which is definitely a substantial advantage because Gunn, IMPATT, and varactor multiplier sources at these frequencies are readily available.

The subharmonically pumped mixer shown schematically in Fig. 29 consists of a full-height waveguide section with a backshort with capacitive coupling of the signal to a suspended strip transmission line. Two diodes are shunt-mounted in the stripline on opposite sides of the stripline center conductor. A low-pass planar filter is used to block the signal and to transmit the pump frequency, which is approximately equal to one-half of the signal frequency. The pump is injected capacitively into the stripline from a pump waveguide, which is terminated by a backshort for tuning purposes.

Figure 30 shows two subharmonically pumped mixers developed at the Institute of Applied Physics at the University of Bern (Kunzi and Berger, 1978). Figure 30a displays a subharmonically pumped mixer built with a WR-10 (75–110 GHz) signal waveguide. The curved waveguide in the upper left is used to transmit the subharmonic pump frequency to the diode, which is located inside the mixer block. The smaller waveguide behind the pump waveguide is used for the signal injection, and the co-axial connector on top of the mixer block is the IF output. The mixer shown in Fig. 30b is a subharmonically pumped circuit built with a WR-4 signal waveguide (170–260 GHz) with a feed horn on the right front, a local oscillator input on the front, and a co-axial connector for the IF output on top of the mixer block.

The harmonic mixer shown in Fig. 31 uses the same diode as a harmonic generator and mixer (Goldsmith and Plambeck, 1976). This circuit is relatively easy to construct; however, good noise performance and con-

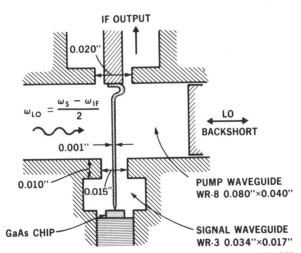

FIG. 31 Cross-sectional view of harmonically pumped mixer for 230-GHz receiver with crossed signal and pump waveguide (Goldsmith and Plambeck, 1976). The diode chip is mounted in the signal waveguide and contacted by a whisker that extends through the pump and signal waveguide.

version characteristics are more difficult to achieve compared with the other designs. The circuit has the advantage that it can be used either as downconverter or as a frequency multiplier. A cross-sectional view of such a multiplier including the exact dimensions with an input frequency of 100 GHz and output frequencies of 200, 300, 400, and 500 GHz is shown in Figs. 32 and 33. This frequency multiplier is useful as a local oscillator for millimeter and submillimeter-wave receivers and as a source for laboratory spectroscopy (Schneider and Phillips, 1981).

Quasi-optical mixers are of interest because the fabrication of extremely precise waveguide parts or the photofabrication of stripline circuits is not necessary. Various realizations of these mixers, which are

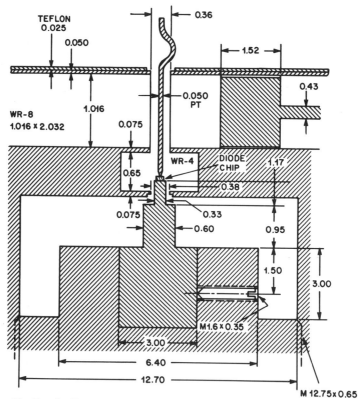

Fig. 32   Detail of harmonic generator with input frequency at 100 GHz and output frequencies at 200, 300, 400, and 500 GHz. The diode chip is mounted on a post in the high-frequency waveguide. The junction is contacted with a spark-eroded Pt whisker (Schneider and Berger, 1979). All dimensions are given in millimeters. (From Schneider and Phillips, 1981. *International Journal of Infrared and Millimeter Waves* **2,** 15–25.)

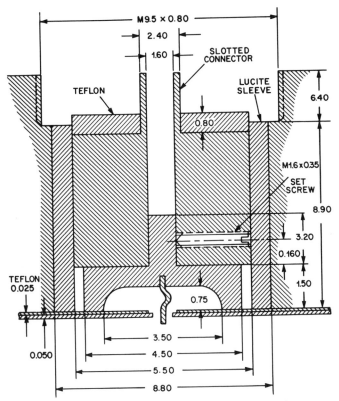

FIG. 33   Cross-sectional view of upper part of harmonic generator showing the insulated central conductor assembly with a slotted connector for applying the dc bias to the diode. All dimensions are given in millimeters. (From Schneider and Phillips, 1981. *International Journal of Infrared and Millimeter Waves* **2**, 15–25.)

mainly used in the submillimeter-wave region, have been reviewed by Clifton (1979) and Wrixon and Kelly (1979).

G.   PERFORMANCE

Low-noise mixers and complete receivers showing good performance have been built at many laboratories (Kerr, 1975; Wilson, 1977; Zimmermann and Haas, 1977; Kerr *et al.*, 1977; Linke *et al.*, 1978; Carlson *et al.*, 1978; Keen *et al.*, 1979; Carlson and Schneider, 1979; Erickson, 1980; Keen, 1980; Ralsanen *et al.*, 1980). The single-sideband noise temperatures achieved for room-temperature mixers in the frequency range from 40 to 400 GHz are displayed in Fig. 34. Typical single-sideband mixer noise temperatures that can be consistently reached at room temperature

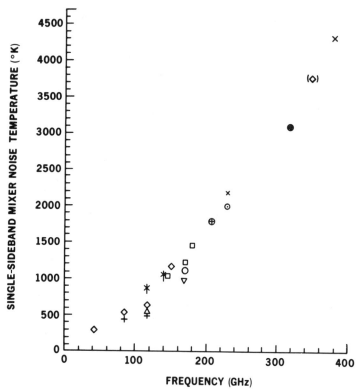

FIG. 34   Single-sideband noise temperatures of room-temperature mixers from 40 to 400 GHz measured at various laboratories: *, Linke *et al.* (1978); +, Kerr (1975); ◇, Zimmermann and Haas (1977); □, Kerr *et al.* (1977); △, Hagstrom and Lidholm (1978); ○, Vizard *et al.* (1979); ●, Erickson (1979); ×, Lidholm (ESTEC); (◇), Zimmermann; ⊙, Vizard; ⊕, Carlson and Schneider (1979); ▽, Carlson and Schneider. The references with no date specified are unpublished results.

at 100 GHz are about 500 K. The noise temperatures that can be attained for cryogenically cooled mixers are substantially lower, as recognized early by Weinreb and Kerr (1973), and are in the range of 100 K at 100 GHz. The conversion loss that has been reported for cooled mixers with a physical circuit temperature of 15 to 20 K is about 6 dB ($L_S = 4$) at a signal frequency of 100 GHz.

A measurement of the single-sideband mixer noise temperature as a function of the physical mixer temperature is shown in Fig. 35. The diode is an MBE junction with the doping profile of Fig. 17 (Schneider *et al.*, 1977). The mixer circuit was designed and tested by Weinreb (1981). It is to be noted that the required local oscillator pump power of 150 μW is relatively small compared with the pump power of 2 to 4 mW needed for

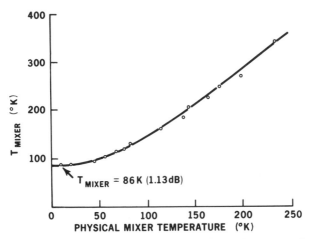

Fig. 35 Measurement of single-sideband mixer noise temperature as a function of the physical mixer temperature for signal frequency of 90.8 GHz for GaAs (N280-92). The minimum mixer temperature is 86 K (1.13-dB noise figure) with an associated conversion loss of 5.2 dB and an LO power of 150 μW. (Weinreb, 1981).

room-temperature mixers. This small power requirement is caused by the much steeper current–voltage characteristics of the diode at low temperatures compared with room temperature, i.e., a much smaller voltage swing is needed to switch the diode from a conducting to a nonconducting state. The same effect has been reported by Keen et al. (1979).

The measured mixer noise temperatures reported for the best mixers are close to the temperatures calculated from Eq. (76). This can be attributed to the fact the current transport mechanism in the diode is nearly ballistic and that there is an absence of parametric effects since the diodes show very little change in capacitance as a function of bias voltage (Linke et al., 1978).

Low-noise temperatures are relatively difficult to achieve at signal frequencies that are much higher than 100 GHz. The rapid degradation of mixer performance in the submillimeter-wave region is caused by the increased skin resistance of the diode, the increased diode spreading resistance for a reduced junction size, and the larger losses in the much smaller waveguides and coupling circuits. This results in an increased conversion loss and also in a relatively higher resistive noise contribution from the series resistance of the diode (Johson noise).

The performance of a recently developed mixer in the 200–230-GHz frequency range is shown in Fig. 36 (Carlson and Schneider, 1979). The mixer is subharmonically pumped and the IF output is connected to a 1.4-GHz GASFET amplifier with a noise temperature of 72 K. The re-

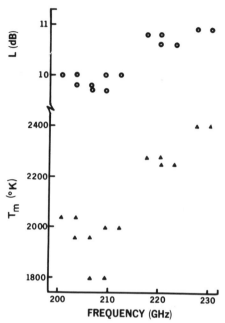

Fig. 36   Single-sideband mixer noise temperature and conversion loss for a WR-4 sub-harmonically pumped mixer as a function of frequency (Carlson and Schneider, 1979).

sulting single-sideband receiver temperature is 2500–3000 K over the 200–300 K band. The receiver has an instantaneous bandwidth of 500 MHz and it requires about 10-dBm local oscillator power at 100 to 115 GHz. Similar performances can also be achieved for conventional single-diode mixers as reported by Erickson (1980). This particular receiver is built with a quasi-optical diplexer to combine the local oscillator and the signal. The local oscillator source is a klystron, which is followed by an efficient frequency doubler made with a vapor-phase-grown diode developed by Schneider and Phillips (1981). This diode is used as a harmonic generator instead of a downconverter. The single-sideband system noise temperature of this receiver is 3100 K at 270 GHz.

VI.   Related Circuits

A.   Direct Detection

Electromagnetic radiation can be detected by direct rectification of the signal at a nonlinear metal–semiconductor contact if the cutoff frequency of the contact is larger or comparable to the signal frequency. The sensitivity obtained for direct detection is much lower compared with super-heterodyne reception. The detector circuits, however, are much simpler, which makes direct detection an inexpensive and very attractive method for many laboratory experiments.

The direct conversion of a signal to a modulated direct current, commonly known as rectification, baseband detection, or video detection, has been treated by many authors (Burrus, 1966; Cowley and Sorenson, 1966; Anand and Moroney, 1971). A dc bias is used to set the operating point of the diode so that a low-level signal is rectified by a highly nonlinear portion of the current–voltage characteristic. An equivalent circuit of a metal–semiconductor contact used as a baseband detector is shown in Fig. 37. The current source, which generates the video signal across the video resistance R_V, is given by

$$I_S = \beta P_{rf}, \tag{82}$$

where β is the current sensitivity and P_{rf} the incident rf power. The current sensitivity can be expressed in terms of the diode parameter as (Torrey and Whitmer, 1948; Cowley and Sorenson, 1966)

$$\beta = \frac{q}{2nkT}\left(1 + \frac{R_S}{R_B}\right)\left\{1 + \left(\frac{f}{f_C}\right)^2\right\}^{-1}, \tag{83}$$

where the dynamic resistance R_B associated with the barrier is given by

$$R_B = nkT/q(I_0 + I_S). \tag{84}$$

The quantity I_0 is the dc bias and I_S the saturation current of the diode. A typical value for the sensitivity β of a good detector diode at microwave frequencies is 1 $\mu A/\mu W$ of rf power with a corresponding video resistance $R_V = R_B + R_S$ of about 5000 Ω. The minimum detectable signal power is given by $i_N = i_S$ (see Fig. 35), which gives

$$P_{rf,min} = (1/\beta)(4kT\,\Delta f/R_V)^{1/2}, \tag{85}$$

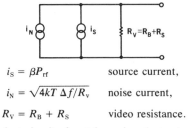

$$i_S = \beta P_{rf} \qquad \text{source current,}$$

$$i_N = \sqrt{4kT\,\Delta f/R_v} \qquad \text{noise current,}$$

$$R_V = R_B + R_S \qquad \text{video resistance.}$$

Fig. 37 Video equivalent circuit of metal–semiconductor junction used for direct detection with signal current source $i_S = \beta p_{rf}$ and noise current source $i_N = (4kT\,\Delta f/R_V)^{1/2}$.

which, for a detector at room temperature, results in

$$P_{rf,min} = 1.8 \times 10^{-12}(\Delta f)^{1/2} \quad \text{W.} \tag{86}$$

This minimum detectable signal power (noise equivalent power) is many orders of magnitudes larger than the corresponding value for a heterodyne receiver, which can be calculated from Eq. (58) as

$$P_{mixer,min} = k \, \delta T_{min} \, \Delta f = (\Delta f/t_0)^{1/2} \, kT_r. \tag{87}$$

For a receiver noise temperature of 500 K, which can be readily achieved for a room-temperature receiver at 100 GHz, one obtains

$$P_{mixer,min} = 6.9 \times 10^{-21}(\Delta f/t_0)^{1/2} \quad \text{W.} \tag{88}$$

Suitable devices for direct detection are metal–semiconductor junctions with a small barrier height or back diodes. A small barrier height can be achieved with a Schottky barrier on an $n/n^+ - In_x Ga_{1-x} As$ double layer, which is heteroepitaxially grown on an $n^+ - GaAs$ substrate. A moderately high band gap of 700 mV is obtained for $x \approx 0.16$, which results in a barrier height of 300 mV as determined by Kajiyama et al. (1979). Another specially developed device, the zero-bias Schottky diode, has a barrier height of only 150 mV, which is substantially less than the height of 800 to 900 mV obtained for most metal contacts on GaAs (Anand, 1976). The low barrier height is attained by forming a metal-silicide on silicon through heat treatment. The resulting current sensitivities are comparable to that of the best available back diodes, which are substantially more complex and more difficult to fabricate.

B. RESISTIVE AND REACTIVE UPCONVERTERS

Upconverters are an essential part in microwave and millimeter-wave radio relay systems where a low-frequency signal must be converted to a much higher frequency for efficient transmission. Both resistive and reactive converters can be used. The resistive upconverter is preferred for low-level applications and for systems that require small intermodulation distortion (Tucker, 1954; Perlow, 1974). The circuit is identical with that of a resistive downconverter; it is built with the same circuit elements and metal semiconductor diodes described in Section V, and it has the same conversion loss if the effect of the variable diode capacitance can be neglected. The only obvious difference is that the input and output terminals are reversed, i.e., the flow of information is from the IF terminal to the signal terminal. The converter can be readily constructed as a subharmonically pumped upconverter by using a single diode with a symmetrical current–voltage characteristic or a pair of two diodes in an antiparallel connection, as shown in Fig. 24.

Reactive upconverters use the nonlinear diode capacitance to convert the low-frequency signal to an output frequency that is either below (lower sideband upconverter) or above (upper sideband upconverter) the pump frequency. Both circuits fulfill the Manley–Rowe (1956) equations and exhibit a gain given by the ratio of the output to the signal frequency. The converters are described in detail in a number of fundamental papers and textbooks (Penfield and Rafuse, 1962; Uenohara and Gewartowski, 1969; Osborne, 1969). An important property of the upper sideband up-converter (USUC) is that its gain is positive and stable, whereas the lower sideband upconverter (LSUC) exhibits negative gain, which is indicative of negative resistance amplification. The lower sideband upconverter has properties that are similar to the nondegenerate parametric amplifier with the idler representing the output frequency.

Reactive upconverters can also be built with a pair of varactor diodes in an antiparallel connection. The diode pair has an even capacitance–voltage characteristic with a resulting differential capacitance that varies twice with the pump frequency (Marquardt and Schieck, 1969). This allows the construction of reactive upconverters or parametric amplifiers with a subharmonic pump. A further development proposed by the authors is to use several varactors with different bias voltages in parallel in order to obtain a parametric amplifier with a differential capacitance variation of four to six times as high as the pump frequency.

C. HARMONIC GENERATORS

Harmonic generators are useful local oscillator sources for mixer circuits in the millimeter and submillimeter-wave frequency range. Typical uses for these multipliers and the associated mixer circuits are the measurement of molecular rotational spectra (Gordy et al., 1966), microwave radiometry applications to remote sensing (Schanda, 1978), and the detection of trace molecules in the atmosphere (Kunzi, 1980). Other examples are radio astronomical observations such as the mapping of carbon monoxide and its isotopes in nebulas and interstellar clouds (Wilson et al., 1970; Penzias and Burrus, 1973) and the detection of the ground-state fine-structure splitting of the carbon atom at 492 GHz (Phillips et al., 1980).

A cross-sectional view of a harmonic generator built with a GaAs Schottky diode contacted by a spark-eroded platinum whisker is shown in Figs. 32 and 33. As in previously described circuits by Lee and Burrus (1968) and by Schneider and Snell (1971), pump power is injected through a rectangular waveguide, coupled to a co-axial probe or thin-film circuit, and transmitted to the varactor diode, which is located near or inside the high-frequency waveguide. Similar structures are also described in detail by Takada and Ohmori (1979).

The diode used in the multiplier of Figs. 32 and 33 is fabricated from a commercially available GaAs slice. The carrier substrate is a heavily doped n^+ layer with a sulfur doping concentration of 2×10^{18} cm^{-3}. The epitaxial layer grown by vapor-phase deposition has a thickness of 0.15 μm and a sulphur doping concentration of 2.4×10^{17} cm^{-3}. The SiO$_2$ layer with a thickness of 0.5 μm is grown on top of the epitaxial layer and an array of junctions (honeycomb diodes) is fabricated by etching holes with a diameter of 2.5 μm in the SiO$_2$ layer and by electroplating Pt and Au on the exposed GaAs surface at the bottom of the holes. The ohmic contact is applied to the back of the slice by electroplating Sn and Au and by alloying the metals into the GaAs at 400 °C for 10 sec in a forming gas atmosphere. The processing of the devices is similar to the fabrication of notch-front millimeter-wave diodes used in subharmonically pumped mixers as described by Carlson *et al.* (1978). The series resistance of the junction is 6.8 Ω and the ideality factor is 1.12. The capacitance as a function of back bias is shown in Fig. 38. The junction is punched through at a reverse voltage of 3 V, i.e., the epitaxial layer is fully depleted at voltages above 3 V. The intersection of the $1/C^2$ versus voltage plot with the horizontal axis should give a barrier height of 0.86 V. Because of the parasitics obtained for such a small junction the intersection is at 1 V. The breakdown of the device measured at a current of 10 μA is 8.5 V. There is no observable hysteresis, which means that there are no hole traps present in

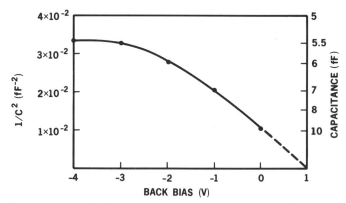

FIG. 38   Capacitance of Schottky varactor diode as a function of back bias. The $1/C^2$ plot shows a straight line at small back bias and a saturation for a back bias above 3 V because of complete depletion of the epitaxial GaAs layer. (From Schneider and Phillips, 1981. *International Journal of Infrared and Millimeter Waves* **2**, 15–22.)

TABLE VI

PERFORMANCE OF HARMONIC GENERATORS

	Input		Output					
Diode	Freq. f(GHz)	Power P(W)	f (GHz)	P (W)	f (GHz)	P (W)	f (GHz)	P (W)
GaAs[a]	100	10^{-2}	200	2×10^{-3}	300	3×10^{-5}	400	10^{-5}
GaAs	115	10^{-2}	230	10^{-3}	345	10^{-5}	460	4×10^{-6}
Si	115	10^{-2}	230	10^{-5}	345	3×10^{-6}	460	3×10^{-7}

[a] For $f_{in} = 100$ GHz and $P_{in} = 10^{-2}$ W, $f_{out} = 500$ GHz with $P_{out} = 4 \times 10^{-6}$ W.

the epilayer at the interface between the GaAs and the electroplated metal. The power output of the multiplier circuit measured at the harmonic frequencies of 100 and 115 GHz is listed in Table VI. (Schneider and Phillips, 1981.) Results obtained with a previously used commercially available Si point contact multiplier are also given. The power measurements are made with an InSb bolometer as described by Phillips and Jeferts (1973). The responsivity of this bolometer was calibrated by comparison of direct current and microwave heating effects.

It is to be noted that the junction is biased at a direct forward current of approximately 4 mA to obtain the maximum output at the harmonic output frequencies. The total direct current flowing with an incident pump power of 10 mA is about 8 mA. The output power obtained for self-bias and back-bias operation is somewhat smaller than the power levels given in Table V. The circuit is therefore most efficient if it is used a resistive multiplier.

VII. Conclusions

Metal–semiconductor junctions have been used to investigate the physical properties of interfaces and carrier transport mechanisms over surface barriers since the early discovery of nonlinear electrical characteristics in small contacts. The junctions have been of interest to surface physicists as tools for probing the bulk and surface properties of semiconductors, and they have been found to be sensitive detectors for electromagnetic radiation, both as quantum detectors and as nonlinear electrical elements.

A practical application of the junctions is their use as frequency downconverters in many areas of applied physics and communications, such as spectroscopy, remote sensing, radar, terrestrial radio links, and satellite communication. Very low noise temperatures and good conversion loss have been developed in the last few decades using conventional components and readily available semiconductor materials. New advances in the fields of microfabrication and molecular engineering make it possible to devise and construct devices and circuits that are superior in comparison to previously available structures. The molecular layers are specifically tailored to achieve high cutoff frequencies and the electrical embedding circuits are photoetched compact circuits showing improved characteristics up to the submillimeter-wave frequency range. These advanced and improved converters are universally attractive as communication circuits and as sensitive tools for probing our environment.

REFERENCES

Allyn, C. L., Gossard, A. C., and Wiegmann, W. (1980). *Appl. Phys. Lett.* **34,** 373–376.
Anand, Y. (1976). U. S. Patent 3,968,272.
Anand, Y., and Moroney, W. J. (1971). *Proc. IEEE* **59,** 1182–1190.
Archer, R. J., and Atalla, M. M. (1963). *Ann. N.Y. Acad. Sci.* **101,** 697–709.
Armstrong, E. H. (1924). *Proc. IRE* **12,** 540–552.
Badertscher, R., Salathe, R. P., and Luthy, W. (1980). *Electron. Lett.* **16,** 113–114.
Ballamy, W. C., and Cho, A. Y. (1976). *IEEE Trans. Electron Devices* **ED-23,** 481–484.
Barber, M. (1967). *IEEE Trans. Microwave Theory Tech.* **MTT-15,** 629–625.
Bayaraktaroglu, B., and Hartnagel, H. L. (1978). *Int. J. Electron.* **45,** 449–463.
Bell, A. E. (1979). *RCA Rev.* **40,** 295–338.
Bethe, H. A. (1942). *Radiat. Lab. Rep.* **43-12,** 1–25.
Braun, F. (1874). *Ann. Phys. (Leipzig)* [2] **153,** 3, 556–563.
Braun, F. (1967). "Nobel Lectures in Physics." Am. Elsevier, New York.
Burrus, C. A., Jr. (1966). *Proc. IEEE* **54,** 575–587.
Calviello, J. A. (1979). *IEEE Trans. Electron Devices* **ED-26,** 1273–1281.
Calviello, J. A., Wallace, J. L., and Bie, P. R. (1979). *Electron. Lett.* **15,** 509–510.
Carlson, E. R., and Schneider, M. V. (1979). *Int. Conf. Infrared Millimeter Waves Their Appl., Tech. Dig., 4th, 1979* pp. 82–83.
Carlson, E. R., Schneider, M. V., and McMaster, T. F. (1978). *IEEE Trans. Microwave Theory Tech.* **MTT-26,** 706–715.
Champlin, K. S., and Eisenstein, G. (1978). *IEEE Trans. Microwave Theory Tech.* **MTT-26,** 31–34.
Cho, A. Y. (1979). *J. Vac. Sci. Technol.* **16,** 275–284.
Cho, A. Y., and Arthur, J. R. (1975). *Prog. Solid State Chem.* **10,** 157–191.
Cho, A. Y., and Dernier, P. D. (1978). *J. Appl. Phys.* **49,** 3328–3332.
Cho, A. Y., E. L. Kollberg, Snell, W. W., Jr., and Schneider, M. V. (1982).
Clifton, B. J. (1977). *IEEE Trans. Microwave Theory Tech.* **MTT-25,** 457–462.
Clifton, B. J. (1979). *Radio Electron. Eng.* **49,** 333–346.
Cohn, M., Degenford, J. E., and Newman, B. A. (1975). *IEEE Trans. Microwave Theory Tech.* **MTT-23,** 667–673.
Copeland, J. A. (1970). *IEEE Trans. Electron Devices* **ED-17,** 404–407.

Cowley, A. M., and Sorenson, H. O. (1966). *IEEE Trans. Microwave Theory Tech.* **MTT-14,** 588–602.

Cunningham, J. A., Fuller, C. R., and Haywood, C. T. (1970). *IEEE Trans. Reliab.* **R-19,** 182–187.

D'Angelo, R., and Verlangieri, P. A. (1981). *Electron. Lett* **17,** 290–291.

De Loach, B. C., Jr. (1964). *IEEE Trans. Microwave Theory Tech.* **MTT-12,** 15–20.

De Loach, B. C., Jr. (1969). *In* "Microwave Semiconductor Devices and Their Applications" (H. A. Watson, ed.), pp. 464–496. McGraw-Hill, New York.

Denlinger, E. J. (1980). *IEEE Trans. Microwave Theory Tech.* **MTT-28,** 513–522.

Dickens, L. E. (1967). *IEEE Trans. Microwave Theory Tech.* **MTT-15,** 101–109.

Dingle, R. (1975). *Festkoerperprobleme* **15,** 21–48.

Dozier, J. W., and Rodgers, J. D. (1964). *IEEE Trans. Microwave Theory Tech.* **MTT-12,** 360.

Dragone, C. (1968). *Bell Syst. Tech. J.* **47,** 1883–1902.

Duchemin, J. P., Bonnet, M., and Huyghe, D. (1977). *Rev. Tech. Thomson-CSF* **9,** 685–716.

Duchemin, J. P., Bonnet, M., Koelsch, F., and Huyghe, D. (1978). *J. Cryst. Growth* **45,** 181–186.

Duke, C. B. (1969). *Solid State Phys., Suppl.* **10,** pp. 1–353.

Eastman, L. F., Stall, R., Woodward, D., Dandekar, N., Wood, C. E. C., Shur, M. S., and Board, K. (1980). *Electron. Lett.* **16,** 524–525.

Eng, S. T. (1961). *IRE Trans. Microwave Theory Tech.* **MTT-9,** 419–425.

Erickson, N. R. (1980). *IEEE/MTT-S Int. Microwave Symp., Tech. Dig.* pp. 19–20.

Esaki, L. (1958). *Phys. Rev.* **109,** 603–604.

Feynman, R. P. (1961). *In* "Miniaturization" (H. Gilbert, ed.), pp. 282–296. Van Nostrand-Reinhold, Princeton, New Jersey.

Ghate, P. B., Blair, J. C., Fuller, C. R., and McGuire, G. E. (1978). *Thin Solid Films* **53,** 117–128.

Goldsmith, P. F., and Plambeck, R. L. (1976). *IEEE Trans. Microwave Theory Tech.* **MTT-26,** 859–861.

Gordy, W., Smith, W. V., and Trambarulo, R. F. (1966). "Microwave Spectroscopy." Dover, New York.

Gruber, S. (1973). *Proc. IEEE* **61,** 237–238.

Gupta, K. C., Garg, R., and Bahl, I. J. (1979). "Microstrip Lines and Slotlines." Artech House, Dedham, Massachusetts.

Gysel, U. (1971). *Mitt. Arbeitsgemeinsch. Elektr. Nachrichtentech. Stift. Hasler-Werke Bern (AGEN)* No. 12, pp. 3–34.

Hagstrom, C., and Lidholm, S. (1978). On mm-wave frequency downconverters. Res. Rep. No. 130. Research Laboratory of Electronics, Onsala Space Observatory, Chalmers Institute of Technology, Gotherburg, Sweden.

Hagstrom, C. E., and Kollberg, E. L. (1980). *IEEE Trans. Microwave Theory Tech.* **MTT-28,** 899–904.

Haitz, R. H., and Smits, F. M. (1966). *IEEE Trans. Nucl. Sci.* **NS-13,** 198–207.

Hannay, N. B. (1976). "Treatise on Solid-State Chemistry." Plenum, New York.

Harris, J. J., and Woodcock, J. M. (1980). *Electron. Lett.* **16,** 317–319.

Harris, J. M., Lau, S. S., and Nicolet, M. A. (1976). *J. Electrochem. Soc.* **123,** 120–124.

Held, D. N., and Kerr, A. R. (1978). *IEEE Trans. Microwave Theory Tech.* **MTT-26,** 49–61.

Henisch, H. K. (1957). "Rectifying Semi-Conductor Contacts." Oxford Univ. Press, London and New York.

Henry, P. S., Glance, B. S., and Schneider, M. V. (1976). *IEEE Trans. Microwave Theory Tech.* **MTT-24,** 254–257.

272 M. V. SCHNEIDER

Hewlett, F. W. (1975). *IEEE J. Solid-State Circuits* **10**, 343-348.
Hill, M. (1980). *Solid State Technol.* **23**, 53-59.
Holm, R. (1967). "Electric Contacts, Theory and Application." 4th ed. Springer-Verlag, Berlin and New York.
Horton, R., Easter, B., and Gopinath, A. (1971). *Electron. Lett.* **7**, 490-491.
Howe, H., Jr. (1974). "Stripline Circuit Design." Artech House, Dedham, Massachusetts.
Ilegems, M. (1977). *J. Appl. Phys.* **48**, 1278-1287.
Immorlica, A. A., Jr., and Wood, E. J. (1978). *IEEE Trans. Electron Devices* **ED-25**, 710-713.
Itoh, T. (1981). *In* "Infrared and Millimeter Waves." (K. J. Button and J. C. Wiltse, eds.), Vol. 4, pp. 199-273. Academic Press, New York.
Joyce, B. A. (1979). *Surf. Sci.* **86**, 92-101.
Kahng, D., and Lepselter, M. P. (1965). *Bell Syst. Tech. J.* **44**, 1525-1528.
Kajiyama, K., Ida, M., Sakata, S., and Mizushima, Y. (1979). *IEEE Trans. Electron Devices* **ED-26**, 244-245.
Keen, N. J. (1977). *Electron. Lett.* **13**, 282-284.
Keen, N. J. (1980). *IEE Proc. I, Solid-State and Electron Devices* **127**(4), 188-198.
Keen, N. J., Kelly, W. M., and Wrixon, G. T. (1979). *Electron. Lett.* **15**, 689-690.
Kelly, W. M., and Wrixon, G. T. (1978). *Electron. Lett.* **14**, 80.
Kelly, W. M., and Wrixon, G. T. (1979). *IEEE Trans. Microwave Theory Tech.* **MTT-27**, 665-672.
Kelly, W. M., and Wrixon, G. T. (1980). *In* "Infrared and Millimeter Waves" (K. J. Button, ed.), Vol. 3, pp. 77-110. Academic Press, New York.
Kerns, V. D., Jr., Kull, J. M., and Ruwe, V. W. (1979). *IEEE Trans. Components, Hybrids, Manuf. Technol.* **CHMT-2**, 218-221.
Kerr, A. R. (1975). *IEEE Trans. Microwave Theory Tech.* **MTT-23**, 781-787.
Kerr, A. R. (1979a). *IEEE Trans. Microwave Theory Tech.* **MTT-27**, 135-140.
Kerr, A. R. (1979b). *IEEE Trans. Microwave Theory Tech.* **MTT-27**, 938-950.
Kerr, A. R., Mattauch, R. J., and Grange, J. A. (1977). *IEEE Trans. Microwave Theory Tech.* **MTT-25**, 399-401.
Kollberg, E. L., and Staehlberg, R. (1982). To be published.
Korwin-Pawlowski, M. L., and Heasell, E. L. (1975). *Solid-State Electron.* **18**, 849-852.
Kraus, J. D. (1966). "Radio Astronomy." McGraw-Hill, New York.
Kunzi, K. (1982). *In* "Infrared and Millimeter Waves" (K. J. Button, ed.). To be published. Academic Press, New York.
Kunzi, K., and Berger, H. (1978). *Proc. AGARD Meet., Munich, 1978.*
Lacombe, J., Duchemin, J. P., Bonnet, M., and Huyghe, D. (1977). *Electron. Lett.* **13**, 472-473.
Lee, T. P., and Burrus, C. A. (1968). *IEEE Trans. Microwave Theory Tech.* **MTT-16**, 287-296.
Lepselter, M. P. (1966). *Bell Syst. Tech. J.* **45**, 233-253.
Liechti, C. A. (1975). *IEEE Trans. Microwave Theory Tech.* **MTT-24**, 279-300.
Linke, R. A., Schneider, M. V., and Cho, A. Y. (1978). *IEEE Trans. Microwave Theory Tech.* **MTT-26**, 935-938.
Malik, R. J., AuCoin, T. R., Ross, R. L., Board, K., Wood, C. E. C., and Eastman, L. F. (1980). *Electron. Lett.* **16**, 836-838.
McColl, M., and Millea, M. F. (1976). *J. Electron. Mater.* **5**, 191-207.
McColl, M., Hodges, D. T., and Garber, W. A. (1977). *IEEE Trans. Microwave Theory Tech.* **MTT-25**, 463-467.

McMaster, T. F., Schneider, M. V., and Snell, W. W., Jr. (1976). *IEEE Trans. Microwave Theory Tech.* **MTT-24**, 948–952.

Maloney, T. J. (1980). *Electron Device Lett.* **EDL-1**, 54.

Manley, J. M., and Rowe, H. E. (1956). *Proc. IRE* **44**, 904–913.

Marquardt, J., and Schiek, B. (1969). *Proc. IEEE* **57**, 2076–2077.

Melchior, H. (1977). *Phys. Today* **30**, 32–39.

Millea, M. F., McColl, M., and Silver, A. H. (1976). *J. Electron. Mater.* **5**, 321–340.

Miller, G. L. (1972). *IEEE Trans. Electron Devices* **ED-19**, 1103–1108.

Miller, G. L., Lang, D. V., and Kimerling, L. C. (1977). *Annu. Rev. Mater. Sci.* **7**, 377–476.

Mittra, R., and Itoh, T. (1974). *Adv. Microwaves* **8**, 67–143.

Mott, H., Raburn, W. D., and Webb, W. E. (1972). *Proc. IEEE* **60**, 899–900.

Mott, N. F. (1938). *Proc. Cambridge Philos. Soc.* **34**, 568–572.

Mott, N. F. (1939). *Proc. R. Soc. London Ser. A* **171**, 27–38.

Mumford, W. W., and Scheibe, E. H. (1968). "Noise Performance Factors in Communication Systems." Horizon House, Dedham, Massachusetts.

Murphy, A. R., Bozler, C. O., Parker, C. D., Fetterman, H. R., Tannenwald, P. E., Clifton, B. J., Donnelly, J. P., and Lindley, W. T. (1977). *IEEE Trans. Microwave Theory Tech.* **MTT-25**, 494–495.

Murphy, R. A., and Clifton, B. J. (1978). *Tech. Dig.—Int. Electron Devices Meet.* pp. 124–128.

Oliver, J. D., Jr. (1980). Ph.D. Thesis, Cornell University, Ithaca, New York.

Nowicki, R. S., Harris, J. M., Nicolet, M. A., and Mitchell, I. V. (1978). *Thin Solid Films* **53**, 195–205.

Osborne, T. L. (1969). *Bell Syst. Tech. J.* **48**, 1623–1649.

Padovani, F. A. (1971). *In* "Semiconductors and Semimetals" (R. K. Willardson and A. C. Beer, eds.), Vol. 7A, pp. 75–146. Academic Press, New York.

Panish, M. G. (1980). *Science* **208**, 916–922.

Panish, M. G., and Cho, A. Y. (1980). *IEEE Spectrum* **17**, 18–23.

Pearsall, T. P., Capasso, F., Nahory, E. M., Pollak, A., and Chelikowsky, J. R. (1978). *Solid-State Electron.* **21**, 297–302.

Pellegrini, B., and Di Leo, T. (1977). *Alta Freq.* **46**, 345–353.

Penfield, P., Jr., and Rafuse, R. P. (1962). "Varactor Applications." MIT Press, Cambridge, Massachusetts.

Penzias, A. A., and Burrus, C. A. (1973). *Annu. Rev. Astron. Astrophys.* **11**, 51–72.

Penzias, A. A., and Schneider, M. V. (1976). *Bell Lab. Rec.* **54**, 45–49.

Penzias, A. A., and Wilson, R. W. (1965). *Astrophys. J.* **142**, 419–421.

Perlow, S. M. (1974). *RCA Rev.* **35**, 35–47.

Phillips, T. G., and Jefferts, K. B. (1973). *Rev. Sci. Instrum.* **44**, 1009–1014.

Phillips, T. G., Huggins, P. J., Kniper, T. H. B., and Miller, R. E. (1980). *Astrophys. J.* **238**, L103–L105.

Pound, R. V. (1948). "Microwave Mixers," MIT Radiat. Lab. Ser., Vol. 16. McGraw-Hill, New York.

Pucel, R. A., Masse, D. J., and Hartwig, C. P. (1968). *IEEE Trans. Microwave Theory Tech.* **MTT-16**, 342–350.

Ralsanen, A., Predmore, R., Parrish, P., Kot, R., Goldsmith, P., and Schneider, M. V. (1980). *Proc., Eur. Microwave Conf., Tech. Dig., 10th, Warsaw, Poland, September 1980.*

Rhoderick, E. H. (1978). "Metal–Semiconductor Contacts." Oxford Univ. Press (Clarendon), London and New York.

Rideout, V. L. (1975). *Solid-State Electron.* **18**, 540–550.

Rideout, V. L. (1978). *Thin Solid Films* **48**, 261–291.

Rideout, V. L., and Crowell, C. R. (1970). *Solid-State Electron.* **13**, 993–1009.

Salardi, G., Pellegrini, B., and Di Leo, T. (1979). *Solid-State Electron.* **22**, 435–441.

Salathe, R., Badertscher, G., Luthy, W., Reinhart, F. K., and Logan, R. A. (1979). *Appl. Phys. Lett.* **35**, 439–440.

Saleh, A. A. M. (1971). "Theory of Resistive Mixers." MIT Press, Cambridge, Massachusetts.

Schanda, E. (1978). *In* "Surveillance of Environmental Pollution and Resources by Electromagnetic Waves" (T. Lund, ed.), pp. 253–273. Reidel, Dordrecht, Netherlands.

Schilder, D. (1976). *Nachrichtentech-Elektron.* **26**, 64–65.

Schneider, M. V. (1969a). *Bell Syst. Tech. J.* **48**, 1421–1444.

Schneider, M. V. (1969b). *Bell Syst. Tech. J.* **48**, 2325–2332.

Schneider, M. V., and Berger, M. (1979). *Electron. Lett.* **15**, 450–451.

Schneider, M. V., and Carlson, E. R. (1977). *Electron. Lett.* **13**, 745–747.

Schneider, M. V., and Phillips, T. G. (1981). *Int. J. Infrared Millimeter Waves* **2**, 15–22.

Schneider, M. V., and Snell, W. W., Jr. (1971). *Bell Syst. Tech. J.* **50**, 1933–1942.

Schneider, M. V., and Snell, W. W., Jr. (1975). *IEEE Trans. Microwave Theory Tech.* **MTT-23**, 271–275.

Schneider, M. V., Glance, B., and Bodtmann, W. F. (1969). *Bell Syst. Tech. J.* **48**, 1703–1726.

Schneider, M. V., Linke, R. A., and Cho, A. Y. (1977). *Appl. Phys. Lett.* **31**, 219–221.

Schottky, W. (1923). *Z. Phys.* **14**, 63–106.

Schottky, W. (1925). *Elektr. Nachrichten-Tech.* **2**, 454–456.

Schottky, W. (1939). *Naturwissenschaften* **26**, 843.

Schottky, W. (1941). *Schweiz. Arch.* **7**, 20–29, 82–86.

Schottky, W., and Spenke, E. (1939). *Wiss. Veroeff. Siemens-Werken* **18**, 225–291.

Shur, M. S., and Eastman, L. F. (1979). *IEEE Trans. Electron Devices* **ED-26**, 1677–1683.

Shur, M. S., Eastman, L. F. (1980). *Electron. Lett.* **16**, 522–523.

Shurmer, H. V. (1964). *Proc. Inst. Electr. Eng.* **111**, 1511–1516.

Siegel, P. H., and Kerr, A. R. (1979). *NASA Tech. Memo.* **NASA TM-X-80324.**

Smythe, W. R. (1950). "Static and Dynamic Electricity," 2nd ed. McGraw-Hill, New York.

Sobol, H., and Caulton, M. (1974). *Adv. Microwaves* **8**, 12–66.

Spenke, E. (1958). "Electronic Semiconductors." McGraw-Hill, New York.

Stall, R., Wood, C. E. C., Board, K., and Eastman, L. F. (1979). *Electron. Lett.* **15**, 800–801.

Sze, S. M. (1981). "Physics of Semiconductor Devices." Wiley, New York. pp. 1–868.

Sze, S. M., and Gibbons, G. (1966). *Appl. Phys. Lett.* **8**, 111–113.

Tada, K., and Laraya, J. L. R. (1967). *Proc. IEEE* **55**, 2064–2065.

Takada, T., and Ohmori, M. (1979). *IEEE Trans. Microwave Theory Tech.* **MTT-27**, 519–523.

Torrey, H. C., and Whitmer, C. A. (1948). "Crystal Rectifiers," MIT Radiat. Lab. Ser., Vol. 15. McGraw-Hill, New York.

Tsui, D. C. (1981). *In* "Handbook of Semiconductor Physics" (W. Paul, ed.), Vol. 1. North-Holland Publ., Amsterdam.

Tucker, D. G. (1954). *Wireless Eng.* **31**, 145–152.

Ueonohara, M., and Gewartowski, J. W. (1969). *In* "Microwave Semiconductor Devices and Their Applications" (H. A. Watson, ed.), pp. 194–270. McGraw-Hill, New York.

Uhlir, A., Jr. (1958). *Bell Syst. Tech. J.* **37**, 951–988.

Viola, T. J., Jr., and Mattauch, R. J. (1973a). *J. Appl. Phys.* **44**, 2805–2808.

Viola, T. J., Jr., and Mattauch, R. J. (1973b). *Proc. IEEE* **61**, 393.

Vizard, D. L., Keen, N. J., Kelly, W. M., and Wrixon, G. T. (1979). *IEEE/MTT-S Int. Microwave Symp., Tech. Dig.* pp. 81–83.

Weinreb, S. (1981). Unpublished.

Weinreb, S., and Kerr, A. R. (1973). *IEEE J. Solid-State Circuits* **SC-8,** 58–63.

Wheeler, H. A. (1965). *IEEE Trans. Microwave Theory Tech.* **MTT-13,** 172–185.

Wilson, A. H. (1932). *Proc. R. Soc. London Ser. A* **136,** 487–498.

Wilson, R. W., Jefferts, K. B., and Penzias, A. A. (1970). *Astrophys. J.* **161,** L43–L44.

Wilson, W. J. (1977). *IEEE Trans. Microwave Theory Tech.* **MTT-25,** 332–335.

Wolf, E. D. (1979). *Phys. Today* **32,** 34–36.

Wolf, E. L. (1975). *Solid State Phys.* **30,** 1–91.

Wrixon, G. T. (1974). *IEEE Trans. Microwave Theory Tech.* **MTT-22,** 1159–1165.

Wrixon, G. T., and Kelly, W. M. (1978). *Infrared Phys.* **18,** 413–428.

Wrixon, G. T., Langley, J. B., and Campbell, J. S. (1979). *Conf. Dig., Int. Conf. Millimeter-Waves Their Appl., 4th, 1979* pp. 80–81.

Young, D. T., and Irvin, J. C. (1965). *Proc. IEEE* **53,** 2130–2131.

Zimmermann, P., and Haas, R. W. (1977). *Nachrichtentech. Z.* **30,** 721–722.

INFRARED AND MILLIMETER WAVES, VOL. 6

CHAPTER 5

Quasi-Optical Techniques at Millimeter and Submillimeter Wavelengths

Paul F. Goldsmith

Five College Radio Astronomy Observatory
Department of Physics and Astronomy
University of Massachusetts,
Amherst, Massachusetts

Introduction

This chapter is a review of quasi-optical techniques used in receivers and telescope feed systems at millimeter and submillimeter wavelengths. The emphasis will be placed on explaining the principles of operation and limitation of devices that have found practical application in radioastronomical systems. We shall first discuss the theory of Gaussian beams; that is to say, the propagation of fairly directed radiation with Gaussian intensity distribution perpendicular to the axis of propagation. This theory will be used to analyze the behavior of and the coupling among focusing optics, receiver, and antenna (telescope). The fundamental behavior is determined by the diffraction of the finite-sized beams; in the correct limit, the formulas of Gaussian optics will be seen to reduce to those of geometrical optics.

The range of functions for which quasi-optical components have been developed is so large that a comprehensive review would not be possible. The present discussion will concentrate on devices relevant for radioastronomical systems. Frequency diplexers and filters will be analyzed in detail, as these have been a major impetus for the development of quasi-optical systems. Also discussed will be calibration, phase modulation, and beamswitching, which are functions now generally performed quasi-optically at millimeter and submillimeter wavelengths.

I. Propagation of Gaussian Beams

The propagation of a beam of electromagnetic radiation with a Gaussian intensity distribution was investigated in detail in the mid-1960s as a result of interest in sequences of lenses or mirrors for low-loss transmission ("beam waveguide") and investigations of the behavior of resonant cavities to be used as components of laser oscillators. The initial investigations approached this problem from a variety of points of view; for references to earlier work, see the reviews by Kogelnik and Li (1966), Goubau (1963), and Martin and Lesurf (1978). Complete discussions are currently available in texts (Arnaud, 1976; Marcuse, 1975); Siegman, 1971), and these can be consulted for details concerning the very brief outline that will be given here.

To start, we wish to find solutions for the wave equation representing radiation not too different from a plane wave. The wave equation in free space for a single component of the electric field ψ can be written as

$$\nabla^2\psi + k^2\psi = 0, \tag{1}$$

where $k = 2\pi/\lambda$ and we have suppressed the time-dependent terms $e^{i\omega t}$. If we consider a wave traveling along the z axis and assume that it is basically a plane wave, we can write

$$\psi = u(x, y, z)e^{-ikz}, \tag{2}$$

where u is assumed to be a slowly varying function of z. Substitution in Eq. (1) then yields

$$\frac{\partial^2 u}{\partial x^2} + \frac{\partial^2 u}{\partial y^2} + \frac{\partial^2 u}{\partial z^2} - 2ik\frac{\partial u}{\partial z} = 0.$$ (3a)

The assumption that the primary dependence on z is contained in the exponential term of Eq. (2) lets us drop the third term with respect to the fourth, and we obtain the simplified equation

$$\frac{\partial^2 u}{\partial x^2} + \frac{\partial^2 u}{\partial y^2} - 2ik\frac{\partial u}{\partial z} = 0.$$ (3b)

A solution of this equation can be written as a product of a Hermite polynomial, a Gaussian transverse amplitude variation, and a phase factor. If we had written the wave equation (1) in cylindrical coordinates instead of rectangular, the solutions would appear essentially the same except with Laguerre polynomials. The orthogonality of either of these sets of polynomials allows one, in principle, to define an arbitrary field distribution in terms of these basic configurations and suggests the identification of each of them with a mode of propagating radiation.[1] We shall be concerned primarily with the propagation of the fundamental mode (as discussed next), and conversion of energy to higher modes is regarded essentially as a loss.

The lowest order, or fundamental Gaussian mode, is the same in rectangular or cylindrical coordinates and can be written as

$$\psi = A\,\frac{w_0}{w(z)}\,\exp\!\left(\frac{-r^2}{w^2(z)}\right)\exp(-ikz)\,\exp\!\left(\frac{-i\pi r^2}{\lambda R(z)}\right)\exp\!\left(i\arctan\frac{\lambda z}{\pi w_0^2}\right),$$ (4)

where $r = (x^2 + y^2)^{1/2}$, A and w_0 constants, and w and R slowly varying functions of z, which will be discussed later. The variation of field amplitude transverse to the direction of propagation is seen to have a purely Gaussian form with the half-width to the $1/e$ point being equal to w, which is called the *beam radius*. The variation of w along the axis of propagation is given by the equation[2]

$$w(z) = w_0[1 + (\lambda z/\pi w_0^2)^2]^{1/2},$$ (5)

---

[1] The validity of the Gaussian mode solution to the wave equation depends, of course, on the validity of the simplified equation (3b) and hence on the relative rate of transverse and longitudinal variation of the electric field. In terms of the diameter-to-wavelength ratio of the beam waist, it has been found that $\lambda/2w_0 < 1$ ensures that the errors in the phase and amplitude distributions resulting from the use of (3b) will be small (van Nie, 1964). This corresponds crudely to a system with $f/D > 0.6$.

[2] The form of w as well as that of R is determined by substitution of a general form for u in the wave equation (3b) (cf. Marcuse, 1975).

where the minimum beam radius w_0, taken to be located at $z = 0$, is called the *beam-waist radius*, or simply the waist radius. The contours of w are hyperboloids of revolution symmetric about the axis of propagation; the region of minimum size is called the *beam waist*. At large distances from the beam waist, the asymptotic angle of growth of the beam radius is given by

$$\theta_{w_0} = \lambda/\pi w_0. \tag{6}$$

The propagation of a Gaussian beam in the vicinity of its waist is shown in Fig. 1.

Some authors (cf. Arnaud, 1976) express the Gaussian distribution as that of the energy flux or power density. In this system, the power distribution perpendicular to the axis of propagation is given by $P(r) = P(o)$ $\exp(-r^2/\xi^2)$ and the variation of ξ along the axis of propagation is given by

$$\xi(z) = \xi_0[1 + (z/k\xi_0^2)^2]^{1/2}. \tag{7}$$

Recalling that $k = 2\pi/\lambda$, we see that $\xi(z) = w(z)/\sqrt{2}$ and $\xi_0 = w_0/\sqrt{2}$. Both systems are encountered in published analyses of quasi-optical devices, which will be discussed in the following sections; we shall utilize the field amplitude system in the present work.

The surfaces of constant phase of the solutions to Eq. (4) are spherical and have radius of curvature $R(z)$. The dependence of $R(z)$ on the distance z from the beam waist is given by

$$R(z) = z[1 + (\pi w_0^2/\lambda z)^2]. \tag{8}$$

The wave fronts are plane at the beam waist, and for large distances from the waist, they have a radius of curvature equal to the distance from the waist, as geometrical optics wave fronts have a radius of curvature equal to the distance from their focus. The radius of curvature of a Gaussian

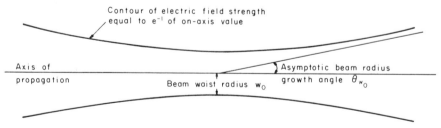

FIG. 1   Contours of Gaussian beam in the vicinity of the beam waist. The contours are of electric field strength equal to $1/e$ of the axis value, which itself varies with distance from the beam waist as $w(z)^{-1}$, as given in Eqs. (4) and (5).

beam attains its minimum value at a distance $z_c = \pi w_0^2 / \lambda$ (sometimes called *confocal distance;* see Section IV.D), where it has a value of $\sqrt{2} z_c$. At this distance, the beam radius is equal to $\sqrt{2} w_0$. From these considerations, it is seen that z_c is a natural division between the near field and far field of the Gaussian beam waist. In what follows the beam radius and beam radius of curvature will be denoted as w and R, respectively, it being understood that they are both functions of the distance from the beam waist.

The Gaussian beam formulas previously discussed assume that whereas the beam waist radius is of finite size, the beam itself is of infinite transverse extent, a situation that clearly cannot be realized in practice. The effects of truncation of a Gaussian beam have been investigated by a number of authors who calculate, in general, the diffraction of a wave of the form given in Eq. (4) after passage through a circular aperture of radius a. For $a/w_0 \leq 1$, there is nonmonotonic variation of the on-axis field strength away from the beam waist (Campbell and DeShazer, 1969; Holmes *et al.*, 1970), and considerable structure in the near-field and far-field angular radiation patterns (Campbell and DeShazer, 1969; Schell and Tyras, 1971). The detailed results are obtained only by numerical evaluation of the diffraction integrals; Campbell and DeShazer indicate that for $a/w_0 = 2$ there is no more than a 4% discrepancy between the computed near-field angular pattern from a truncated Gaussian beam and the pure Gaussian shape with the same value of w_0. Schell and Tyras show that the far-field patterns are less sensitive to truncation showing significant non-Gaussian behavior only at relatively low levels (sidelobes). The main lobe of the radiation pattern, although quite Gaussian, will, in general, have a width considerably in excess of that predicted on the basis of a nontruncated Gaussian (see Section III). The ratio of the widths is only a factor of 1.03 for $a/w_0 = 2$ and thus a ratio of aperture diameter to beam radius of 4 [which produces an electric field at the edge equal to 0.018 of the on-axis value and a relative power density 3.4×10^{-4} (-35 dB)] appears to be a reasonable value to ensure propagation of essentially pure Gaussian beams and proper performance of focusing elements (Section II.C).

II. Production and Focusing of Gaussian Beams

Within an approximation analogous to the ''thin lens'' approximation of geometrical optics, there exists a simple and elegant theory for the transformation of Gaussian beams by focusing elements; its results will be summarized in Section II.C. However, for practical utilization, there are certain useful caveats and guidelines that will also be mentioned.

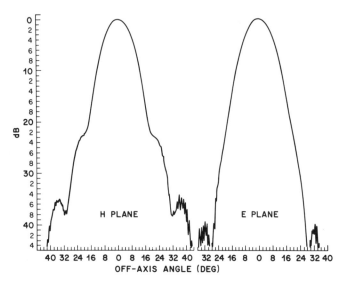

FIG. 2   *E* plane and *H* plane patterns of dual-mode feedhorn operating at 100 GHz. The horn aperture diameter is 1.58 cm. The shoulders on the *H* plane curve are characteristic of dual-mode feedhorn patterns.

A.   GAUSSIAN BEAMS AND MICROWAVE FEEDHORNS

The validity of an analysis based on the fundamental Gaussian mode clearly depends on the extent to which such a mode can be produced. Since it would not be acceptable to remove unwanted modes by filtering, it is fortunate that at microwave frequencies there are efficient devices for coupling a fundamental-mode Gaussian beam to a waveguide. A Gaussian beam pattern is not necessarily optimal for all purposes; illuminating an antenna to achieve highest aperture efficiency obviously demands a different distribution, but for applications requiring the use of focusing optics in addition to the antenna itself, a feedhorn producing a Gaussian radiation pattern is desirable.

A relatively simple horn, which gives nearly Gaussian radiation patterns, is the dual-mode feed (Potter, 1963; Turrin, 1967).[3] In this device, the TE_{11} and TM_{11} modes in cylindrical waveguide are combined with appropriate amplitude and phase to produce an aperture distribution whose radiation pattern is nearly symmetric and highly gaussian. The 100-GHz patterns in the principle planes of such a feed having an aperture of 1.58 cm and mode-transducer angle β (Turrin, 1967) equal to 27.5° are shown in Fig. 2. To levels down to 20 dB below the on-axis power, the

[3] A comprehensive collection of papers on feedhorns is that edited by A. W. Love (1976).

beamwidths in the two principal planes differ by less than 5% (the E plane always being wider). A major disadvantage of such horns is that the phasing between the two propagating modes is critical for proper performance. Particularly for larger aperture horns having a longer distance between the mode transducer and the aperture, this severely limits the bandwidth; for the horn just discussed, as well as a version having an aperture diameter of 3.0 cm, the instantaneous bandwidth is ~5 GHz at frequencies between 80 and 150 GHz. This type of horn can be constructed in a "trombone" type of design allowing tuning over a broad range, however. From the point of view of Gaussian optics the beam-waist radius for measured patterns of dual-mode horns is approximately 0.3 times the aperture diameter and is essentially frequency-independent.

A somewhat more elaborate feedhorn is the hybrid mode or scalar feed (Simmons and Kay, 1966; Thomas, 1978; see also references in Love, 1976). In this device, corrugated walls impose identical boundary conditions for both polarizations of the electric field, this producing a symmetric pattern and, ideally, one with very low sidelobes. The patterns of typical scalar feeds are well fit by Gaussian distribution to a level approximately 20 dB below the on-axis power. Achieving the optimum radiation pattern requires launching the proper hybrid mode in the horn, which can be done in a number of ways (cf. Dragone, 1977). A scalar feedhorn with a mode-launching-section design based on the work of Dragone (1977) is shown in Fig. 3.

An important quantity for characterizing scalar feedhorns is the difference (measured in wavelengths) between the path from the horn apex to the edge of the aperture and that from the apex to the center of the aperture. This is the aperture phase error and is given by[4]

$$\Delta = \frac{R}{\lambda} \sin \theta_0 \tan \frac{\theta_0}{2}, \tag{9a}$$

where R is the slant length of the horn and $2\theta_0$ the opening angle of the horn. For angles that are not too large, this can be approximated by

$$\Delta \simeq a^2/2\lambda R, \tag{9b}$$

where a is the radius of the horn aperture. It can be seen that Δ is also the ratio of the geometrical horn-opening angle to a characteristic beam dif-

[4] The article by Thomas (1978) gives a very clear discussion of the requirements for frequency-independent operation of scalar feedhorns as well as useful design curves. The nomenclature for the different types of feedhorns utilizing hybrid modes in corrugated waveguides is both confusing and generaly inconsistent from author to author. If one ignores the name and considers only the aperture phase error, it will be clear in what regime the feedhorn is operating.

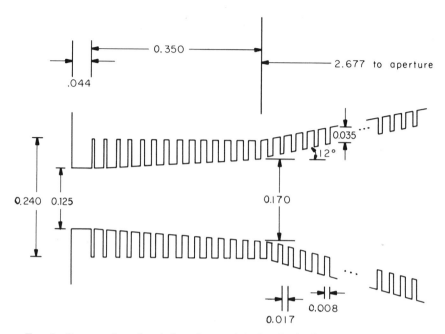

FIG. 3   Cross section of mode-launcher section of scalar feedhorn built at FCRAO (dimensions are in inches), with an aperture diameter of 1.308 in. Both the groove width and depth are tapered, starting from 0.005 and .0575 in., respectively, to achieve a broad-band match from the circular to the corrugated guide. The horn, together with rectangular to circular transition has a VSWR less than 1.1 between 75 and 115 GHz and radiation patterns that agree quite well with theory.

fraction angle for the aperture. Large values of Δ correspond to the geometrical angle exceeding the diffraction angle, producing the frequency-independent limit. In this situation, the flare angle θ_0 corresponds closely to the -17-dB contour of the power pattern and thus $w_0 = 0.45(\lambda/\theta_0)$. For certain applications, however, it may be preferable to have the beam-waist radius independent of frequency. Narrow flare-angle scalar feeds can be constructed for this purpose. In the limit of low-aperture phase error, one finds that the beam-waist radius w_0 is equal to 0.33 times the aperture diameter, quite similar to the case of the dual-mode horn discussed previously. Scalar feeds are widely used to illuminate parabolic reflectors, but they obviously make excellent Gaussian beam launchers as well. One complication is that the phase center of the horn (the center of curvature of the far-field spherical wave, which is equivalent to the location of the Gaussian beam waist) can be located well behind the horn aperture and can move as the frequency is changed as well. This effect is discussed in several articles in the collection by Love (1976) and by Ohtera and Ujiie (1975) and Thomas (1978).

Pyramidal feedhorns have the important advantage of simplicity of construction and low loss, which are particulary important at shorter wavelengths. Their drawbacks include significant sidelobe content and dissimilar E and H plane patterns. The exact patterns depend on the horn length as well as the aperture dimensions due to the phase error produced by the horn flare. Using an approximate expression given by Kraus (1950) for the gain of a pyramidal waveguide horn

$$G(\text{dB}) = 10 \log 4.45 A_{E\lambda} A_{H\lambda}, \tag{10}$$

where $A_{E\lambda}$ and $A_{H\lambda}$ are the E and H plane aperture dimensions in wavelengths, and the ratio $A_{E\lambda}/A_{H\lambda} = 1.3$ commonly used in optimum horns, one finds for the beam-waist radius

$$w_0 = (\lambda/2\pi)10^{G(\text{dB})/20}. \tag{11}$$

This expression is useful for analyzing the approximate Gaussian beam behavior of a pyramidal horn, but the relatively large sidelobe content (10–15% of the radiated power) makes the coupling to a Gaussian beam somewhat dependent on the exact application.

B. SUBMILLIMETER FEEDS

As one approaches shorter wavelengths, fabrication of the more elaborate feeds discussed in the previous section with the required accuracy becomes a major obstacle. The simple feeds such as pyramidal horns have thus been those most often used at frequencies above 200 GHz with rectangular waveguide systems (Goldsmith and Plambeck, 1976; Erickson, 1978). Circular waveguides appear to have significant advantages in construction (Blaney, 1978) at these frequencies, which should be an incentive to fabricate scalar feedhorns. If this is not done, conical horns can still be utilized, though they have the same disadvantages as pyramidal horns as discussed previously.

For wavelengths less than 0.5 mm, metallic guided-wave structures of any type become exceedingly difficult to fabricate and are highly lossy as well. Because micron-sized diodes are still employed as mixing elements for heterodyne systems, a coupling structure (antenna) is required and numerous designs have been proposed, and quite a few have been evaluated. Not surprisingly, a number of these are simply scaled versions of devices that are familiar at much longer wavelengths. One such device is the bioconical antenna, developed for submillimeter wavelengths by Gustincic (1977a; see Kraus, 1950, for discussion). With suitable passive reflecting elements, this device is indicated to have a relatively well defined radiation pattern, although few details are available. A number of antenna designs for use at submillimeter wavelengths are discussed by Rutledge *et al.* (1978).

One design that has been relatively successful throughout the submillimeter region is the cube-corner antenna (Krautle *et al.*, 1977; Fetterman *et al.*, 1978), which is discussed by Kraus (1950). A diagram of this device is given in Fig. 4, together with measured radiation patterns. These patterns do appear to have a fairly Guassian main lobe and modest-amplitue sidelobes. Laboratory measurements of the patterns of a number of examples, as well as a main-lobe efficiency determination of approximately 0.5 (Goldsmith *et al.*, 1981) at 692 GHz, indicate less ideal performance. This may be due to variations in the long-wire–vertex distance, which is only one-quarter wavelength ($\sim$100 μm) and which is extremely difficult to control accurately. Further development is required to increase the efficiency of this type of feed, which, in other respects, including low ohmic loss, ease of fabrication, and convenient driving point impedance, is highly advantageous.

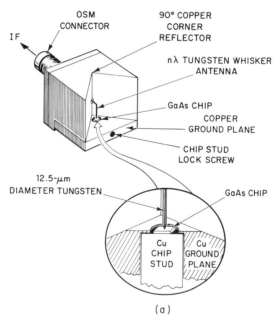

(a)

FIG. 4   Ninety-degree corner reflector mixer for submillimeter wavelengths. (a) Construction of the mixer showing long-wire antenna (generally 2 or 4 wavelengths long) and an inset of the diode chip mounting; the GaAs Schottky diodes have a capacitance of $\sim$1.5 fF and are 1 μm in diameter.

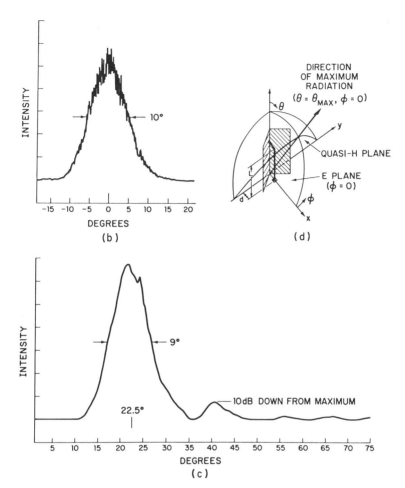

FIG. 4 (*Continued*)   Parts (b) and (c) show radiation patterns in the principal planes *H* (part b) and *E* (part c) taken at $\lambda = 394\ \mu$m. The largest amplitude sidelobes are found not to lie in either of the two principal planes. (From Fetterman *et al.*, 1978.) Also shown (part d) is the corner-reflector coordinate system.

Other types of feeds have been developed with the goal of combining a well-behaved radiation pattern and the elimination of the delicate whisker contact to the diode. Controlling the radiation pattern from an antenna integrated on a high-dielectric-constant substrate (Murphy *et al.*, 1977) appears to be difficult. A technique of sandwiching the diode and metallic conducters between two sheets of dielectric results in a relatively clean pattern (Schwartz and Rutledge, 1980), but data that can be compared with other devices such as the cube-corner antenna are not yet available.

C. Focusing of Gaussian Beams

For paraxial rays in geometrical optics, a thin lens acts simply as a phase transformer, providing a phase advancement approximately proportional to the square of the distance r of a ray from the axis of propagation:

$$\Delta\phi = \pi r^2/\lambda f, \tag{12}$$

where f is the focal length of the lens. This phase pertubation acting on a spherical wave changes the wave front curvature (reciprocal of the radius of the curvature) by an amount equal to the reciprocal of the focal length of the lens. This approximation can also be applied to the spherical waves of a Gaussian beam [cf. Eq. (4)].

Let us consider the effect of a thin lens on the propagation of a Gaussian beam with beam-waist radius w_{01} propagating to the right, as shown in Fig. 5. We place the lens a distance d_1 from this waist and assume that the transformed beam has waist radius w_{02}, located a distance d_2 to the right of the lens. Then, because we assume that a thin lens does not affect the amplitude distribution of the field in the plane of the lens, we obtain

$$w_{01}[1 + (\lambda d_1/\pi w_{01}^2)^2]^{1/2} = w_{02}[1 + (\lambda d_2/\pi w_{02}^2)^2]^{1/2}. \tag{13a}$$

The phase-transforming properties of the lens yield

$$\frac{1}{d_1[1 + (\pi w_{01}^2/\lambda d_1)]} + \frac{1}{d_2[1 + (\pi w_{02}^2/\lambda d_2)]} = \frac{1}{f} \tag{13b}$$

[see Kogelnik (1964) and Chu (1966) for a discussion of sign convention; the one generally adopted is that R is positive when an observer from the left sees a concave wave front for waves in Fig. 5]. The two preceding

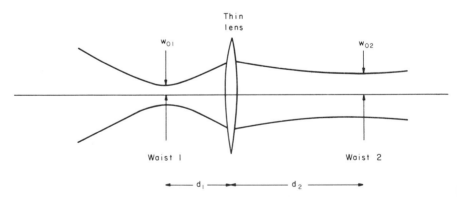

Fig. 5 Schematic illustration of the imaging of Gaussian beam waist by a thin lens. The relative sizes and locations of the waists are given by Eqs. (14a) and (14b).

equations can be solved for the output beam parameters as a function of those of the input beam, which yields

$$\frac{d_2}{f} = 1 + \frac{(d_1/f) - 1}{[(d_1/f) - 1]^2 + (\pi w_{01}^2/\lambda f)^2}, \tag{14a}$$

$$\left(\frac{w_{02}}{w_{01}}\right)^2 = \frac{1}{[(d_1/f) - 1]^2 + (\pi w_{01}^2/\lambda f)^2}. \tag{14b}$$

These relations between input and output waists can also be obtained directly from diffraction theory (Dickson, 1970). The transformation equations for waist location and radius are shown graphically in Fig. 6. The location of the waists with respect to the lens, given by Eq. (14a), differs from the analogous geometrical optics imaging relationship because of the second term in the denominator. Thus in Fig. 6 there are regions in which d_2 increases as d_1 is increased, quite unlike the situation in geometrical optics.

A particularly important case of the Gaussian beam transformation equations is that occurring when the waist of the input beam is located at a

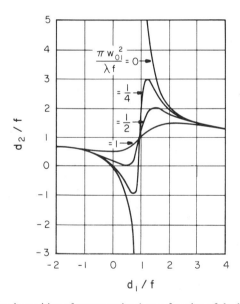

FIG. 6 Variation in position of output waist d_2 as a function of the location of the input waist d_1 for a Gaussian beam imaged by a lens of focal length f. The different curves correspond to different values of the parameter $\pi w_{01}^2/\lambda f$. The curve for the zero value corresponds to the geometrical optics limit with the input beam a spherical wave from a point source at distance d_1 from the lens.

distance equal to the focal length of the lens or mirror ($d_1 = f$). Then we see that

$$d_2 = f,$$

$$w_{02}/w_{01} = \lambda f/\pi w_{01}^2.$$

(14c)

The location of the output waist is thus seen to be independent of the wavelength of the radiation. A pair of lenses separated by the sum of their focal lengths with the waist of the input beam at the focal point of one of the lenses thus produces an output beam whose waist location is independent of frequency and whose radius is given by

$$w_{\text{out}}/w_{\text{in}} = \frac{f_2}{f_1}.$$

(14d)

These features of a "Gaussian beam telescope" are of particular importance for designing systems that must operate over a large range of frequencies.

The transformation equations for a Gaussian beam can also be written in a useful form as a solution to the following problem. If we wish to transform waist radius w_{01} to w_{02} with a thin lens of focal length f, what input and output distances are required? One finds that

$$d_1 = f \pm (w_{01}/w_{02})[f^2 - (\pi w_{01} w_{02}/\lambda)^2]^{1/2},$$

(15a)

$$d_2 = f \pm (w_{02}/w_{01})[f^2 - (\pi w_{01} w_{02}/\lambda)^2]^{1/2},$$

(15b)

where the signs must be either both positive or both negative. The focal length f must be greater than the minimum characteristic length $\pi w_{01} w_{02}/\lambda$.

Using Eq. (14a,b), a Gaussian beam propagating through a sequence of focusing elements can be analyzed. It is assumed that the focusing elements are large enough to introduce negligible truncation and diffraction; these effects will be analyzed in more detail in the discussion of feed systems presented next.

The beam parameters w and R can be combined into a complex parameter q defined by

$$\frac{1}{q} = \frac{1}{R} - \frac{i\lambda}{\pi w^2}.$$

(16a)

The complex beam parameter can equivalently be defined as

$$q = z + (i\pi w_0^2/\lambda).$$

(16b)

If z is measured from the beam waist, q is purely imaginary at the waist, and as a beam propagates in free space, only its real part varies. A focusing element is seen to affect only the real part of q. These relationships can be put in a useful form by letting q after a particular operation (length of free space, passage through lens, etc.) be given by

$$q_{out} = (Aq_{in} + B)/(Cq_{in} + D). \tag{17}$$

If we write the four coefficients in a 2×2 matrix, then for a distance of d of free space we have a matrix

$$\begin{pmatrix} A & B \\ C & D \end{pmatrix} = \begin{pmatrix} 1 & d \\ 0 & 1 \end{pmatrix}$$

and for a lens of focal length f

$$\begin{pmatrix} A & B \\ C & D \end{pmatrix} = \begin{pmatrix} 1 & 0 \\ -1/f & 1 \end{pmatrix}.$$

The effect on the complex beam parameter of a sequence of operations is then given by matrix multiplication of the individual operations and insertion of the result in Eq. (17). This formulation, known as the ABCD law, is especially useful for the analysis of resonators where one is often interested in a sequence of operations that returns the beam to its original state.

The ABCD law shows that the effect of an optical element (or system) on a Gaussian beam can be calculated once its transformation coefficients are known. The coefficients that are required are identical to those for the transformation of ray position and slope in geometrical optics. Thus, once the transformation coefficients for a geometrical optics device are known, it is straightforward to calculate the effects of the device on a Gaussian beam. In this approximation then, a thin lens is also eqivalent to a parabolic mirror having focal length f.[5] An ellipsoidal mirror can be thought of as two cascaded simple focusing elements separated by zero distance, with the focal length of each being the distance from the ellipsoid focus to the center of the section of the surface being used. From the ABCD law we see that the equivalent focal length of the system is given by

$$f_{ellip} = d_1 d_2/(d_1 + d_2), \tag{18}$$

d_1 and d_2 being the two distances in question.

D. GEOMETRICAL OPTICS LIMIT OF GAUSSIAN BEAM FORMULAS

It is instructive to examine the geometrical optics limit of the Gaussian beam propagation and focusing formulas. In order to avoid confusion, one must keep clear which quantities are being fixed and which are allowed to

[5] More realistically, the equivalence would be to an offset parabolic mirror because we are assuming that there is no blockage.

vary. Consider first the free-space propagation formulas (5) and (8). If we let $\lambda \to 0$ and keep w_{01} fixed, we see that $w \to$ const and $R \to \infty$, which corresponds to a plane wave. If we let $\lambda \to 0$ but keep the beam divergence angle $\theta_{w_{01}}$ [Eq. (6)] fixed, we see that $w \to \theta_{w_{01}}z$ and $R \to z$, which represents a spherical wave expanding from the location of the beam waist at $z = 0$. The focusing formulas (14) give, for the first case, $d_2 \to f$ and $w_{02} \to 0$, which is what one expects because in geometrical optics a lens focuses a plane wave to a point. The second limit of the waist location formula (14a) yields the geometrical optics thin-lens formula

$$(1/d_1) + (1/d_2) = 1/f.$$

Holding the beam divergence angle fixed in Eq. (14b) with $d_1 = f$ yields $w_2 = \theta_{w_{01}}f$, which gives the beam diameter produced by a collimated point source with divergence angle $\theta_{w_{01}}$.

E. LIMITATIONS AND DIFFICULTIES OF GAUSSIAN OPTICS

To dispel the notion that the thin-lens formulas with Gaussian beams furnish a universal solution to all focusing problems, some of the outstanding limitations should be indicated.

For the case of the microwave lenses, it has been recognized that the proper shape is not one that consists of spherical surfaces (Silver, 1949). Most designs for horn–lens combinations produce a wave front which has the beam waist located in the plane of the lens. The general design procedure is to calculate the lens shape necessary to produce the change in radius of curvature required to correct the phase errors produced by the feedhorn. In many practical situations, the lens will perturb the amplitude as well as the phase distribution. The deviations from spherical surfaces need to be determined for a particular case, but it is clear that systems with small f/D ratio present the greatest problems. The design of dielectric lenses is discussed by Silver (1949), Jasik (1961), and Wolff (1966). Lenses have been combined with scalar feedhorns in the microwave region; this is discussed in the articles by Clarricoats and Saha (1969) and Padman (1978).

For reflecting optics, there are three effects that must be considered; they are all dependent on the extent to which the system is not axisymmetric and also on the f/D ratio. The first problem arises from the breakdown of the thinness condition. This results in the radius of curvature being different for the intersection of different portions of the mirror surface with the beam. Since, in general, the radius of curvature of a Gaussian beam is not equal to the distance from the waist, one can obtain differences from the usual ellipsoidal or parabolic shapes. This problem can be minimized by keeping the focusing optics in the far field of the waist (where the behavior is more nearly geometrical) and by using a large f/D ratio, which results in mirrors of minimum extent parallel to the axis of propagation. In extreme cases, a second problem arises because of the change in power density due to the variation in the distance of different portions of mirror from the waist. This effect will distort the output beam. In certain cases, a system of multiple offset reflectors can be used, which avoids these difficulties (Dragone 1978).

An additional problem is that the curvature of off-axis mirrors can introduce cross-polarization effects. The magnitude of this effect increases with decreasing f/D ratio and offset angle. Calculations have been carried out for offset parabolic reflectors (Chu and Turrin, 1973); the restrictions thus imposed depend on the system application. For example, for many of the frequency-selective devices discussed in Section IV, cross-polarized radiation is not transmitted efficiently, and thus even if polarization purity per se is not required, efficiency will be degraded if the optics affects the polarization state. The absorption in reflecting optics is generally not a significant problem, as will be shown in Appendix III.

III. Gaussian Beams and Antenna Feed Systems

Illumination of an antenna aperture by a Gaussian amplitude distribution is not optimum for achieving highest aperture efficiency, but as we have seen, it accurately represents the pattern produced by several important types of feedhorns as well as more complex feed systems. In this section we shall discuss some of the results that can be obtained by simple Gaussian beam theory that are useful for assessing the effect of feed system parameters on the performance of the antenna.

We assume that the feed system produces a Gaussian beam and that the antenna f/D ratio is large enough to ensure the validity of the Gaussian beam and optics formulas. (This will generally not be a problem for any Cassegrain system.) The ABCD law lets us treat such an antenna as a system with a specified effective focal length f_E and focal point. We assume an unblocked aperture, which is strictly valid only for offset antennas but introduces relatively small effects for the blockage found in most radioastronomical systems.

The edge taper of the main reflector, defined as the ratio of the power density at its center compared with that at the edge, is a useful parameter for analyzing the tradeoffs between different aperture illuminations. The edge taper T_E and aperture beam radius are related by

$$w_A = (D_M/2)[20/(T_E(\text{dB}) \ln 10)]^{1/2}, \qquad (19)$$

where D_M is the main reflector diameter. In most cases, the subreflector is in the far field of the beam waist, and we can take $w_A = f_E\lambda/\pi w_{CASS}$, w_{CASS} being the Cassegrain beam-waist radius. Thus we find

$$w_{CASS} = 0.22[T_E(dB)]^{1/2}(f_E/D_M)\lambda. \tag{20}$$

The Cassegrain waist radius is thus proportional to the wavelength, the f/D ratio at the Cassegrain focus, and to the square root of the edge taper in decibels. The fraction of power spilling past the edge of the subreflector is given by

$$\mathscr{F}_{spill} = \exp[-2(D_M/2w_A)^2] = T_E^{-1}. \tag{21}$$

The radiation pattern of the aperture with a truncated Gaussian amplitude distribution will not be purely Gaussian but the computation can be done numerically [see references given in Section I; the paper by Chai and Wertz (1965) contains some useful mathematical suggestions for the computations]. The location and amplitude of the sidelobes depends in a complex fashion on the edge taper. For a 10-dB edge taper, the first side-lobe level is 24 dB down from the on-axis power; for a 15-dB edge taper, 30 dB down; for a 20-dB edge taper, 37 dB down. These values are for an unblocked aperture; the sidelobe levels for a centrally blocked antenna will be increased (Dragone and Hogg, 1974). A graph of the beam effi-ciency (fraction of total power radiated contained between the first minima of the pattern) and spillover as a function of edge taper is given in Fig. 7a. In Fig. 7b, we show the FWHM beamwidth as a function of edge taper. An edge taper of ~ 14 dB (a factor of 25) appears to be a reasonable compromise between beam efficiency and beamwidth; in this case, $w_{CASS} = 0.82(f_E/D_M)\lambda$ and the Cassegrain beam diameter at the waist is $4w_{CASS} \simeq 3(f_E/D_M)\lambda$.

As discussed in Section I, the truncation of a Gaussian beam broadens the size of the central lobe of the far-field radiation pattern. If we had an infinite-sized aperture with waist radius w_A, the beam radius growth angle would be given by Eq. (6), and the full width to one-half maximum beam-width is related to θ_{w_0} by

$$\theta_{FWHM} = 1.18\theta_{w_0}. \tag{22}$$

If we assume for the moment that the edge taper and antenna diameter serve only to define together the beam-waist radius through Eq. (19), we can calculate for an "infinite" antenna with edge taper T_E that

$$\theta^*_{FWHM} = 0.80[T_E(dB)]^{1/2}(\pi D_M/\lambda)^{-1}. \tag{23}$$

These values are compared with those obtained for truncated Gaussian illumination (Fig. 7b) in Table I. We see that for $T_E < 25$ dB the actual truncated beam is significantly broader than that obtained using an infinite antenna.

The axial focusing curve for an antenna illuminated by a Gaussian beam-waist can be calculated using Eq. (14b) for the waist size produced by an optical system. The telescope system produces a second waist of radius w_2 on the far side of the main reflector and the on-axis gain is proportional to w_2^2. In most cases, the change in gain owing to the change in spillover resulting from the displacement of the Cassegrain waist is small

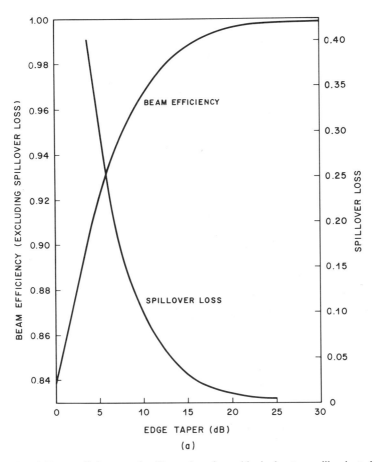

FIG. 7 (a) Beam efficiency and spillover loss for unblocked antenna illuminated by a Gaussian pattern as a function of the edge taper (power density at the center divided by power density at edge).

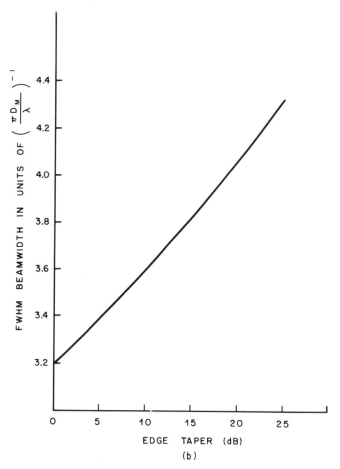

FIG. 7 (*Continued*)   (b) Beamwidth of power pattern (full width to one-half maximum) as a function of edge taper.

compared with the variation produced by the change in w_2. Then, we determine that

$$G/G_{\max} = [1 + (\lambda/\pi w^2_{\text{CASS}})^2(\delta)^2]^{-1}, \tag{24}$$

where δ is the displacement of w_{CASS} from the nominal focal position at the geometrical focus of the Cassegrain system. From this relationship, one finds that for a 1-dB decrease in gain, δ is given by

$$\delta(-1 \text{ dB}) = 0.075 T_{\text{E}}(\text{dB})(f_{\text{E}}/D_{\text{M}})^2\lambda, \tag{25}$$

which, for a typical edge taper of 14 dB, is very close to $(f_{\text{E}}/D_{\text{M}})^2\lambda$.

TABLE I

EFFECT OF TRUNCATION ON BEAMWIDTH

Edge taper (dB)	D_M/w_A	θ_{FWHM}(actual)/θ_{FWHM}(untruncated)
5	1.52	1.89
10	2.15	1.42
15	2.63	1.24
20	3.03	1.13
25	3.40	1.08
30	3.72	1.05

The ratio of the beam radius at the antenna to the antenna beam waist w_2 [located approximately at a distance f_E on the far side of the antenna, as indicated by Eq. (14)] is given by $w_A/w_2 \simeq 1 + 2[(f_E/D_M)(\lambda/D_M)]^2$, which is close to unity for large antennas of any reasonable focal ratio. Thus the beam from the telescope remains essentially parallel for a distance $\simeq z_c \simeq \pi w_A^2/\lambda$ and then diverges, becoming a normal spherical wave for $z \gg z_c$.

The simple Gaussian optics theory dealt with here does not provide information on lateral defocusing, cross-polarization, and other higher order effects. These must be analyzed using more rigorous electromagnetic calculations, as described in several of the articles in the collection of Love (1978).

IV. Quasi-Optical Frequency-Selective Devices

One of the areas in which quasi-optical techniques play an important role is that of frequency-selective devices. The most common form of receiver at millimeter wavelengths and, to an increasing extent, at submillimeter wavelengths is the frequency downconverter. At these high frequencies, these devices, generally based on Schottky diode mixers, require a system for combining the local oscillator and signal that normally must also provide filtering action for the former. For high calibration accuracy and for optimum system tuning in spectroscopic use, a single-sideband filter is highly advantageous. A variety of quasi-optical devices have been developed to perform these functions, and these will be analyzed and compared in this section. As the majority of quasi-optical frequency-selective devices utilize single- or multibeam interferometers, the discussion will be introduced by an analysis of the coupling between Gaussian beams.

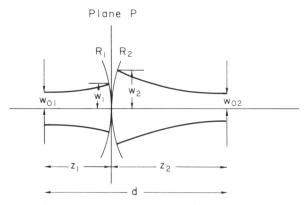

FIG. 8 Coupling between two Gaussian beams with waist radii w_{01} and w_{02} and radii of curvature R_1 and R_2 at distances z_1 and z_2 from their respective waists.

A. GAUSSIAN BEAM-COUPLING FORMULAS

Consider two fundamental-mode Gaussian beams of infinite transverse extent with different waist sizes and locations, as shown in Fig. 8. The power coupling coefficient K is defined as the fraction of the power incident in the one beam that would be extracted in the second beam (Kogelnik, 1964). The coupling is derived from a two-dimensional integral over the electric field distribution of the two beams[6]:

$$C_{12} \equiv \langle E_1 \mid E_2 \rangle = \int_0^\infty \int_0^{2\pi} d\phi \, r \, dr \, \psi_1 \psi_2^*, \qquad (26)$$

and $K_{12} = |C_{12}|^2$. If the integral is carried out at reference plane P, we find

$$C_{12} = \frac{w_1 w_2}{2} \left[\frac{1}{w_1^2} + \frac{1}{w_2^2} + \frac{i\pi}{\lambda} \left(\frac{1}{R_1} - \frac{1}{R_2} \right) \right]^{-1}, \qquad (27)$$

where the parameters all refer to those at plane P.[7] Note that R_2, as shown in Fig. 8, is negative by the sign convention. It is generally more convenient to express the coupling coefficients in terms of the parameters at their beam waists. Then, the power coupling coefficient is given by

$$K_{12} = 4 \bigg/ \left[\left(\frac{w_{01}}{w_{02}} + \frac{w_{02}}{w_{01}} \right)^2 + \left(\frac{\lambda}{\pi w_{01} w_{02}} \right)^2 d^2 \right], \qquad (28)$$

where d is the separation of the waists.

[6] For the fundamental mode, this integral is the same whether expressed in cylindrical or in rectangular coordinates.

[7] The definition of q used by Kogelnik in his discussion of coupling formulas differs from that employed in most other work and we have thus not used it here.

In terms of Eq. (27), one can see that for perfect coupling one must have equal beam radii and radii of curvature in the reference plane. From Eq. (28) (which also shows that the coupling efficiency is independent of the choice of reference plane), we see that this is equivalent to saying that the beams must have the same waist radius and location. A fundamental Gaussian mode can couple to higher order modes of a beam with a different waist radius; this is discussed in detail by Kogelnik (1964), but we consider power not coupled to the fundamental mode as "lost."

A result of particular importance for the analysis of interferometers is the coupling of a fundamental Gaussian beam to the same beam, except that it has a displaced displaced waist location. In this case, we define ψ as the beam having waist radius w_0, and ψ_Δ as the beam with waist displaced by a distance Δ; then

$$\langle \psi \mid \psi_\Delta \rangle = [1 + (i\lambda\Delta/2\pi w_0^2)]^{-1}. \tag{29}$$

B. DUAL-BEAM INTERFEROMETERS

In this section we shall discuss the operation of and some design criteria for dual-beam interferometers. These devices have found use as LO injection systems for wavelengths between 3 mm and 120 μm and have also been employed as single-sideband filters. In what follows, we shall analyze the dual-beam amplitude-division interferometer in terms of diffractionless and realistic operating conditions. The limitations imposed by the growth of the beam radius within the interferometers will also be discussed. A variant of this type of device, which uses the principle of polarization rotation, will then be described.

1. Analysis of the Amplitude-Division Dual-Beam Interferometer

We shall first discuss the operation of the amplitude-division dual-beam interferometer neglecting diffraction and shall then determine the effect of diffraction on the fundamental-mode system transmission, assuming excitation by a single fundamental Gaussian mode.

Considering the system shown in Fig. 9, we assume the two beam splitters to be identical and let the complex electric field reflection coefficient of each be r and the transmission coefficient be t. Because of the symmetry (under interchange of input and output ports) of almost any quasi-optical beam splitter, we know that the phase angles of r and t differ by 90° (Levy, 1966), assuming that the beam splitter is lossless, which is a good approximation. We then also have $|t|^2 + |r|^2 = 1$. The amplitude of the wave transmitted from port 1 to port 3, assuming unit incident electric field amplitude and neglecting diffraction, is

Port 1

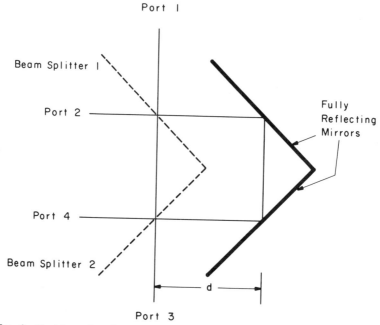

Beam Splitter 1

Port 2

Fully
Reflecting
Mirrors

Port 4

Beam Splitter 2

|← d →|

Port 3

FIG. 9   Dual-beam interferometer employing two beam splitters (1 and 2) and two fully reflecting mirrors. This design is described by Payne and Wordeman (1978). The path difference between the two paths through the device is equal to $2d$.

$$A_{1\to3} = |t|^2 - |r|^2 \exp(2\pi i\Delta/\lambda), \qquad (30)$$

where Δ is the path-length difference and we have omitted any overall phase factor. The power transmission given by $A_{1\to3}A_{1\to3}^*$ is found to be

$$P_{1\to3} = 1 - 2R(1 - R)[1 + \cos(2\pi\Delta/\lambda)], \qquad (31a)$$

where we have defined $R = |r|^2$. Power from port 1 not exiting via port 3 does so through port 4. Similarly, we find

$$P_{2\to3} = 2R(1 - R)[1 + \cos(2\pi\Delta/\lambda)]. \qquad (31b)$$

It can be seen that $P_{1\to3}$ is equal to unity irrespective of R for $\lambda = \Delta/(M - \frac{1}{2})$ and $P_{1\to3} = (1 - 2R)^2$ $(=0$ for $R = 0.5)$ for $\lambda = \Delta/(M - 1)$, where M is $1, 2, 3, \ldots$; $P_{2\to3}$ is equal to zero for $\lambda = \Delta/(M - \frac{1}{2})$ (again irrespective of R), whereas $P_{2\to3} = 4R(1 - R)$ for $\lambda = \Delta/(M - \frac{1}{2})$ $(=1$ for $R = 0.5)$. Considering these results, we conclude that $1 \to 3$ is a good path for the signal because losses will be low at the appropriate resonances. For diplexing, we choose

$$\Delta_{DIP} \simeq (2K - 1)(\lambda_{IF}/2), \qquad K = 1, 2, 3, \ldots. \qquad (32a)$$

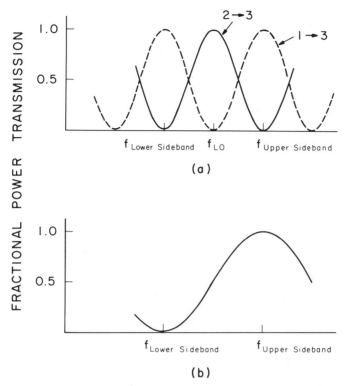

FIG. 10 Response of dual-beam interferometer when tuned for local oscillator injection (a) and as a single-sideband filter (b).

The resulting transmission function is shown schematically in Fig. 10a. We see that both sidebands are coupled to port 3. Choosing K equal to 1 gives the largest bandwidth and makes satisfying Eq. (32a) (approximately) relatively easier, while maintaining the resonance condition. For the systems being considered, $\lambda/\lambda_{IF} \simeq 0.05$, so that this is not a significant problem. With $\lambda_{IF} \simeq 6$ cm, the resulting small value of Δ makes the configuration of Fig. 11 or of Erickson (1977) preferable to the more straightforward configuration shown in Fig. 9 and described by Payne and Wordeman (1978). The latter design is essentially equivalent, however, when low intermediate frequencies are employed, necessitating a large value of Δ.

If the mixer is attached to port 3, power is fed to it at the signal and image frequencies from ports 1 and 2 according to Eqs. (31a) and (31b). For single-sideband filtering, we halve the path-length difference used for

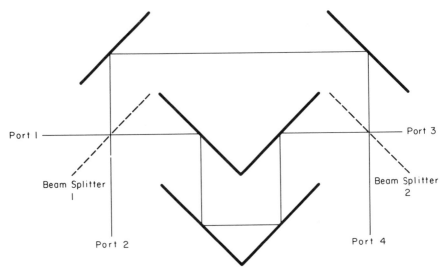

FIG. 11   Dual-beam interferometer built at FCRAO. In this version, the path-length difference can be reduced to near zero (as shown above), permitting double-sideband operation, if desired, in addition to normal filtering functions.

diplexing and obtain the response shown in Fig. 10b. In this case we require

$$\Delta_{\text{SSB}} = (2K - 1)(\lambda_{\text{IF}}/4), \qquad K = 1, 2, 3, \ldots . \qquad (32b)$$

We see from Eqs. (31a) and (31b) that to obtain high efficiency for LO injection in the diplexer ($P_{2\rightarrow3,\,\text{max}} \approx 1$) and high image rejection for the SSB filter ($P_{1\rightarrow3,\,\text{min}} \approx 0$) we require that $R \approx 0.5$. In Table II we tabulate the values of the reflectivity-dependent parameters as a function of the beam-splitter reflectivity. As mentioned before, $P_{1\rightarrow3,\,\text{max}} = 1$ and $P_{2\rightarrow3,\,\text{min}} = 0$ irrespective of R. In the absence of diffraction, then, the signal-transmission, noise-rejection, and image-suppression performances can all be made close to perfect with the proper choice of beam splitter. Metallic mesh (discussed in Section IV.E) can be made to have $R = 0.5$ but, in general, will be relatively frequency-sensitive. Dielectric beam splitters can also be used; these will be discussed in Appendix II.

2. Diffraction Loss

Let us assume that we have a fundamental-mode Gaussian beam of infinite transverse extent denoted by ψ_0. We assume that the coupling has been optimized either without the interferometer or with it present but set

TABLE II

EFFECT OF BEAM-SPLITTER REFLECTION
ON DIPLEXER/FILTER PERFORMANCE

R	$P_{1 \to 3, \, min}$	$P_{2 \to 3, \, max}$
0.1, 0.9	0.64	0.36
0.2, 0.8	0.36	0.64
0.3, 0.7	0.16	0.84
0.4, 0.6	0.04	0.96
0.45, 0.55	0.01	0.99
0.50	0.00	1.00

to zero path-length difference if it adds a fixed path length (the design of Fig. 11 does, whereas that of Fig. 9 does not). When the interferometer is in place and set to a nonzero path-length difference Δ, the fraction of power collected in the original mode is

$$K_{t0} = |\langle \psi_0 | \psi \rangle|^2, \tag{33}$$

where ψ denotes the total electric field distribution after the two beams have been recombined. Considering the path from port 1 to port 3 (in Fig. 9) and recalling the definition of beam-splitter characteristics previously discussed, we have

$$\psi_{1 \to 3} = |t|^2 \psi_0 - |r|^2 \exp(2\pi i \Delta / \lambda) \psi_\Delta, \tag{34}$$

where ψ_Δ denotes the delayed beam (see Section A) and we have again omitted any overall phase factors. We then find

$$K_{t0} = \left| |t|^2 - |r|^2 \exp(2\pi i \Delta / \lambda) \langle \psi_0 | \psi_\Delta \rangle \right|^2. \tag{35}$$

The coupling coefficient between the original and delayed beams has been discussed previously, and insertion of Eq. (29) into (35) yields

$$K_{t0, \, 1 \to 3} = (1 - R)^2 + \frac{R^2}{1 + \alpha^2} - 2R(1 - R) \frac{\cos \theta + \alpha \sin \theta}{1 + \alpha^2}, \tag{36a}$$

where $\alpha = \lambda \Delta / 2\pi w_0^2$ and $\theta = 2\pi \Delta / \lambda$. Note that for $\alpha \to 0$ (no diffraction) the result reduces to Eq. (31a). If we use the same values of θ for the extrema of K_{t0} as for the diffractionless case, we obtain

$$K_{t0, \, 1 \to 3, \, max, \, min} = (1 - R)^2 + \frac{R^2 \pm 2R(1 - R)}{1 + \alpha^2} \tag{36b}$$

where the $+$ and $-$ signs give the maximum and minimum values, respectively. However, from the form of Eq. (36a), we see that these are not the optimum values of θ when there is diffraction; solving for the new extrema of K yields

$$K'_{t0,\,1\to3,\,\text{min, max}} = \left[1 - R\left(1 \mp \frac{1}{(1 + \alpha^2)^{1/2}}\right)\right]^2. \qquad (36c)$$

For the case $R = 0.5$ and $\alpha \ll 1$, the "retuned" loss at the transmission is two-thirds as large as the "unretuned" loss. At the minima, the loss (in decibels) is twice as large for the "retuned" case as for the "unretuned" case. The shift between the diffractionless and actual extremal resonator tuning is given by $\delta\theta \simeq \alpha$ for $\alpha \ll 1$, and hence the retuning represents a small change for a low-loss device.

In the situation where $R = 0.5$, we obtain

$$K'_{t0,\,1\to3,\,\text{min, max}} = \frac{1}{4}\left(1 + \frac{1}{(1 + \alpha^2)^{1/2}}\right)^2, \qquad (36d)$$

and assuming $\alpha \ll 1$, the resulting fractional power loss is just equal to $\alpha^2/2$. An upper limit of 2% on the power loss requires

$$w_0 \geq (5\,\Delta\lambda/2\pi)^{1/2}. \qquad (37)$$

Diplexing with an intermediate frequency of 5 GHz ($\Delta = 3$ cm) at a frequency of 100 GHz requires, then, $w_0 \geq 0.85$ cm. The minimum size for the waist radius is smaller for single-sideband filtering.

These calculations are conservative in the sense that, first, not all of the power lost from the fundamental mode is lost to the $1 \to 3$ path, but some will emerge at a higher mode at the desired port of the interferometer; this may or may not be ultimately useful. Second, one can reoptimize the optics system to allow for the presence of two beams with different path lengths; by designing for a compromise between the two path lengths, the coupling loss can be significantly reduced. This analysis has been carried out by Brewer (private communication, 1981). In practice, the beam growth within the diplexer often sets larger lower limits to the waist size than does the coupling loss calculation we have carried out here.

3. Restrictions on Beam Diameter Due to Beam Growth within the Interferometer

It is necessary to avoid significant beam truncation within the interferometer in order to ensure low loss as well as the validity of the formulas given previously. If we specify that the available diameter for the beam should be equal to βw_{max}, where w_{max} is the maximum beam radius within the device, and that the interferometer extends to a distance $\varepsilon[\beta w_{\text{max}}]$

from the beam waist, then we obtain the following relationship between w_{max} and w_0:

$$w_{max} = w_0/[1 - (\lambda\varepsilon\beta/\pi w_0)^2]^{1/2} \qquad (38)$$

For any λ, ε, β, and w_0, this then gives the maximum beam diameter βw_{max} and hence the size of the interferometer. We see that there is a minimum waist radius ($= \varepsilon\beta\lambda/\pi$), which makes the device physically realizable, but that increasing w_0 to a value equal to $\sqrt{2}$ times this, with a beam-waist radius

$$w_0^* = \sqrt{2}\,\varepsilon\beta\lambda/\pi \qquad (39a)$$

and a beam diameter

$$D^* = 2\varepsilon\beta^2\lambda/\pi, \qquad (39b)$$

results in the minimum interferometer size. In this situation, $w_{max} = w_0\sqrt{2}$. Of course, the value of w_0 must be examined to consider the diffraction loss implied. In all cases considered to date, w_0^* is much larger than any reasonable criterion in terms of diffraction loss would dictate. As discussed by Erickson (1977), to obtain the smallest distance for a given waist radius, one puts the waist in the center of the interferometer. Neglecting the path-length difference, the interferometer shown in Fig. 11 has $\alpha = 3$, and taking $\beta = 4$ as a reasonable value for ensuring low spillover, we obtain $D^* = 30.5\lambda$. The interferometer shown in Fig. 9 has $\alpha = 2$. However, it has a minimum path-length difference of $2\beta w_{max}$, which is a severe handicap for a single-sideband filter as one cannot approach zero path-length difference. The tradeoffs among the different interferometer configurations must be analyzed for the particular function, frequencies, and bandwidths required, together with any size restrictions, in order to choose the optimum device.

4. Results Achieved in Practice

The signal insertion loss of the dual-beam amplitude-division interferometer depends on $\alpha = \Delta\lambda/2\pi w_0^2$, assuming that the beam truncation and ohmic or dielectric losses are neglible. Results in the 1–3-mm-wavelength region reported by Erickson (1977) and Payne and Wordeman (1978) indicate that losses <0.1 dB are attainable. At much shorter wavelengths (~400 μm), the amplitude-division interferometer discussed by Fetterman et al. (1978) (identical to the configuration shown in Fig. 9 and employing Mylar beam splitters) has an insertion loss <0.5 dB, with this fairly large upper limit being set by non-Gaussian components in the submillimeter laser's output radiation pattern and measurement difficulties. A single-sideband filter using quartz beam splitters constructed at the

FCRAO using the design in Fig. 11 has shown low signal insertion loss (<0.25 dB at 100 GHz) and image rejection greater than 17 dB over a 2.25-GHz bandwidth with a path-length difference of 1.5 cm.

5. *Polarization-Rotating Dual-Beam Interferometer*

The polarization-rotating dual-beam interferometer is a variation of the type of interferometer discussed previously, but several distinctive features make a separate discussion worthwhile. The principle of operation as an interferometer is discussed by Martin and Puplett (1969), Lambert and Richards (1978), and Burton and Akimoto (1980). The basic elements are shown in Fig. 12. If we consider a wave horizontally polarized on the 45° grid, it will be divided equally between the transmitted and reflected beams. Each of these travels a certain distance before being reflected by a roof mirror. Because a mirror of this type reflects the direction of the polarization vector with respect to the line of intersection of its plane surfaces, the beam initially reflected by the 45° grid will be returned with its polarization rotated 90° and hence will be transmitted by the grid, whereas the beam first transmitted will subsequently be reflected. The two beams thus propagate co-axially toward the output. Let the unit vectors in the horizontal and vertical directions be represented by $\hat{\mathbf{H}}$ and $\hat{\mathbf{V}}$, respectively, and the vectors in the directions transmitted and reflected by the 45° be $\hat{\mathbf{T}}$ and $\hat{\mathbf{R}}$. Because $\hat{\mathbf{H}} = (\hat{\mathbf{R}} - \hat{\mathbf{T}})/\sqrt{2}$ and $\hat{\mathbf{V}} = (\hat{\mathbf{R}} + \hat{\mathbf{T}})/\sqrt{2}$, we see that the electric field at the output of the interferometer for unit horizontal input amplitude is given by

$$\mathbf{E}_{\text{out}} = (1/\sqrt{2})[\hat{\mathbf{R}} \exp(2\pi i d_{\text{R}}/\lambda) - \hat{\mathbf{T}} \exp(2\pi i d_{\text{T}}/\lambda)], \qquad (40a)$$

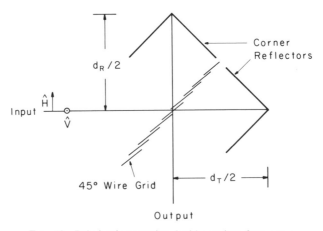

FIG. 12   Polarization-rotating dual-beam interferometer.

where d_R and d_T are, respectively, the distances traveled by the beams initially reflected and transmitted by the 45° grid between encounters with the grid. Again, taking Δ as the path-length difference ($= d_T - d_R$), we obtain

$$\mathbf{E}_{out} = \tfrac{1}{2}\{\hat{\mathbf{H}}[1 + \exp(2\pi i\Delta/\lambda)] + \hat{\mathbf{V}}(1 - \exp(2\pi i\Delta/\lambda)]\}, \quad (40b)$$

where we have neglected an overall phase factor. We see that the polarization state of the output beam depends on Δ, being horizontally polarized for $\Delta = 0$, vertically polarized for $\Delta = \lambda/2$, and circularly polarized for $\Delta = \lambda/4, 3\lambda/4$. The output intensity is always unity because we are neglecting any ohmic losses. This device thus may be useful as a polarization analyzer.

If we place a horizontal polarizer at the output, the power transmission function is given by

$$P_{\hat{\mathbf{H}}\to\hat{\mathbf{H}}} = \tfrac{1}{2}[1 + \cos(2\pi\Delta/\lambda)], \quad (41a)$$

whereas a vertical polarizer at the output yields

$$P_{\hat{\mathbf{H}}\to\hat{\mathbf{V}}} = \tfrac{1}{2}[1 - \cos(2\pi\Delta/\lambda)]. \quad (41b)$$

We see that the $\hat{\mathbf{H}} \to \hat{\mathbf{H}}$ path has a transmission function identical to that of path $2 \to 3$ of the amplitude-division interferometer for the case $R = 0.5$, and the $\hat{\mathbf{H}} \to \hat{\mathbf{V}}$ path is equivalent to path $1 \to 3$. If the grid is not tilted at 45°, the analysis is slightly more involved, but one finds that the efficiency of the device is slightly reduced; this effect is discussed in detail by Lambert and Richards (1978). An arrangement for using the interferometer as a diplexer is shown in Fig. 13, where the addition of the two polarizing grids makes four ports available. The use of the polarization-rotating interferometer as a local oscillator injector has been described by Wrixon and Kelly (1978) and Erickson (1980). In this application, the output polarizing grid can be omitted if the mixer structure is sensitive to a single polarization. The path-length difference criterion is the same for the amplitude-division interferometer in this function: $\Delta \simeq (2K - 1)\lambda_{IF}/2$ [Eq. (32a)].

Again, for use as a single-sideband filter, we require $\Delta \simeq (2K - 1)\lambda_{IF}/4$ [Eq. (32b)]. Then a mixer accepting horizontal polarization at output port $\hat{\mathbf{H}}$ can be coupled to input port $\hat{\mathbf{H}}$ at the signal frequency and to input port $\hat{\mathbf{V}}$ at the frequency of the image. In this application, it may be advantageous to retain the output polarizing grid and terminate output port $\hat{\mathbf{V}}$ with an absorbing load. This will have the effect of suppressing possible resonances between the optics or antenna structure and the mixer because energy at the image frequency incident from input port $\hat{\mathbf{H}}$ will be ro-

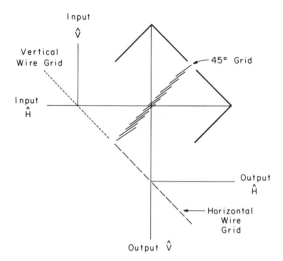

FIG. 13   Polarization-rotating dual-beam interferometer in a configuration suitable for use as a diplexer. The assignment of the input and output ports is arbitrary; the vertical and horizontal wire grids could be interchanged or have the same orientation.

tated in polarization and would otherwise be relfected by the mixer after reaching the output.

6. *Instantaneous Bandwidth of Dual-Beam Interferometer*

If we restrict operation to $K = 1$, which yields the lowest diffraction loss and greatest bandwidth, then we cannot exactly satisfy the resonance requirement for the signal and the antiresonance requirement for the image simultaneously for a single-sideband filter. Similarly, for a local oscillator injector, we cannot achieve optimum double-sideband operation and noise rejection. These effects are important primarily for relatively high intermediate frequencies, and we thus should examine the question of the device bandwidth to see what loss or local oscillator noise leakage will be incurred. The symmetry of the transmission function for the amplitude-division interferometer with $R = 0.5$ or for the polarization-rotating version lets us derive an approximate expression for the bandwidth for loss less than L (near maxima) or transmission less than L (near minima) given by

$$\delta\nu = (\tfrac{1}{2})(8/\pi)L^{1/2}[\nu_{IF}/(2K - 1)], \qquad (42)$$

which is valid for $\delta\nu/\nu_{IF} < 1$. The factor of $\tfrac{1}{2}$ applies to the local oscillator injection tuning. Results quoted by Erickson (1977) and Payne and

Wordeman (1978) agree with this result for LO injection. For $L = 0.02$, we find $\delta\nu = 1.75$ GHz for single-sideband filter tuning, which agrees fairly well with the FCRAO filter described previously. If one is tuned so that the signal frequency ν_s is resonant in Mth order giving a particular value of Δ and one changes to the next highest order, then the shift in frequency of the adjacent transmission minima is approximately $\nu_s/2M^2 = 2$ GHz for $\nu_s = 100$ GHz and $M = 5$. Thus, for a SSB filter, there does not appear to be a significant tuning problem even for relatively low M.

C. MULTIPLE-BEAM INTERFEROMETERS

Multiple-beam interferometers, based on repeated reflections between two partially reflecting surfaces, are widely used at optical and infrared wavelengths and are generally referred to as Fabry–Perot interferometers. These devices generally employ a pair of plane-parallel mirrors. A number of variations have been developed for use at millimeter and submillimeter wavelengths where they have been used for local oscillator injection and single-sideband filtering. Some of these employ internal focusing optics to reduce the effect of diffraction and are often referred to as resonators, although the distinction in terminology is by no means definite. In the discussion that follows, we shall first give some results for the diffractionless case and then analyze the results for plane-parallel interferometers when diffraction is included. Finally, we shall give a brief discussion of resonators. The partially reflecting mirrors for Fabry–Perot interferometers at millimeter and submillimeter wavelengths are generally made of metallic mesh, which will be discussed in Section E.

1. *Response in Absence of Diffraction*

We shall consider here only interferometers consisting of two identical partially reflecting plane mirrors. Discussions of multiple-mirror systems, including surfaces with different reflectivites, can be found in papers by Ulrich (1967), Saleh (1974), Garg and Pradhan (1978), and Chen (1979). We shall also assume that there are no losses in the reflecting surfaces or in the intervening medium. If each mirror has a fractional power reflection R (and fractional power transmission $T = 1 - R$), the interferometer transmission function is given by

$$|\tau|^2 = \{1 + [4R/(1 - R)^2] \sin^2(\delta\phi/2)\}^{-1} \tag{43a}$$

and the reflection function by

$$|\mathcal{R}|^2 = \{[4R/(1 - R)^2] \sin^2(\delta\phi/2)\}/\{1 + [4R/(1 - R)^2] \sin^2(\delta\phi/2)\} \tag{43b}$$

where $\delta\phi$ is the round-trip phase delay.[8] If the partially reflecting surfaces are separated by a distance d, if they are tilted from normal incidence with respect to the incident beam by angle θ_i, and if the material between them has index of refraction n, then we obtain

$$\delta\phi = (4\pi d/\lambda)(n^2 - \sin^2 \theta_i)^{1/2}, \tag{43c}$$

where λ is the free-space wavelength. The above formula omits any phase shift due to the reflections. This will be dependent upon the nature of the mirrors, and is most important for wire grids (cf. Ulrich *et al.*, 1963; Ulrich, 1967). If we assume that the phase shift is frequency-independent, then for normal incidence the spacing between the transmission maxima (called the free spectral range) is equal to $c/2nd$. For the lossless Fabry–Perot, $|\mathscr{R}|^2 + |\tau|^2 = 1$, and the maximum and minimum values of $|\mathscr{R}|^2$ are $4R/(1 + R)^2$ and 0, respectively. If loss is included, the reflection coefficent must be calculated separately and the absorption of the interferometer is given by $1 - |\mathscr{R}|^2 - |\tau|^2$.

2. Diffraction and Walkoff Effects in Multiple-Beam Interferometers

The simplest type of Fabry–Perot interferometer, as previously discussed, consists of two identical plane-parallel mirrors. Because, in practice, we do not have an incident beam of infinite transverse extent, the radiation on successive bounces will have a different beam radius and hence, as shown in Fig. 14, the various beams will not be able to interfere perfectly as assumed in deriving Eq. (43). The output of the interferometer will thus consist of a sum of Gaussian beams with the same beam-waist radius but having traveled different distances from the waist. Assuming no beam truncation, the total electric field transmitted by the device can be written, utilizing the notation of Section B, as

$$\psi = t^2 \sum_{s=0}^{\infty} \rho^{2s}\psi_s, \tag{44}$$

where ρ and t are the (complex) electric field reflection and transmission coefficients, respectively, and ψ_s represents the Gaussian beam after s

---

[8] If each mirror has a fractional power absorption A ($A + R + T = 1$, where T is the fractional power transmission) and the one-way power transmission through the intervening medium is ζ, then

$$|\tau|^2 = \{\zeta[(1 - R - A)/(1 - \zeta R)]^2\}/\{1 + [4\zeta R/(1 - \zeta R)^2] \sin^2(\delta\phi/2)\}. \tag{42c}$$

Fabry–Perot resonators with loss are discussed by Ulrich *et al.* (1963) and Garnham (1969).

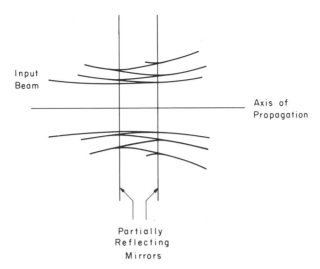

Input
Beam

Axis of
Propagation

Partially
Reflecting
Mirrors

FIG. 14   Diffraction and growth of a beam within a Fabry–Perot interferometer. The reflected and transmitted outputs are seen to be the sum of beams having traveled different distances from the waist.

round trips through the interferometer. The fraction of power transmitted in the fundamental mode is given by Eq. (33) as $K_{t0} = \langle \psi_0|\psi \rangle^2$, where

$$\langle \psi_0|\psi \rangle = t^2 \sum_{s=0}^{\infty} \rho^{2s} \langle \psi_0|\psi_s \rangle. \tag{45}$$

If we are at normal incidence, the coupling factor between the original beam and that after s round trips is obtained from Eq. (29)[9] as

$$\langle \psi_0|\psi_s \rangle_{\text{NI}} = e^{is\delta\phi}/(1 + \tfrac{1}{2}is\,TD), \tag{46}$$

with

$$\delta\phi = (4\pi nd/\lambda) + 2\sphericalangle\rho \quad \text{and} \quad D = 2\lambda d/\pi w_0^2 nT,$$

where $\delta\phi$, as before, is the phase delay per round trip, d the mirror spacing, and n the index of refraction between the reflectors. Substitution of Eq. (46) into (45) unfortunately does not lead to an analytic expression for the transmission as for the dual-beam interferometer. Again, as discussed by Arnaud *et al.* (1974), the diffraction of the beam shifts the fre-

---

[9] The discussion of diffraction loss follows that of Arnaud *et al.* (1974), but using the electric field rather than power-density distribution. The quantity TD is the difference in the distances propagated by the two beams, measured in units of the confocal distance in whatever medium the radiation is propagating. Because $z_c = \pi w_0^2/\lambda_{\text{eff}}$, we see that it is larger in a medium with $n > 1$, and thus D is smaller.

quencies of maximum transmission from those expected from plane-wave illumination. The computations are best carried out numerically. Curves are given by Arnaud *et al.* for total power transmission as a function of D; certain of these can be modified to be applicable to fundamental-mode transmission because, as these authors discuss, for $T \ll 1$, the loss in the fundamental mode (in decibels) will be twice as large as the total power-transmission loss. In the limit of small loss, we find that the fractional power transmitted at resonance is given by the expression

$$K_{t0} \simeq 1 - D^2. \tag{47}$$

In the case of nonnormal incidence, we find that $\delta\phi$ is given by (43c) (with the addition of the round-trip reflection phase shift) and

$$D = 2\lambda d(1 - \sin^2 \theta_i)/\pi w_0^2 T(n^2 - \sin^2 \theta_i)^{1/2}, \tag{48}$$

where θ_i is the angle from normal incidence. In addition, the overlap between the beams is complicated by the fact that there is a lateral displacement after each successive transit. This effect is present even with finite-sized "plane-wave" illumination and, in this case, is called *geometrical walkoff*. Again, curves for the power transmission are given by Arnaud *et al.* (1974). For $T \ll 1$ and low loss, the expressions for walkoff and diffraction loss can be combined yielding

$$K_{t0} \simeq 1 - (G^2 + D^2), \tag{49}$$

where

$$G = 2\sqrt{2}d \sin \theta_i \cos \theta_i/w_0 T(n^2 - \sin^2 \theta_i)^{1/2}. \tag{50}$$

Note that for interferometers filled with material $n \gg 1$, both G and D are proportional to n^{-1} for fixed d; but because d is proportional to n^{-1} for operation in the same order (same $\delta\phi$), the loss in this case will be reduced by a factor of approximately n^4. Thus dielectric-filled Fabry–Perot interferometers can have substantially lower loss than their air-filled counterparts, as discussed in detail by Goldsmith (1982).

3. Configurations for Multiple-Beam Interferometers

Fabry–Perot interferometers can operate as local oscillator injectors, as single-sideband filters, and as devices that can perform both functions simultaneously. Various designs have been developed with these uses in mind as well as the minimization of the effects of diffraction and walkoff as previously discussed. We shall describe a number of devices that have proven to be of practical interest and also mention some possibilities for designs that offer additional attractive features.

The form of the Fabry–Perot response function [Eq. (43a)] in the absence of diffraction sets basic limits on its performance as a frequency-selective device. For example, if used in transmission as a single-sideband filter, the maximum image discrimination is given by the minimum transmission $[(1 - R)/(1 + R)]^2$ and will set a minimum value required for the mirror reflectivity R. The finesse of the Fabry–Perot is defined by

$$F = \pi\sqrt{R}/(1 - R), \tag{51}$$

and the fractional bandwidth $(\Delta\lambda_{FWHM}/\lambda)$ when the interferometer is operated in order q $(\delta\phi/2 \simeq q\pi)$ is given by $1/qF$ (Ulrich et al., 1963). Thus, the requirement of large bandwidth will impose a limitation on the maximum R that can be employed. The tradeoffs for a particular practical single-sideband filter application at $\nu \simeq 100$ GHz are discussed by Goldsmith (1977), where $R \simeq 0.75$ is found to be the optimum value.

The most straightforward use of the Fabry–Perot interferometer is as a single-sideband filter in transmission, which can be accomplished with plane-parallel mirrors close to normal incidence. An example of this application is described by Wannier et al. (1976). The interferometer mirrors consist of two-dimensional etched copper mesh whose spacing can be varied to transmit the signal frequency in the range 100–115 GHz, with the free spectral range approximately equal to $4\nu_{IF}$ to keep the frequency of minimum transmission coincident with that of the image. With a Cassegrain f/D ratio of 13.6, the waist radius (Section III) is approximately 3 cm for $\lambda = 0.27$ cm, and because $|\tau|_{min}^2 \simeq 0.1$, we can conclude that $R \simeq 0.5$ for the mirrors. With the spacing used of 0.8 cm, we find $D = 0.03$ [Eq. (46)], so that the diffraction loss will be given by (47) and should be negligible. These authors report a transmission loss of 0.4 dB (0.09), which is presumably due to ohmic losses and imperfections in the grids. This device configuration is obviously convenient for "add-on" applications and can achieve excellent performance if the waist radius is large enough to keep the diffraction loss small. A significant drawback is that the reflected image is terminated in a relatively ill defined manner, which will generally lead to an increase in the system temperature.

The use of the Fabry–Perot as a diplexer requires access to three separate ports of the system. Two types of designs that accomplish this while retaining freedom from geometrical walkoff have been developed. The folded Fabry–Perot described by Gustincic (1976, 1977a,b) is shown schematically in Fig. 15. The beam returns upon itself after a round trip through the interferometer and thus there is no walkoff despite the 45° inclination of the mirrors. This arrangement has been used for local oscillator injection by transmitting this power while reflecting the signal into the mixer. It could also be used, however, as a single-sideband transmission filter or as a sideband dropping filter with the desired frequency being

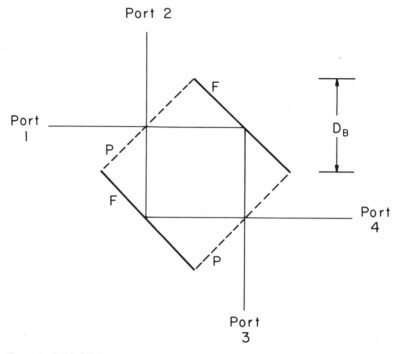

Fig. 15   Folded Fabry–Perot interferometer as described by Gustincic (1976, 1977a,b). The mirrors P are partially reflecting and those indicated as F are completely reflecting. The round-trip path delay is $4D_B$.

rejected. A disadvantage is that the minimum mirror spacing d is equal to $2D_B$, where D_B is the clear-beam diameter, which must be large enough to account for the growth in beam radius after successive passes through the interferometer, in order to avoid effects of truncation. This can make the diffraction loss unnecessarily large for high intermediate frequencies, which could otherwise advantageously use a smaller value of d. Gustincic (1976) reports a minimum local oscillator (transmitted) loss of 3 to 6 dB and a minimum transmission at least 15 dB below this value, although these numbers include the losses in the horns and lenses coupling to the interferometer.

A polarization-rotating Fabry–Perot has been described by Watanabe and Nakajima (1978) and is shown in Fig. 16. The vertically polarized input is converted to circular polarization by quarter-wave plate B, transmitted by the Fabry–Perot, and reconverted to vertical linear polarization by quarter-wave plate A since its fast axis is perpendicular to that of B. This beam is then transmitted by the horizontal grid to the output. The horizontally polarized input is initially reflected by the grid, converted to circular polarization, and then reflected by the Fabry–Perot. Owing to the reversal of the handedness of the circular polarization by the reflection, it is converted to vertical polarization on its second passage through quarter-wave plate A and is also transmitted through the grid to the output. Again, the transmission function for input with the electric field vertical is given by Eq. (43a) and the reflection function for by Eq. (43b). This configuration has the advantage of freedom in the choice of interferometer spacing but suffers from the loss and bandwidth limitations of the quarter-wave plates. Watanabe and Nakajima (1978) report an insertion loss of ≤ 1 dB for a device designed to have a flat-topped transmission response.

The simplicity of a single plane-parallel Fabry–Perot can be combined with the diplexing function if the interferometer is inclined from normal incidence. The disadvantage suffered is that of walkoff loss, which is found to exceed the diffraction loss for practical inclination angles. The general criteria for diplexing with this device configuration are discussed by Arnaud *et al.* (1974); they conclude that the minimum inclination angle is

$$\theta_{min} \simeq 2\sqrt{2}\lambda/\pi w_0 \tag{52}$$

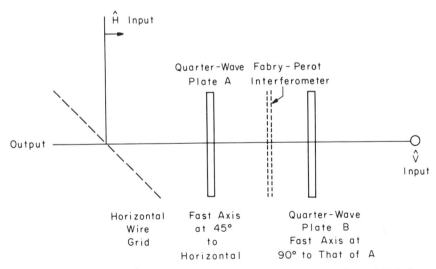

FIG. 16   Polarization-rotating Fabry–Perot interferometer (Watanabe and Nakajima, 1978). The two quarter-wave plates have their axis at $\pm 45°$ to the horizontal; the interferometer operates for circular polarization and hence must be capable of linear polarization-independent operation.

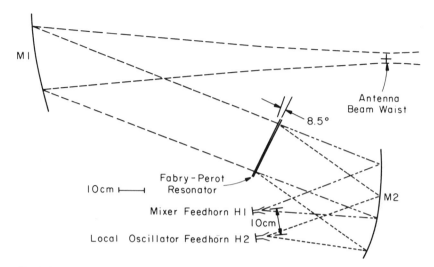

Fig. 17   Fabry–Perot interferometer operating at oblique incidence used as a combination LO injector and single-sideband filter (Goldsmith, 1977). The angle of the interferometer from normal incidence is 8.5° and the waist radius of the beam between M1 and M2 is 9.2 cm.

in order that the various beams be separable when their diffraction is taken into account. Using Eq. (49) and assuming $D \ll G$, we find that for an air-filled interferometer the minimum loss for small angles is given by

$$K_{t0,\,min} = 1 - (8\lambda d/\pi w_0^2 T)^2. \tag{53}$$

Since T is fixed by considerations of bandwidth and minimum transmission and d by the required periodicity of the response function, Eq. (53) implies that a relatively large waist radius is required for low loss. In a system designed for the $\lambda = 2$–4-mm range, Goldsmith (1977) utilized a waist radius of 9.2 cm and an inclination angle of 8.5°, which was set by constraints other than the separability condition [Eq. (52)]. The insertion loss of the Fabry–Perot was measured to be 0.25 dB, with the 1-dB bandwidth near maximum 780 MHz at 100 GHz and the minimum transmission 19 dB below the maximum.[10] The system shown in Fig. 17 functions simultaneously as a single-sideband filter and local oscillator injector by transmitting the signal while reflecting the local oscillator into the mixer feedhorn. The outstanding advantage is that a single adjustment allows tuning the system for single-sideband response as well as diplexing the os-

[10] The loss calculated from Eqs. (49b) and (50) is 0.13 dB, and the excess value measured is presumably due to ohmic loss in the grids.

cillator with low loss.[11] A drawback is that the local oscillator filtering is only ~3 dB because noise at the image frequency is reflected into the mixer along with the signal. This does not appear to present a problem with the relatively high (4.75 GHz) intermediate frequency used; this would also presumably be the case for a lower frequency source and a frequency multiplier used as the local oscillator.

D. RESONATORS

It has long been recognized that diffraction losses can be a significant problem for interferometers at microwave and millimeter wavelengths. The addition of focusing elements to reduce diffraction has been extensively studied, both theoretically and practically, as a result of the work on resonant cavities for optical lasers carried out in the early 1960s.[12] The great number of different designs that have been developed, together with the relatively limited application seen to date at millimeter wavelengths, preclude an extensive discussion here. A general analysis and review of resonators is given by Kogelnik and Li (1966) and Siegman (1971).

We shall describe here one device used as a single-sideband filter at 100 GHz in a Cassegrain system having a waist radius $w_0 \simeq 1$ cm. Since the IF is only 1.4 GHz, the diffraction loss for a plane-parallel interferometer would be approximately 3 dB. The filter developed by Goldsmith and Schlossberg (1980) is shown in Fig. 18a. It consists of a semiconfocal cavity[13] and dielectric film, which serves to couple power into and out of the resonant structure. The plane mirror and spherical mirror are separated by a distance $d \simeq$ half the radius of curvature ρ_M of the spherical mirror; this is called a *semiconfocal resonator* and (in the case where the diffraction loss is a small effect) has as its field configuration a Gaussian mode with waist radius (located at the plane mirror):

$$w_0^{\text{res}} = (\lambda \rho_M / 2\pi)^{1/2} = (\lambda d / \pi)^{1/2}. \tag{54a}$$

---

[11] The quoted loss of 2.7 dB is dominated by aberrations due to a relatively fast optical system, which is off-axis for the local oscillator beam. This is not an inherent problem for a Fabry–Perot used in this manner; the dips in the reflection function (43b) are sufficiently narrow that the local oscillator leakage through the interferometer should be negligible.

[12] In the article by Auston *et al.* (1964), a number of references to work on optical and microwave resonators and interferometers are given; two important early works on resonators are those by Fox and Li (1961) and Boyd and Gordon (1961).

[13] A confocal cavity is one having two (identical) spherical mirrors, each with its center of curvature located at the other mirror. The beam waist is thus located halfway between the two mirrors. As seen from Eq. (8), at a distance $z_c = \pi w_0^2 / \lambda$ from the waist, the radius of curvature $R(z_c) = 2z_c$ and hence if the mirror separation is equal to $2z_c$, which is also equal to the radii of curvature, then the wave front of the beam at the mirrors will have the same curvature as the reflectors and will be reflected in its original mode.

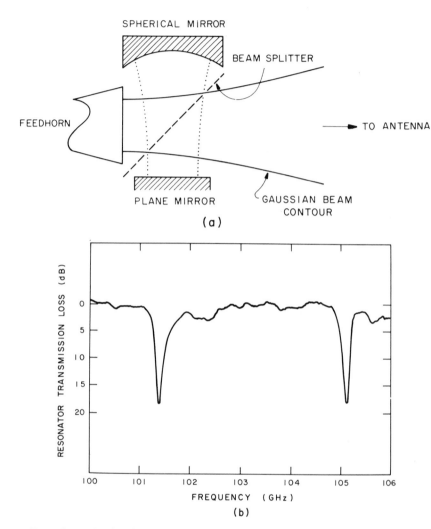

FIG. 18 (a) Semiconfocal resonator sideband dropping filter. (Goldsmith and Schloss-berg, 1980. Reprinted from *IEEE Transactions on Microwave Theory and Techniques* **28,** 1136–1139. © 1980 IEEE.) The beamsplitter is a Mylar film. (b) Transmission curve of the filter showing two resonances. The drop in the transmission at 102.5 GHz may be due to the excitation of higher order modes.

The resonant frequencies are determined by the requirement that the phase shift $\delta\phi$ between the mirrors be equal to $(q + 1)\pi$ rad for $q = 0, 1,$ 2 . . . ; using Eq. (4) for the phase shift of a Gaussian beam as a function of the distance from its waist, we determine that

$$\nu^{\text{res}} = (c/2d)(q + \tfrac{5}{4}), \tag{54b}$$

and we see that the separation of the resonances is given by $c/2d$. The transmission function for this resonator is obtained from the normal interferometer-reflection function but with the dielectric film power-transmission coefficient T replacing the mirror-reflection coefficient. If we define β as the fraction of the electric field surviving each pass, then $\beta = (1 - \alpha)^{1/2}$, where α is the fraction of the power lost per round trip due to diffraction. [The definition of β in Goldsmith and Schlossberg (1980) is incorrect.] Assuming no loss in the dielectric film, we then find

$$|\tau|^2 = \frac{[4\beta T/(1 - \beta T)^2] \sin^2(\delta\phi/2) + T[(1 - \beta)/(1 - \beta T)]^2}{1 + [4\beta T/(1 - \beta T)^2] \sin^2(\delta\phi/2)}, \tag{55}$$

where $\delta\phi$ is again the round-trip phase shift within the resonator. The transmission function is near unity except in the vicinity of the resonances, and the resonator thus functions as a sideband dropping filter. We also see that the maximum resonator transmission increases as the dielectric film transmission increases but that the rejection bandwidth decreases. The measured transmission curve shown in Fig. 18b was obtained with a film transmission of 0.75. The power loss per round trip for the semiconfocal resonator is given by Li (1965) as

$$\alpha = 16\pi^2 N e^{-4\pi N}, \tag{56a}$$

where

$$N = a^2/2\lambda d, \tag{56b}$$

a being the radius of the spherical mirror.[14] The loss is lowest for the exactly semiconfocal mirror spacing but only increases slightly for modest increases of d. A value of $d \simeq \lambda_{\text{IF}}/8$ is again optimum for a single-sideband filter, but we see that for $d = 5\text{–}6$ cm and $a = 1.5$ cm, as used by Goldsmith and Schlossberg (1980), the loss is still very small, justifying the previous calculations. The device shown in Fig. 18a was measured to have a minimum loss of <1 dB and a maximum loss of 17 dB, thus exhibiting a useful level of performance. The loss was found to be dominated by ohmic and dielectric losses, rather than by diffraction, however. One difficulty with microwave resonators as compared with plane-parallel Fabry–Perot interferometers is the critical alignment relative to the inci-

---

[14] It is assumed that the plane mirror is large enough compared with the waist radius that diffraction there is negligible to that occurring at the spherical mirror.

dent beam required for optimum performance; this is the price paid, however, for the lower diffraction loss that can be obtained.

E. METALLIC MESH

Mirrors made of metal mesh are used to perform a number of functions in quasi-optical systems at millimeter and submillimeter wavelengths. These include partial reflection, particularly important for Fabry–Perot interferometers, and polarization discrimination. We shall give here a brief summary of some basic results together with references to more extensive treatments of the subject.

The problem of a plane wave incident on a plane metallic system is an electromagnetic field problem of considerable difficulty if the metal is not entirely uniform (e.g., a grid). An extensive historical review and comparison of different results is given by Larsen (1962). In the discussions of Marcuvitz (1951), Casey and Lewis (1952), and Wait (1954, 1957) this problem is treated in considerable detail. More recent discussions are those by Ulrich *et al.* (1963, 1970) and Ulrich (1967). A variety of techniques have been employed to determine the amplitude and phase of the reflected and transmitted waves for a beam incident from a particular direction with a specified polarization. The results are conveniently expressed in terms of an equivalent transmission-line problem; the grid is represented by an impedance Z_g shunting the transmission line; this situation is shown schematically in Fig. 19a. The reflection coefficient for the electric field is given by

$$r = E_{ref}/E_{inc} = - [1 + (2Z_g/Z_0)]^{-1} \qquad (57a)$$

where Z_g is the shunt impedance of the the grid, Z_0 the wave impedance of the medium, and E_{ref} and E_{inc} are the reflected and incident electric field amplitudes, respectively. The shunt impedance can be complex: $Z_g = \mathrm{Re}(Z_g) + i\,\mathrm{Im}(Z_g)$. From this expression, the phase of the electric field amplitude-reflection coefficient is found to be

$$\phi = \pi + \arctan\left(\frac{\mathrm{Im}(Z_g)/Z_0}{1 + 2[\mathrm{Re}(Z_g)/Z_0]}\right), \qquad (57b)$$

and the power reflection coefficient is

$$R \equiv |r|^2 = [|1 + (2Z_g/Z_0)|^2]^{-1} \qquad (57c)$$

The most simple, as well as an important practical utilization of metal mesh, is the situation in which a plane wave is incident on a one-dimensional array of thin wires. The electric field is taken to be parallel to the direction of the conductors, which have a spacing g. Defining the

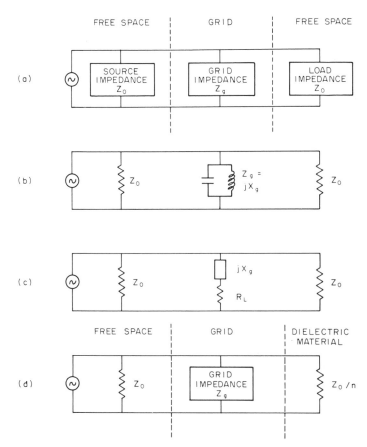

FIG. 19 (a) Treatment of behavior of grid in terms of transmission line shunted by grid impedance Z_g, which, in the simplest models, is either an inductance or a capacitance, depending on the type of grid. (b) Equivalent circuit for lossless inductive grid suggested by Ulrich (1967), having no reflection at the grid resonant frequency. (c) Addition of loss resistance R_L to account for the effect of ohmic dissipation in the conductors. (d) Equivalent circuit for grid mounted on dielectric surface. The absolute grid impedance Z_g is not affected by the presence of the dielectric but the grid transmission is different.

impedance of the grid relative to free space as Z_g/Z_0, for wavelengths $\lambda \gg g$ (Marcuvitz, 1951):

$$Z_g/Z_0 = i \ln \csc(\pi a/g)(g/\lambda) \qquad (58a)$$

for thin strips of width $2a$ and

$$Z_g/Z_0 = i \ln(g/2\pi a)(g/\lambda) \qquad (58b)$$

for round wires having radius a. The form of these equations, with an imaginary impedance increasing in proportion to the frequency suggests an analogy with an inductor shunting a transmission line, hence grids with wires parallel to the electric field are often called inductive grids. These formulas are only the limiting cases for grid dimensions that are very small compared with a wavelength; more complete expressons are given by MacFarlane (1946), Marcuvitz (1951), and Wait (1954).

For nonnormal incidence but with electric field parallel to the wires, MacFarlane (1946) shows that for $g \ll \lambda$ the impedance must be multiplied by $\cos \theta$, where θ is the angle from normal incidence. Marcuvitz (1951) gives expressions that treat grids with finite thickness for which the equivalent circuit is necessarily more complex than that indicated by Eq. (58). The finite thickness will increase the reflectivity of inductive grids.

Ulrich (1967) suggests modification of the simple equivalent circuit in order to fit measured data in the vicinity of $g = \lambda$, which indicate that $R \rightarrow 0$ for $g \simeq \lambda$. The addition of a capacitance in parallel with the inductance as shown in Fig. 19b results in a resonance with $Z \rightarrow \infty$ as $R \rightarrow 0$ as required, with the proper choice of the capacitance. This behavior of $R \rightarrow 0$ for $g/\lambda \approx 1$ has also been measured at microwave frequencies by Pursley (1956); the equivalent circuits of Marcuvitz or Ulrich appear to be able to reproduce these results reasonably well. Mok *et al.* (1979) report submillimeter measurements on inductive grids consistent with the formulas previously discussed. Their measurements extend to $g/\lambda \simeq 3$, and they find maxima in the transmission for $g/\lambda \simeq 1$ and 2. These authors also give results of theoretical calculations of the grid transmission based on a rather different method than that used previously, which gives a reasonable agreement throughout the very wide frequency range measured; the disagreement in the vicinity of the transmission maxima is attributed to nonuniformity in the spacing of the wires.

The general formulas for grid shunt impedance given by Wait (1954) simplify, for round wires with $g \ll \lambda$, to Eq. (58b) reduced by a factor dependent on the two angles from normal incidence; the reflectivity of the grid thus increases as the incident beam departs from normal incidence.

A useful approach for the analysis of certain types of grids is the principle of complementarity (Ulrich, 1967). This is an extension of Babinet's principle (Jackson, 1962) and, when applied to the thin, perfectly conducting pattern making up a grid, indicates that there is a simple relationship between the transmission function of the original grid and that for a complementary grid—one made by interchanging the open and the conducting portions of the grid. The relationship is that

$$t + t_c = 1, \tag{59}$$

where t and t_c are the field-transmission coefficients of the original and complementary grids, respectively. The direction of polarization of the incident wave must be rotated by 90° to make this relationship apply. As we are assuming thin lossless grids, we also have, letting r and r_c be the reflection coefficients, $1 + r = t$ and $1 + r_c = t_c$ (from the continuity of the electric field) and $r^2 + t^2 = r_c^2 + t_c^2 = 1$ (from conservation of energy). From these relationships, it is straightforward to derive that, for each grid, ρ and t have a 90° relative phase shift and also that

$$t_c = -r \tag{60a}$$

and that

$$r_c = -t. \tag{60b}$$

Thus the properties of one grid determine the behavior of its complement (under restriction of 90° polarization rotation). The complement of a grid consisting of strips narrow compared with their spacing, for example, is a grid with wide strips nearly touching. This discussion indicates that the properties of such a grid for the electric field perpendicular to the gaps is simply related to the narrow strip grid with **E** parallel to the strips. A calculation of the equivalent shunt impedance [using Eqs. (57a), (58a), and (60)] of a grid with gaps having width $2a$ and spacing g yields

$$Z_g/Z_0 = -i/[4 \ln \csc(\pi a/g)(g/\lambda)] \tag{61}$$

The complementary grid behaves like a shunt capacitor and hence is called a *capacitive* grid.

The inductive grid previously discussed is often called a "one-dimensional" grid because the current flow is in one dimension only. If we consider the effect of a grid of thin strips with the electric field perpendicular to the strips, we see that it acts like a capacitive grid (but not the complementary grid) with gap width almost equal to gap spacing. From Eq. (61) we see that the shunt impedance is extremely large and the grid is essentially invisible. Thus, to first order, a "two-dimensional" grid of an array of perpendicular wires or strips should behave as two simple grids acting on the two principal polarizations independently. This problem has been treated by the method of "average boundary conditions" by Astrakhan (1964, 1968), who finds that this is the case for normal incidence. A two-dimensional wire grid thus acts like a polarization-independent inductive grid and a two-dimensional array of square conductor patterns acts like a polarization-independent capacitive grid.

A grid pattern that is invariant under a 90° rotation is polarization-indepenent for radiation incident at normal incidence. The results of Astrakhan (1964, 1968) have been used by Saleh and Semplak (1976) to derive the requirements for a two-dimensional grid to be polarization-independent when the radiation is not normally incident. For a rectangular grid pattern, the result is approximately that the pattern appears square when viewed from the direction of the incident radiation.

Two-dimensional grids having special frequency-dependent properties have been developed for particular diplexing applications (Arnaud and Pelow, 1975; Anderson, 1975b). These generally employ conductor patterns that are resonant at a particular frequency to give a maximum or a minimum transmission there and are thus called *self-resonant* grids; when used in a Fabry–Perot interferometer, for example, these grids can select a particular order of interference and reduce the likelihood of confusion from transmission at unwanted frequencies.

The performance of wire grids as partially reflecting mirrors or polarizers is limited by the loss arising from the finite conductivity of the metal. Assuming that the wire thickness is much greater than a skin depth, the loss resistance for a grid under normal incidence will be approximately proportional to the surface resistance R_s of the metal (see Appendix III) multiplied by a factor η, which is determined by the fraction of the surface area available for carrying current (Ulrich, 1967); for one-dimensional inductive grids, for example, this factor is $g/2a$.[15] The loss resistance $R_L = \eta R_s$ appears in series with the shunt reactance in the equivalent circuit for the grid, as shown in Fig. 19c. The fraction of incident power absorbed by the grid is

$$A = P_{abs}/P_{inc} = R_L Z_0/[(\tfrac{1}{2}Z_0 + R_L)^2 + X_g^2], \tag{62a}$$

where X_g is defined from the lossless grid impedance by $Z_g = iX_g$. Comparing this with the expression for the fraction of the power reflected obtained from Eq. (57c)

$$R = \tfrac{1}{4}Z_0^2/[(\tfrac{1}{2}Z_0 + R_L)^2 + X_g^2], \tag{63}$$

we see that

$$A = R(4R_L/Z_0). \tag{62b}$$

For inductive grids under nonnormal incidence, the loss for fixed reflectivity is multiplied by a factor $\cos\theta \cos\psi$, where θ and ψ are the angles from normal incidence (Wait, 1954). Thus, even allowing for surface resistance somewhat greater than that expected from dc values (see Appendix II), the loss in the grids will by typically only a fraction of a percent at millimeter wavelengths. At submillimeter wavelengths, the measured ab-

---

[15] For inductive grids, Ulrich *et al.* (1963) give $\eta = 2g/u$, where u is the circumference of the conducting wires having arbitrary cross section.

sorption exceeds the predicted loss (Mok *et al.*, 1979), but nonuniformity in the wire spacing may be making a significant contribution to the loss in this case. The effect of nonuniform wire spacing is discussed by Ulrich *et al.* (1963).

It is often desirable for practical reasons (and necessary for two-dimensional capacitive grids) to support the conductors on a sheet of dielectric material. The presence of the dielectric changes the reflectivity of the grid; this problem has been analyzed in detail by Wait (1957). A discussion of lossless grids at normal incidence with $g \ll \lambda$ is given by Ulrich *et al.* (1963). For these conditions, the two treatments yield the result that the absolute impedance of the grid is unaffected by the presence of the dielectric. However, the impedance of the dielectric is Z_0/n, and hence the transmission is different. Using the equivalent circuit shown in Fig. 19d, the transmission and reflection can be calculated. The result can be put into the following form, which relates the power reflectivity R' of the grid on the boundary of a dielectric film to that of the grid in free space R:

$$R' = [4R + (1 - R)(n - 1)^2]/[4R + (1 - R)(n + 1)^2]. \tag{64}$$

This expression gives the reflectivity of the grid on the air–dielectric interface; to calculate accurately the transmission and reflection on a thin dielectric substrate, the latter must be treated as a dielectric-filled Fabry–Perot interferometer with one mirror being the grid and the other just the air–dielectric interface. Because in most cases the reflectivity of the latter will be small compared with that of the grid, it can be neglected.

V. Practical Realizations of Quasi-Optical Antenna Feed Systems

The purpose of this section is to give an overview of the quasi-optical feed systems that have been developed for use at millimeter and submillimeter wavelengths. It is impractical to discuss all designs and their variations; rather, we shall use a few examples to illustrate how the constraints of antenna illumination and the use of frequency-selective devices can be satisfied by systems designed using Gaussian beam theory.

The quasi-optical feed system described by Goldsmith (1977) is designed to provide constant antenna illumination over a wide range of frequencies (60–150 GHz) and also to provide the large-diameter beam-waist radius required by the tilted Fabry–Perot diplexer/local oscillator injector described above. The two parabolic mirrors (shown in Fig. 17) have offset angles of 20° and 30°, respectively, and were machined on a numerically controlled mill. They are separated by a distance of 140 cm; ideally, their separation should be equal to the sum of their focal lengths (136 + 44 cm) making the perfectly frequency-independent Gaussian telescope described in Section II.C, but using the Gaussian beam imaging formulas and the expression for power coupling loss [Eq. (28)], one can verify that in this case the frequency variations and loss are negligible (owing to the very large size of the waist between the mirrors).

A system described by Erickson (1980) uses 60° off-axis paraboloids for coupling the mixer and local oscillator beams to a polarization-rotating dual-beam interferometer (Fig. 20). For coupling to the $f/D = 0.5$ beam of the antenna, it was felt that an off-axis mirror would be difficult to realize because interference problems of the very divergent output beam with the mirror, as well as possible undesirable effects arising from the extreme curvature of the surfaces required. Thus the Newtonian type of system shown was developed, which produces satisfactory antenna illumination, although the blockage loss is approximately 12%.

The optical system for the FCRAO cryogenically cooled 2–4-mm receiver, which operates with the 14-m-diameter $f/D = 4.15$ antenna (Fig. 21) uses ellipsoidal reflectors to couple beams from the scalar feedhorn

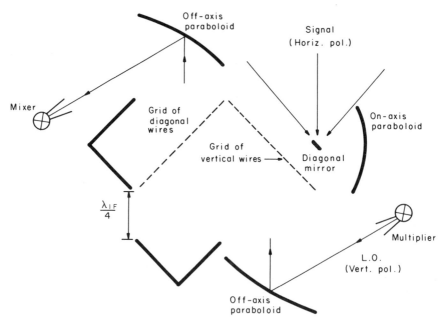

FIG. 20 Antenna feed and local oscillator injection system described by Erickson (1980) and used at frequencies between 200 and 345 GHz. The telescope beam has $f/D = 0.5$; this is reimaged by the diagonal mirror/on-axis paraboloidal Newtonian system to a waist located within the polarization-rotating dual-beam interferometer (described in Section IV.B).

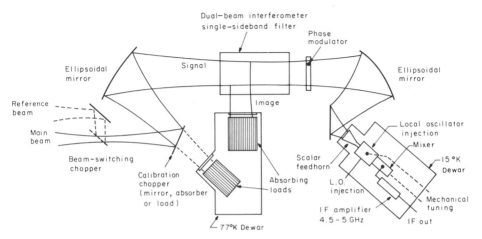

FIG. 21   Quasi-optical feed system for use in the λ-2–4-mm region. The feedhorn waist is reimaged twice by a pair of ellipsoidal mirrors; between them is a dual-beam interferometer, which sends the image sideband to an absorbing load at 77 °K. Calibration and beam-switching functions are provided by the two choppers. A rotating dielectric sheet phase modulator (Section VI.B) reduces baseline ripple due primarily to reflections between the receiver system and telescope structure.

and antenna beam waist through an amplitude-division dual-beam interferometer used as a single-sideband filter. In this case the scalar feedhorn was included in an iterative design procedure, which accounted for the changes in waist sizes and locations as a function of frequency due to motion of the horn phase center as well as the Gaussian beam imaging formulas. This results in minor differences from the Gaussian telescope design; the overall system exhibits very little Cassegrain illumination variation in theory over the 70–115-GHz frequency range despite a feedhorn phase center motion of over 2 cm across this frequency interval.

A system designed for use with a $f/D = 16$, 150-cm-dia optical telescope at a frequency of 345 GHz is described by Erickson (1978). A 90° off-axis parabolic mirror couples the telescope beam into an amplitude-division dual-beam diplexer while 90° off-axis ellipsoids couple in the beams from the local oscillator and the mixer. Each ellipsoidal mirror has significant near-field corrections (see Section II.E) due to the combination of being in the nearfield of its foci and the large offset angle involved. These mirrors were machined using a normal milling machine but employing a technique described by Erickson (1979).

Quasi-optical techniques have also been applied to the design of systems for use at submillimeter wavelengths. An astronomical receiver system operating at a wavelength of 434 μm is described by Goldsmith et *al.* (1981) and is shown in Fig. 22. The laser beam ($w_0 = 0.14$ cm) couples directly to the diplexer while the cube-corner mixer is illuminated by a 90° off-axis ellipsoid. As discussed in Section II.B, the measured pattern has significant side-lobe content so that accurate characterization in terms of a Gaussian beam is not possible; it appears that the devices require a convergent beam with f/D between 1 and 2 for best performance. The telescope beam has $f/D = 120$; to reduce this to the value appropriate for the diplexer, a pair of off-axis parabolic mirrors is employed. The design of this portion of the system was carried out using essentially geometrical optics calculations because of the small effects of diffraction under these conditions. A submillimeter feed system using $f/D = 2$ Teflon lenses has

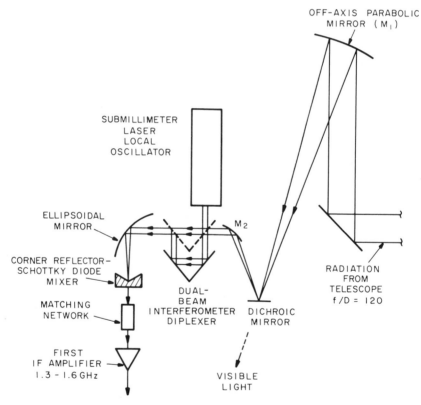

Fig. 22   Submillimeter feed system described by Goldsmith *et al.* (1981) and used at telescope coudé focus ($f/D = 120$) at a wavelength of 434 μm. The offset parabolic mirrors are denoted as M1 and M2; the diplexer is a dual-beam interferometer, which is described in Section IV.B.

also been developed (Lidholm and de Graauw, 1979) and has been used in conjunction with a number of optical and infrared telescopes. A loss of 20% is reported for the pair of lenses and folded Fabry–Perot diplexer at wavelengths of ~1 mm.

One type of antenna feed that is closely related in principle to Gaussian optics systems is the beam waveguide system employing two (or more) focusing elements to transport the beam to a more accessible or larger area than the Cassegrain focus (Mizusawa and Kitsuregawa, 1973). Such systems, which are not restricted to Gaussian beams, have been primarily employed to date at longer wavelengths and over fairly narrow bandwidths (Brain and Gill, 1978). The design for the 45-m and 10-m-diameter antennas of the Radio Telescope Project of Japan includes beam waveguide systems for 20 to 115 GHz, which will allow receivers to be in a room that while rotating in azimuth does not move in elevation (Kaifu, 1980).

VI. Quasi-Optical Components

The limitations of waveguide devices for millimeter and submillimeter wavelengths have been recognized for a considerable time, and there has been, as a result, an ongoing effort to develop devices in other transmission media to perform the system functions for which highly refined devices have been developed in waveguide at longer wavelengths. The use of an oversized waveguide has been one approach, and although not significantly exploited in current systems, devices in this medium bear a close similarity to quasi-optical technology. A review of the developments of oversized waveguide components is given by Taub et al. (1963), but a further discussion of these devices is beyond the scope of this chapter.

An early review of optical techniques at microwave frequencies is that by Harvey (1959) in which several types of components, particulary matched loads, are discussed. Garnham (1969) describes a variety of devices, including interferometers, used at millimeter and submillimeter wavelengths, but the effects of diffraction are not dealt with in detail. Both of these articles do describe a wide range of quasi-optical components.

Our purpose here is to review a number of specialized components other than frequency-selective devices that have been developed for use in conjunction with millimeter and submillimeter radiometer systems. We shall concentrate on devices that have seen practical implementation and that appear to offer attractive performance.

A. Calibration Systems

Radiometers for radioastronomical use must be calibrated absolutely; that is, an output scale factor K/V must be established to fairly high accuracy. For this purpose, blackbody loads are desirable because power standards and attenuator calibration are sufficiently uncertain and frequency sensitive to make the approach using coherent sources undesirable. The most widely used technique at millimeter and submillimeter wavelengths is to fill the input beam of the radiometer with absorbing loads at ambient and liquid-nitrogen temperatures. Satisfactory loads can be made of Eccosorb[16] carbon-loaded foam absorber—the material in pyramidal or hillock-shaped forms has the lowest reflection coefficient. Low-temperature calibration loads using down-looking liquid-nitrogen-filled Dewars are described by Gordon-Smith and Gibbins (1972), Hardy (1973), and Goldsmith (1977). The reflection from the air–liquid interface probably does not exceed 3%. The pieces of absorbing foam can be immersed in liquid nitrogen and viewed through a low-loss dielectric such as expanded polystyrene (Ulich *et al.*, 1980). Alternatively, a piece of liquid-nitrogen-saturated material can be held in the beam without any Dewar; the temperature for reasonable-sized pieces of absorber remains constant for at least 30 sec.

For some applications, a cryogenic load is inconvenient, and heated loads can be used, although radiometer linearity may be a greater problem. One such system described by Goldsmith *et al.* (1979) consists of wedges of high thermal conductivity castable absorber (Emerson and Cuming's type CR 117) tilted with respect to the radiometer beam and heated from the back. This hot load (shown in Fig. 23) provides a radiometric temperature of 410 °K at 87.5 GHz. Blackbody reference loads for the submillimeter region employing multiple reflections are discussed by Carli (1974).

B. Baseline Ripple Reduction

Radiometers used with symmetric antennas often exhibit variations in output power that are periodic in frequency. One cause is multiple reflections betwen the feedhorn/receiver and the elements of the antenna structure blocking portions of the beam (i.e., the subreflector or subreflector support structure); this problem is analyzed by Allen (1969) and Morris (1978). Although these can be minimized by careful design, including

---

[16] Eccosorb material is manufactured by Emerson and Cuming Inc., Canton, Massachusetts. The pyramidal material is type VHP-2 and the hillock-shaped material is type CV-3.

FIG. 23   Heated blackbody calibration load described by Goldsmith *et al.* (1979) for use in the ~3-mm region. The scale included in the photograph is calibrated in inches.

proper shaping of the subreflector edge and of the supporting legs, this remains a significant practical problem for high-sensitivity observations.
The source of the modulated output can be the input noise from the

atmosphere (or source being observed) or power radiated from the radiometer input.

One can improve the isolation of the radiometer from reflections produced by elements of the antenna structure by using circular polarization; rays having undergone an odd number of reflections will return with the opposite sense of polarization and hence will not couple to the radiometer. This approach has been employed by Erickson (1978), who used a quarter-wave plate to convert linear to circular polarization. This technique could be extended by including a linear polarizer between the circular polarizer and the feed (which accepts a single linear polarization) arranged to reflect any returned signal of the unwanted polarization into an absorbing load rather than letting it be reflected once again from the feedhorn.

The baseline ripple (or chromatism) may, however, be produced by a variety of reflected paths. A more general approach to reducing the magnitude of the output power variations is to vary the round-trip phase shift between the radiometer feed and the antenna structure. A variety of methods for achieving this have been developed. If a path-length perturbation with a sinusoidal time dependence is introduced, the magnitude of the perturbation can be chosen to produce a time-averaged reflection coefficient that is independent of wavelength (Gustinic, 1977b). Good results have also been achieved by varying the path length linearly in time over many wavelengths with quasi-randomly selected endpoints (J. M. Payne, private communication, 1980). Alternatively, a step function in the path length can be introduced by insertion of a dielectric having a one-way differential phase delay of 90°.[17] By inclining the dielectric to be near Brewster's angle, its reflection can be minimized for a single polarization. An order-of-magnitude reduction in the magnitude of the baseline ripple has been achieved using this technique on the $f/D = 4.15$ FCRAO Cassegrain telescope system at a frequency of 115 GHz (Goldsmith and Scoville, 1980).

C. BEAM SWITCHING

For observations of small sources, switching between two closely spaced beams on the sky is an effective technique to cancel both sky emission and radiometer drifts. This can be easily accomplished by the introduction of a pair of reflectors producing a lateral shift of the effective feed position in the focal plane of the antenna. One such system operating at rates up to 50 Hz has been described by Goldsmith (1977). The resulting change in gain can be calculated using Eq. (24) and will be quite small for

---

[17] This effect could also be produced by the motion of a reflecting surface.

most Cassegrain systems. A nutating subreflector can also switch between different beams on the sky, but the throw in one system reported is limited to 5 arcmin away from the antenna axis (Payne, 1976). It is also possible to divert the input of the radiometer away from the subreflector in a Cassegrain system and thus have as a reference beam a wide-angle pattern, which facilitates observation of moderately extended sources. This is achieved by the insertion of a dielectric prism in a system described by Payne and Ulich (1978), who report good balance between the directive and diffuse beams.

D. POLARIZATION DIPLEXING

The properties of metallic mesh discussed in Section IV.E indicate that a grid with properly chosen parameters can efficiently separate two linear polarizations in an incident beam. Using an inductive grid, for example, we can make the reflectivity for the electric field parallel to the wires high, whereas the reflectivity for radiation polarized with E perpendicular to the wires will be very low [see the discussion following Eq. (61)]. This technique has been effectively used in receivers at the NRAO 36-ft telescope at wavelengths ≤ 6 mm and has less loss than waveguide polarization diplexers (J. M. Payne, private communication, 1980). The generation of cross-polarized radiation sets one limit to the performance of the quasi-optical diplexers. Chu et al. (1975) investigated a Mylar-supported inductive grid polarization diplexer at 19 and 28.5 GHz and found that, for lowest cross-polarization, the conducting strips should be perpendicular to the plane of incidence of the radiation. In this situation, the maximum level of cross-polarization is $\simeq -28$ dB, whereas the insertion loss for reflection and transmission (of the appropriate polarization) is only 0.1 dB.

E. QUARTER-WAVE PLATES

A quarter-wave plate is a device having a relative 90° phase shift for two perpendicularly polarized components of the incident electric field. An important use is as a circular polarizer (see Sections IV.C and VI.B) converting linearly polarized radiation to circular polarization and vice versa. An overview of quarter-wave plates for use in the submillimeter region is given by van Vliet and de Graauw (1981). Specific designs that have been constructed are discussed by Farrukh and Koepf (1976), Erickson (1978), and Watanabe and Nakajima (1978).

Appendix I. Dielectric Materials

Dielectric materials play a potentially important role in millimeter and submillimeter systems; lenses, beam splitters, and Dewar windows being only a few of their applications. In most situations, both the dielectric constant and loss are important parameters. In this appendix, we shall give a selection of the dielectric properties of some useful materials, collected from the literature. A comprehensive compilation of available data is that by Simonis (1980); some of the shorter wavelength data is summarized by Button (1978). In Table AI, references to other work can also be found.

There is relatively little information concerning the temperature dependence of dielectric properties; all the data in Table AI apply to "room temperature." Lowenstein et al. (1973) have investigated the submillimeter properties of crystalline quartz, sapphire, germanium, and silicon at 1.5 °K. Haas and Zimmermann (1976) have measured the behavior of a number of dielectrics at 77 °K at a frequency of 22 GHz. Both these studies indicate a modest reduction in the dielectric constant and a significant reduction in the absorption factor upon cooling.

Another troublesome point is the variability in the composition of some dielectric materials. The different compositions of alumina–titanate ceramics produce large changes in their dielectric properties (Perry, 1970). In the case of fused quartz, different quantities of H_2O incorporated into the material can produce changes of $\geq 0.1\%$ in the index of refraction, which may be significant for some applications (Parker et al., 1978).

At least two systems are in wide use for reporting the loss of dielectric materials. First, taking the dielectric function as $\varepsilon = \varepsilon' - i\varepsilon''$, the loss tangent is defined by $\tan \delta = \varepsilon''/\varepsilon'$. Second, the power attenuation coefficient is generally denoted as α and defined by $P(z) = P_0 e^{-\alpha z}$, where z is the distance propagated in the dielectric. (Defined thusly, α is equal to twice the real part of the propagation constant, which often is given the same symbol, providing a source of confusion.) The relationship between the power attenuation coefficient and the loss tangent is given by

$$\alpha = (2\pi\sqrt{\varepsilon'}/\lambda) \tan \delta,$$

where ε' is the real part of the dielectric function as previously defined and λ the free-space wavelength.

Appendix II. Dielectric Beam Splitters

A dielectric beam splitter of finite thickness behaves like a Fabry–Perot interferometer. Defining its thickness to be d, index of refraction n, and angle of incidence of the radiation to be θ_i, its reflectivity $|\mathcal{R}|^2$ is given by Eq. (43b). The reflectivity of each individual surface is different for the two principal planes of polarization, making these devices relatively

TABLE AI

DIELECTRIC PROPERTIES OF SELECTED SOLIDS

Material	Dielectric constant	Loss tangent	Frequency (GHz) and reference
Alumina	9.4	1.7×10^{-3}	$(14-50)^c$
	8.63	—	$(245)^d$
Germanium	16.05	1×10^{-4}	$(900)^e$
Polyethylene	2.14	1×10^{-3}	$(891)^f$
Polystyrene	2.54	3.3×10^{-3}	$(35)^g$
	2.53	1.2×10^{-3}	$(90)^h$
	2.56	—	$(143)^i$
	2.57	—	$(343)^i$
Quartz	4.47^a	8×10^{-4}	$(900)^e$
(Crystalline)	4.65^b	5×10^{-5}	$(900)^e$
Quartz	3.81	—	$(245)^d$
(Fused)	3.81	2.5×10^{-3}	$(600)^j$
	3.73	$5-7 \times 10^{-3}$	$(891)^f$
Rexolite	2.54	4.7×10^{-4}	$(10)^k$
	2.34	—	$(128)^l$
	2.44	—	$(143)^i$
	2.54	—	$(343)^i$
Sapphire	9.42^a	2×10^{-3}	$(900)^e$
	11.66^b	3×10^{-3}	$(900)^e$
Silicon	11.42	—	$(245)^d$
	11.67	6×10^{-4}	$(900)^e$
Styrofoam	1.03	—	$(245)^d$
Teflon	2.04	1×10^{-3}	$(35)^g$
	2.06	2×10^{-4}	$(90)^h$
	2.07	—	$(143)^i$
	2.07	—	$(343)^i$
TPX	2.13	—	$(245)^d$
	2.12	1×10^{-3}	$(891)^f$

[a] Ordinary.
[b] Extraordinary.
[c] Laboratory for Insulation Research (1957).
[d] Simonis and Felock (1980).
[e] Lowenstein et al. (1973).
[f] Chantry (1971).
[g] Hakki and Coleman (1960).
[h] Balanis (1971).
[i] Degenford and Coleman (1966).
[j] Parker et al. (1978).
[k] von Hipple (1954).
[l] Goldsmith and Scoville (1980).

unsuitable for dual-polarization use except near normal incidence. For reference, we list the formulas for the single-surface power reflectivity with the electric field perpendicular to the plane of incidence as

$$R = \{[\cos \theta_i - (n^2 - \sin^2 \theta_i)^{1/2}]/[\cos \theta_i + (n^2 - \sin^2 \theta_i)^{1/2}]\}^2 \quad (A1)$$

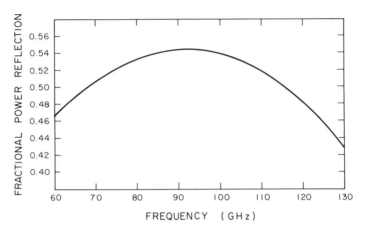

FIG. 24   Response of quartz ($n = 1.95$) dielectric beam splitter. The incident radiation is polarized perpendicular to the plane of incidence, where $\theta_i = 45°$ and the thickness is 0.045 cm.

and for the electric field parallel to the plane of incidence as

$$R = \{[n^2 \cos \theta_i - (n^2 - \sin^2 \theta_i)^{1/2}]/[n^2 \cos \theta_i + (n^2 - \sin^2 \theta_i)^{1/2}]\}^2 \quad (A2)$$

(cf. Garnham, 1969). There is a phase shift of 180° for the wave reflected from the dielectric; this does not affect the phase delay given by Eq. (43c). The Rs are the same for the wave entering or leaving the dielectric.

For dual-beam interferometers (Section IV.B), we wish to have a beam splitter with $|\mathscr{R}|^2 \simeq 0.5$ over a wide range of wavelengths; from the behavior of the beam splitter, we conclude that a choice of $|\mathscr{R}|_{max}$ between 0.52 and 0.54 will keep $|\mathscr{R}|^2$ very close to 0.50 over a bandwidth of an octave. For $\theta_i = 45°$, fused quartz has $|\mathscr{R}|_{max} = 0.54$ for the electric field perpendicular to the plane of incidence. The frequency dependence of the reflectivity of such a beam splitter is shown in Fig. 24.

Appendix III.   Mirror Reflectivity and Loss

One limitation on the performance of quasi-optical systems using reflective optics is the loss due to absorption in the metal mirrors. A metal conductor can be characterized at a frequency f by a surface resistivity given by

$$R_s(\Omega/\text{sq}) = (\pi f \mu_0/\sigma)^{1/2}, \quad (A3)$$

where μ_0 is the permeability of the vacuum ($= 4\pi \times 10^{-7}$ H/m), we have assumed a relative permeability of the conductor of unity, and σ the bulk (dc) conductivity ($\mho/m$). The fractional power loss per reflection depends on the geometry of the mirror and incident beam, but for a plane wave at normal incidence, the fractional power lost is given by

$$f_L = P_{abs}/P_{inc} = 4R_s/Z_0 = R_s/30\pi \tag{A4}$$

(Batt *et al.*, 1977; a cgs formulation can be found in Jackson, 1962). Combining the two previous equations yields

$$f_L = 2.1 \times 10^{-4}[f(GHz)/\sigma(10^7\mho/m)]^{1/2}. \tag{A5}$$

Formulas such as this have been compared with measured absorption coefficients at millimeter and submillimeter wavelengths (Beyer and Scheibe, 1962; Batt *et al.*, 1977). The results of the former authors indicate that at $\lambda = 3.3$ cm, R_s is approximately 30% higher than that predicted from dc characteristics of aluminum, whereas the latter investigation implies that the value of R_s for gold is approximately 2.2 times higher than predicted, for a wavelength of 337 μm. Measurements of waveguide loss can also be related to effective surface resistivity, and studies at wavelengths between 0.9 cm and 0.15 cm (Benson, 1969) indicate a consistent trend, implying an excess surface resistivity of a factor of 1.5 to 2.0 throughout the millimeter range for copper, brass, and silver. Because the dc values of σ for most metals are in the range from 3 to 6 $\times$ $10^7\mho/m$, we see that even using an effective conductivity equal to a quarter of the dc value, the fractional loss per reflection is only 0.004 at 300 GHz. Thus even complex focusing and beam-steering systems should introduce negligible loss at millimeter wavelengths, but losses at submillimeter wavelengths may become significant.

For plane mirrors there is, in general, little difficulty in making the surface finish essentially perfect; because the reflection is largely due to the impedance discontinuity at the conductor boundary, extremely thin metal layers suffice for obtaining the maximum attainable reflectivity (Jackson, 1962) so that evaporation on optical flats is a convenient technique. For nonnormal incidence, the loss differs for the two electric field polarizations, and it is preferable to have the electric field perpendicular to the plane of incidence (Beyer and Scheibe, 1962). For focusing mirrors, the surface finish is more difficult to control and because these are diffraction-limited systems, it must be held within very stringent limits. The effects of surface roughness are analyzed in the literature on antennas (cf. Ruze, 1966; Anderson, 1975a) and suggest that a rms surface deviation of less than 1/50 wavelength is required to closely approach ideal performance.

ACKNOWLEDGMENTS

This work derives in part from discussions with many colleagues. I would particularly like to thank Neal Erickson, Michael Brewer, Jacques Arnaud, Adel Saleh, and Bob Haas for their contributions. E. Twelves provided valuable assistance in converting the manuscript into a legible form. I would also like to thank the Institute for Applied Physics, the University of Bern, and the Institute for Millimeter Radio Astronomy, Grenoble, for their hospitality during the time part of this work was carried out. Research at the Five College Radio Observatory is supported in part by the National Science Foundation under grant AST 80-26702. This is contribution no. 457 of the Five College Observatory.

REFERENCES

Allen, R. J. (1969). The radio spectrum of Virgo A from 1411.7 to 1423.8 MHz. *Astron. Astrophys.* **3**, 316–322.

Anderson, I. (1975a). The effect of small phase errors upon transmission between confocal apertures. *Bell Syst. Tech. J.* **54**, 783–795.

Anderson, I. (1975b). On the theory of self-resonant grids. *Bell Syst. Tech. J.* **54**, 1725–1731.

Arnaud, J. A. (1976). ''Beam and Fiber Optics,'' pp. 50–64. Academic Press, New York.

Arnaud, J. A., and Pelow, F. A. (1975). Resonant grid quasi-optical diplexers. *Bell Syst. Tech. J.* **54**, 263–283.

Arnaud, J. A., Saleh, A. A. M., and Ruscio, J. T. (1974). Walk-off effects in Fabry–Perot diplexers. *IEEE Trans. Microwave Theory Tech.* **MTT-22**, 486–493.

Astrakhan, M. I. (1964). Averaged boundary conditions at the surface of a lattice with rectangular cells. *Radio Eng. Electron. Phys. (Engl. Transl.)* **9**, No. 9, 1446–1454.

Astrakhan, M. I. (1968). Reflecting and screening properties of plane wave grids. *Radio Eng.* **23**, No. 1, 76–83.

Auston, D. H., Primich, R. I., and Hayami, R. A. (1964). Further considerations of the use of the Fabry–Perot resonators in microwave plasma diagnostics. *Proc. Polytech. Inst. Brooklyn Symp. Quasi-Opt., June 8–10, 1964* pp. 273–301.

Balanis, C. A. (1971). Dielectric constant and loss tangent measurements at 60 and 90 GHz using the Fabry–Perot interferometer. *Microwave J.* **14**, (3), 39–44.

Batt, R. J., Jones, G. D., and Harris, D. J. (1977). The measurement of the surface resistivity of evaporated gold at 890 GHz. *IEEE Trans. Microwave Theory Tech.* **MTT-25**, 488–491.

Benson, F. A. (1969). Attenuation of rectangular waveguides. *In* ''Millimetre and Submillimetre Waves'' (F. A. Benson, ed.), Chapter 14. Iliffe, London.

Beyer, J. B., and Scheibe, E. H. (1962). Loss measurement of the beam waveguide. *IEEE Trans. Microwave Theory Tech.* **MTT-11**, 18–23.

Blaney, T. G. (1978). A theoretical study of Josephson frequency mixers for heterodyne reception in the submillimeter wavelength range. *Natl. Phys. Lab. (U.K.), Rep.* pp. 5.7–5.8.

Boyd, G. D., and Gordon, J. P. (1961). Confocal multimode resonator for millimeter through optical wavelength masers. *Bell Syst. Tech. J.* **40**, 489–508.

Brain, J., and Gill, J. (1978). An 11/14 GHz 19 M satellite Earth station antenna. *Proc. Int. Conf. Antennas Propagation, 1978* Part I, pp. 384–388.

Burton, C. H., and Akimoto, Y. (1980). A polarizing Michelson interferometer for the far-infrared and millimeter regions. *Infrared Phys.* **20**, 115–120.

Button, K. J., ed. (1978). "Far Infrared and Submillimeter Waves Technical Group," Newsletter. Optical Society of America, New York.

Campbell, J. P., and DeShazer, L. (1969). Near fields of truncated-Gaussian apertures. *J. Opt. Soc. Am.* **59,** 1427–1429.

Carli, B. (1974). Design of a blackbody reference standard for the submillimeter region. *IEEE Trans. Microwave Theory Tech.* **MTT-22,** 1094–1099.

Casey, J. P., and Lewis, E. A. (1952). Interferometer action of a parallel pair of wire gratings. *J. Opt. Soc. Am.* **42,** 971–977.

Chai, A. S., and Wertz, H. J. (1965). The digital computation of the far-field radiation pattern of a truncated Gaussian distribution. *IEEE Trans. Antennas Propagation* **AP-13,** 994–995.

Chantry, G. W. (1971). "Submillimeter Spectroscopy," p. 341. Academic Press, New York.

Chen, M. H. (1979). The network representation and the unloaded Q for a quasi-optical bandpass filter. *IEEE Trans. Microwave Theory Tech.* **MTT-27,** 357–360.

Chu, T. S. (1966). Geometrical representation of Gaussian beam propagation. *Bell Syst. Tech. J.* **45,** 287–299.

Chu, T. S., and Turrin, R. H. (1973). Depolarization properties of offset reflector antennas. *IEEE Trans. Antennas Propagation* **AP-21,** 339–345.

Chu, T. S., Gans, M. J., and Legg, W. E. (1975). Quasi-optical polarization diplexing of microwaves. *Bell Syst. Tech. J.* **54,** 1665–1680.

Clarricoats, P. J. B., and Saha, P. K. (1969). Radiation pattern of a lens corrected scalar horn. *Electron. Lett.* **5,** 300–301.

Degenford, J. E., and Coleman, P. D. (1966). A quasi-optics perturbation technique for measuring dielectric constants. *Proc. IEEE* **54,** 520–522.

Dickson, L. D. (1970). Characteristics of a propagating Gaussian beam. *Appl. Opt.* **9,** 1854–1861.

Dragone, C. (1977). Reflection, transmission and mode conversion in a corrugated feed. *Bell Syst. Tech. J.* **56,** 835–867.

Dragone, C. (1978). Offset multireflector antennas with perfect pattern symmetry and polarization discrimination. *Bell Syst. Tech. J.* **57,** 2663–2684.

Dragone, C., and Hogg, D. C. (1974). The radiation pattern of offset and symmetrical near-field Cassegrain and Gregorian antennas. *IEEE Trans. Antennas Propagation* **AP-22,** 472–475.

Erickson, N. R. (1977). A directional filter diplexer using optical techniques for millimeter to submillimeter wavelengths. *IEEE Trans. Microwave Theory Tech.* **MTT-25,** 865–866.

Erickson, N. R. (1978). A 0.9 mm heterodyne receiver for astronomical observations. *IEEE MTT-S Int. Microwave Symp. Dig.* pp. 438–439.

Erickson, N. R. (1979). Off-axis mirrors made using a conventional milling machine. *Appl. Opt.* **18,** 956–957.

Erickson, N. R. (1980). A 200–300 GHz heterodyne receiver. *IEEE MTT-S Int. Microwave Symp. Dig.* IEEE Cat. No. 80CH1545-3MTT, pp. 19–20.

Farrukh, U. O., and Koepf, G. A. (1976). FIR Fresnel prism used as a quarterwave plate. *Proc. Int. Conf. Winter Sch. Submillimeter Waves Their Appl., 2nd, December 6–11, 1976* IEEE Cat. No. 76CH1152-8MTT, pp. 52–53.

Fetterman, H. R., Tannenwald, P. E., Clifton, B. J., Parker, C. D., Fitzgerald, W. D., and Erickson, N. R. (1978). Far-IR heterodyne radiometer measurements with quasioptical Schottky diode mixers. *Appl. Phys. Lett.* **33,** pp. 151–154.

Fox, A. G., and Li, T. (1961). Resonant modes in a Maser interferometer. *Bell Syst. Tech. J.* **40,** 453–488.

Garg, R. K., and Pradhan, M. M. (1978). Far-infrared characteristics of multi-element inter-ference filters using different grids. *Infrared Phys.* **18**, 292–298.

Garnham, R. H. (1969). Quasi-optical components. *In* "Millimetre and Submillimetre Waves" (F. A. Benson, ed.), Chapter 21. Iliffe, London.

Goldsmith, P. F. (1977). A quasioptical feed system for radioastronomical observations at millimeter wavelengths. *Bell Syst. Tech. J.* **56**, 1483–1501.

Goldsmith, P. F. (1982). Diffraction loss in dielectric-filled Fabry–Perot interferometers. *IEEE Trans. Microwave Theory Tech.*, in press.

Goldsmith, P. F., and Plambeck, R. L. (1976). A 230-GHz radiometer system employing a second-harmonic mixer. *IEEE Trans. Microwave Theory Tech.* **MTT-24**, 859–861.

Goldsmith, P. F., and Schlossberg, H. R. (1980). A quasi-optical single sideband filter em-ploying a semiconfocal resonator. *IEEE Trans. Microwave Theory Tech.* **MTT-28**, 1136–1139.

Goldsmith, P. F., and Scoville, N. Z. (1980). Reduction of baseline ripple in millimeter radio spectra by quasi-optical phase modulation. *Astron. Astrophys.* **82**, 337–339.

Goldsmith, P. F., Kot, R. A., and Iwasaki, R. S. (1979). Microwave radiometer blackbody calibration standard for use at millimeter wavelengths. *Rev. Sci. Instrum.* **50**, 1120–1122.

Goldsmith, P. F., Erickson, N. R., Fetterman, H. R., Clifton, B. J., Koepf, G. A., Buhl, D., and McAvoy, N. (1981). Detection of the $J = 6 \rightarrow 5$ transition of carbon monoxide. *Astrophys. J.* **243**, L79–L82.

Gordon-Smith, A. C., and Gibbins, C. J. (1972). Simple overall absolute calibrator for millimeter-wavelength horn-radiometer systems. *Electron. Lett.* **8**, 59–60.

Goubau, G. (1963). Optical relations for coherent wave beams. *In* "Electromagnetic Theory and Antennas" (E. C. Jordan, ed.), Part 2, pp. 907–918. Macmillan, New York.

Gustinic, J. J. (1976). A quasi-optical radiometer. *Proc. Int. Conf. Winter Sch. Submilli-meter Waves Their Appl., 2nd, 1976* pp. 106–107. IEEE Catalog No. 76CH1152-8MTT.

Gustinic, J. J. (1977a). Receiver design principles. *Proc. Soc. Photo-Opt. Instrum. Eng.* **105**, 40–43.

Gustinic, J. J. (1977b). A quasi-optical receiver design. *IEEE MTT-S Int. Microwave Symp. Dig.* pp. 99–101. *IEEE* Catalog No. 77CH1295-5MTT.

Haas, R. W., and Zimmermann, P. (1976). 22-GHz measurements of dielectric constants and loss tangents of castable dielectrics at room and cryogenic temperatures. *IEEE Trans. Mi-crowave Theory Tech.* **MTT-24**, 881–883.

Hakki, B. W., and Coleman, P. D. (1970). A dielectric resonator method of measuring induc-tive capacities in the millimeter range. *IRE Trans. Microwave Theory Tech.* **8**, 402–410.

Hardy, W. N. (1973). A precision temperature reference for microwave radiometry. *IEEE Trans. Microwave Theory Tech.* **MTT-21**, 149–150.

Harvey, A. F. (1959). Optical techniques at microwave frequencies. *Proc. Inst. Electr. Eng., Part B* **106**, 141–157.

Holmes, D. A., Avizonis, P. V., and Wrolstad, K. H. (1970). On-axis irradiance of a fo-cused, apertured Gaussian beam. *Appl. Opt.* **9**, 2179–2180.

Jackson, J. D. (1962). "Classical Electrodynamics." Wiley, New York.

Jasik, H., ed. (1961). "Antenna Engineering Handbook," Chapter 14. McGraw-Hill, New York.

Kaifu, N. (1980). Radio telescope project of Japan. *In* "Collection of Abstracts of Invited Papers," URSI Symposium on Millimeter Technology in Radioastronomy, Grenoble.

Kogelnik, H. (1964). Coupling and conversion coefficients for optical modes. *Polytechnic Inst. Brooklyn Symp. Quasi-Opt., 1964* pp. 333–347.

Kogelnik, H., and Li, T. (1966). Laser beams and resonators. *Proc. IEEE* **54**, 1312–1329.

Kraus, J. D. (1950). "Antennas." McGraw-Hill, New York.

Krautle, H., Sauter, E., and Schultz, G. V. (1977). Antenna characteristics of whisker diodes used as submillimeter receivers. *Infrared Phys.* **17**, 477–483.

Laboratory for Insulation Research (1957). "Tables of Dielectric Materials," Vol. 5. Massachusetts Institute Technology, Cambridge.

Lambert, D. K., and Richards, P. L. (1978). Martin–Puplett interferometers: An analysis. *Appl. Opt.* **17**, 1595–1602.

Larsen, T. (1962). A survey of the theory of wire grids. *IRE Trans. Microwave Theory Tech.* **MTT-10**, 191–201.

Levy, R. (1966). Directional couplers. *Adv. Microwaves* **1**, 121–122.

Li, T. (1965). Diffraction loss and selection of modes in Maser resonators with circular mirrors. *Bell Syst. Tech. J.* **44**, 917–932.

Lidholm, S., and de Graauw, T. H. (1979). A heterodyne receiver for submillimeter-wave astronomy. *Proc. Int. Conf. Infrared Millimeter Waves Their Applications, 4th.*

Love, A. W., ed. (1976). "Electromagnetic Horn Antennas." IEEE Press, New York.

Love, A. W., ed. (1978). "Reflector Antennas." IEEE Press, New York.

Lowenstein, E. V., Smith, D. R., and Morgan, R. L. (1973). Optical constants of far infrared materials. 2. Crystalline solids. *Appl. Opt.* **12**, 398–406.

MacFarlane, G. G. (1946). Surface impedance of an infinite parallel-wire grid at oblique angles of incidence. *Proc. Inst. Electr. Eng., Part 3A* **93**, 1523–1527.

Marcuse, D. (1975). "Light Transmission Optics," Chapter 6. Van Nostrand-Reinhold, Princeton, New Jersey.

Marcuvitz, N. (1951). "Waveguide Handbook," Chapter 5. McGraw-Hill, New York.

Martin, D. H., and Lesurf, J. (1978). Submillimeter wave optics. *Infrared Phys.* **18**, 405–412.

Martin, D. H., and Puplett, E. (1969). Polarised interferometric spectrometry for the millimetre and submillimetre spectrum. *Infrared Phys.* **10**, 105–109.

Mizusawa, M., and Kitsuregawa, T. (1973). A beam-waveguide feed having a symmetric beam for cassegrain antennas. *IEEE Trans. Antennas Propagation* **AP-21**, 884–886.

Mok, C. L., Chambers, W. G., Parker, T. J., and Costley, A. R. (1979). The far-infrared performance and application of free-standing grids wound from 5 μm diameter tungsten wire. *Infrared Phys.* **19**, 437–442.

Morris, D. (1978). Chromatism in radio telescopes due to blocking and feed scattering. *Astron. Astrophys.* **67**, 221–228.

Murphy, R. A., Bozler, C. O., Parker, C. D., Fetterman, H. R., Tannenwald, P. E., Clifton, B. J., Donnelly, J. P., and Lindley, W. T. (1977). Submillimeter heterodyne detection with planar GaAs Schottky-Barrier diodes. *IEEE Trans. Microwave Theory Tech.* **MTT-25**, 494–495.

Ohtera, I., and Ujiie, H. (1975). Nomographs for phase centers of conical corrugated and TE_{11} mode horns. *IEEE Trans. Antennas Propagation* **AP-23**, 858–859.

Padman, R. (1978). Lenses for widebead corrugated conical horns. *Electron. Lett.* **14**, 311–312.

Parker, T. J., Ford, J. E., and Chambers, W. G. (1978). The optical constants of pure fused quartz in the far-infrared. *Infrared Phys.* **18**, 215–219.

Payne, J. M. (1976). Switching subreflector for millimeter wave radio astronomy. *Rev. Sci. Instrum.* **47**, 222–223.

Payne, J. M., and Ulich, B. L. (1978). Prism beamswitch for radiotelescopes. *Rev. Sci. Instrum.* **49**, pp. 1682–1683.

Payne, J. M., and Wordeman, M. R. (1978). Quasi-optical diplexer for millimeter wavelengths. *Rev. Sci. Instrum.* **49**, 1741–1743.

Perry, G. S. (1970). Dielectric properties of titanate-alumina ceramics. *IEEE Conf. Publ.* **67**, 63–66.

Potter, P. D. (1963). A new horn antenna with suppressed sidelobes and equal beamwidths. *Microwave J.* **6**, 71–78.

Pursley, W. K. (1956). The transmission of electrmagnetic waves through wire diffraction gratings. Ph.D. Dissertation, University of Michigan, Ann Arbor.

Rutledge, D. B., Schwartz, S. E., and Adams, A. T. (1978). Infrared and submillimetre antennas. *Infrared Phys.* **18**, 713–719.

Ruze, J. (1966). Antenna tolerance theory—a review. *Proc. IEEE* **54**, 633–640.

Saleh, A. A. M. (1974). An adjustable quasi-optical bandpass filter Part I. Theory and design formulas. *IEEE Trans. Microwave Theory Tech.* **MTT-22**, 728–739.

Saleh, A. A. M., and Semplak, R. A. (1976). A quasi-optical polarization-independent diplexer for use in the beam feed system of millimeter-wave antennas. *IEEE Trans. Antennas Propagation* **AP-24**, 780–785.

Schell, R. G., and Tyras, G. (1971). Irradiance from an aperture with a truncated Gaussian field distribution. *J. Opt. Soc. Am.* **61**, 31–35.

Schwartz, S., and Rutledge, D. B. (1980). Moving towards NMM wave integrated circuits. *Microwave J.* **23**, 47–52.

Siegman, A. E. (1971). "An Introduction to Lasers and Masers," Chapter 8. McGraw-Hill, New York.

Silver, S. (1949). "Microwave Antenna Theory and Design," Chapter 11. McGraw-Hill, New York.

Simmons, A. J., and Kay, A. F. (1966). The Scalar feed—a high performance feed for large paraboloid antennas. *IEE Conf. Publ.* **21**, 213–217.

Simonis, G. J. (1980). Private communication.

Simonis, G. J., and Felock, R. D. (1980). Quasi-optical near-millimeter optical measurements. *Proc. Int. Conf. Infrared Submillimeter Waves, 5th, 1980* Suppl., p. 2.

Taub, J. J., Hindin, H. J., Hinckelmann, O. F., and Wright, M. L. (1963). Submillimeter components using oversize quasi-optical waveguide. *IEEE Trans. Microwave Theory Tech.* **MTT-11**, 338–345.

Thomas, B. M. (1978). Design of corrugated conical horns. *IEEE Trans. Antennas Propagation* **AP-26**, 367–372.

Turrin, R. H. (1967). Dual mode small aperture antennas. *IEEE Trans. Antennas Propagation* **AP-15**, 307–308.

Ulich, B. L., Davis, J. H., Rhodes, P. L., and Hollis, J. M. (1980). Absolute brightness temperature measurements at 3.5-mm wavelength. *IEEE Trans. Antennas Propagation* **AP-28**, 367–377.

Ulrich, R. (1967). Far-infrared properties of metallic mesh and its complementary structure. *Infrared Phys.* **7**, 37–55.

Ulrich, R., Renk, K. F., and Genzel, L. (1963). Tunable submillimeter interferometers of the Fabry–Perot type; *IEEE Trans. Microwave Theory Tech.* **MTT-11**, 363–371.

Ulrich, R., Bridges, T. J., and Pollack, M. A. (1970). A variable output coupler for far infrared lasers. *Polytech. Inst. Brooklyn Symp. Submillimeter Waves, 1970* pp. 605–614.

van Nie, A. G. (1964). Rigorous calculation of the electromagnetic field of wave beams. *Philips Tech. Rev.* **19**, 378–394.

van Vliet, A. H. F., and de Graauw, T. H. (1981). Quarter wave plates for submillimetre wavelengths. *Int. J. Infrared Millimeter Waves* **2**, 465–477.

von Hipple, A. (1954). "Dielectric Materials and Applications," pp. 291–433. Wiley, New York.

Wait, J. R. (1954). Reflection at arbitrary incidence from a parallel wire grid. *Appl. Sci. Res., Sect. B* **4,** 393–400.

Wait, J. R. (1957). The impedance of a wire grid parallel to a dielectric interface. *IRE Trans. Microwave Theory Tech.* **MTT-5** 99–102.

Wannier, P. G., Arnaud, J. A., Pelow, F. A., and Saleh, A. A. M. (1976). Quasioptical band-rejection filter at 100 GHz. *Rev. Sci. Instrum.* **47,** 56–58.

Watanabe, R., and Nakajima, N. (1978). Quasi-optical channel diplexer using Fabry–Perot resonator. *Electron. Lett.* **14,** 81–82.

Wolff, E. A. (1966). "Antenna Analysis," Chapter 10. Wiley, New York.

Wrixon, G. T., and Kelly, W. M. (1978). *Infrared Phys.* **18,** 413–428.

CHAPTER 6

Far-Infrared and Submillimeter-Wavelength Filters*

G. D. Holah

Engineering Experiment Station
Georgia Institute of Technology
Atlanta, Georgia

I. Introduction

In most forms of spectroscopy, for example, grating or Fourier transform, the spectrum obtained is not necessarily unique. The wavelengths present at some particular point in space after diffraction by a grating include all of those satisfying the grating equation

$$n\lambda = 2d \cos \theta,$$

* Much of the work discussed in this chapter was performed while the author was with the Physics Department, Heriot-Watt University, Edinburgh.

where n is an integer, d the grating periodicity, and θ an angular term that depends on the particular spectrometer design. It is immediately seen from this equation that for any value of θ there will be wavelengths λ_1 ($n = 1$), $\lambda_1/2$, $\lambda_1/3$, etc., present simultaneously. For conventional black-body sources used in the infrared, the energy of the shorter wavelengths, i.e., the higher orders, is much higher than that of the first-order wavelength, which is usually the one under investigation. This is because the infrared region corresponds to the tail of the Planck curve. To make sure that the information at a particular grating setting is uniquely related to the wavelength λ_1, some means of rejecting all higher orders before they reach the detector is required. In addition to making the data ambiguous, the presence of intense short-wavelength radiation would soon overload most detectors. To reject the higher orders, various techniques of filtering have been used. These range from natural absorption in alkali halides, semiconductors, and plastics to more sophisticated interference filters whose properties can be more easily tailored for any spectral region.

Filters are needed, not only for grating spectroscopy, but also for the fourier transform spectrometers in which finite sampling intervals have the effect of introducing spurious data unless all frequencies above a particular value, the aliasing frequency, are optically removed (Chantry, 1971). The type of filters required in these cases are termed *edge* filters or *low-frequency pass* filters.

These are not the only type of filter used, however, and it is often required that we be able to isolate a small spectral range about some predetermined wavelength. In this case, bandpass filters are needed. For example, our intention may be to map the emission of some particular line source in the galaxy, e.g., the OH radical. This can be done simply by having a narrow bandpass filter centered at the wavelength of interest. Line-isolation filters are also used in selecting just one wavelength from lasers that emit more than one line or whose lasing transition may also be accompanied by nonlasing discharge lines.

Thus various types of filters are defined according to how they respond to radiation—low-frequency pass, high-frequency pass, and bandpass— with either narrow or broad characteristics. Further experiments may be required to enable the rejection of long-wavelength radiation, but this ability is not often required, so it will be considered in somewhat less detail.

It has been possible for quite some time to make use of evaporated dielectric layers, stacked according to well-defined designs, to produce filters for the visible region of the spectrum (McLeod, 1969). These are called *all-dielectric multilayer interference filters*. The extension of these techniques to the infrared has proven to be difficult for wavelengths between 10 and 25 μm and almost impossible for longer wavelengths. This has been a severe limitation in many aspects of infrared investigation because only interference filters can produce high-performance filters at any desired wavelength. The last ten years, however, has seen the application of interference filters that use metallic mesh as the reflecting interfaces, and it is now possible to cover the region $\lambda > 50$ μm completely (Renk and Genzel, 1962; Ulrich, 1967a,b, 1968; Holah and Smith, 1972; Holah and Auton, 1974; Holah and Morrison, 1977; Holah and Smith, 1977; Holah, Davis, and Morrison, 1978; Holah,, Morrison, and Goebel, 1979; and Holah, Davis, and Morrison, 1979).

Interference filters are not the only means of restricting optical bandwidths, and there are a number of other important techniques. In some cases, for example, absorption in direct-gap semiconductors, filters that are as good as interference devices can be made. The problems associated with most of these techniques are the lack of flexibility and the limited range.

We shall begin our discussion with noninterference techniques in Section II and procede to interference techniques in Section III.

II. Noninterference Filters

A. RESTSTRAHLEN FILTERS

One of the interesting phenomena under investigation through the use of infrared radiation is the vibrational dynamics of atoms in solids. The frequency at which atoms vibrate and hence solids absorb, given that an electric dipole transition is allowed by symmetry, occurs in the infrared region for most crystals. The frequency of vibration is a function of the interatomic forces and the masses of the vibrating atoms (Houghton and Smith, 1966).

Under the simplest assumptions, Hooke's law and diatomic crystals, it can readily be shown that the absorption frequency is given by

$$\omega^2 = f/m, \tag{1}$$

where f is some effective force constant between the atoms and m the reduced mass of the atoms, which, for a simple binary compound, is given by

$$1/m = (1/m_1) + (1/m_2). \tag{2}$$

A list of frequencies for a number of common crystals is given in Table I.

From standard dispersion theory, expressions for the reflectivity asso-

TABLE I

Absorption Maxima for Common Materials[a]

Material	λ (μm)	$\bar{\nu}$ (cm^{-1})	Material	λ (μm)	$\bar{\nu}$ (cm^{-1})
CsF	78.6	127	RbBr	112	89.3
	87	115	TlBr	210	47.6
KF	52.1	192	CdBr$_2$	156	64.1
	52.6	190	CsI	163	61.3
LiF	32.9	304	KI	99	101
NaF	40.6	246	LiI	69.4	144
RbF	62.5	160	NaI	85.6	117
	64.1	156	RbI	134	74.6
TlF	67.5	148	TlI	193	51.8
BaF$_2$	54.3	184			
	52.9	189	AlSb	31.9	314
CaF$_2$	38.9	257	GaSb	44.4	225
	37.6	266	InSb	57.6	174
SrF$_2$	45.6	219	InAs	47.8	210
	46.1	217	GaAs	37.5	266
			InP	32.9	304
AgCl	97.1	103	GaP	27.3	366
CsCl	101	99	ZnSe	46.5	215
			CdSe	54.1	185
KCl	70.8	141	CdS(Z)[b]	37.1	270
LiCl	58.5	171	CdS(W)[b]	38.4	260
	62.9	159	PbS	154	64.9
NaCl	61.0	163			
RbCl	84.1	119	CdTe	66.6	150
TlCl	159	62.8	ZnTe	52.6	190
			ZnS(Z)[b]	32.3	310
AgBr	125	80	ZnS(W)[b]	33.3	300
			HgTe	85	118
CsBr	135	74	Mg$_2$Sn	53	189
KBr	86.5	116	Mg$_2$Si	36	278
	84.7	118	Mg$_2$Ge	45	222
LiBr	58.5	171			
	62.8	159	MgO	24.8	403
NaBr	73.9	135	ZnO	24.2	413

[a] Reproduced from Houghton and Smith (1966).
[b] (Z), zinc blende structure; (W), wurzite structure.

ciated with this absorption can be obtained, together with expressions for the complex dielectric constant:

$$\varepsilon(\omega) = \varepsilon'(\omega) + i\varepsilon''(\omega)$$
$$= \varepsilon_\infty + S\omega_0^2/(\omega_0^2 - \omega^2 - i\Gamma\omega_0) \tag{3}$$

and

$$R(\omega) = \{[n(\omega) - 1]^2 + k^2(\omega)\}/\{[n(\omega) + 1]^2 + k^2(\omega)\}, \qquad (4)$$

where

$$\varepsilon'(\omega) = n^2 + k^2, \qquad \varepsilon''(\omega) = 2nk, \qquad (5)$$

$\varepsilon'(\omega)$, $\varepsilon''(\omega)$, $n(\omega)$, and $k(\omega)$ are the real and imaginary parts of the dielectric constants and the refractive index, respectively, and k is sometimes called the *extinction coefficient*.

The reflectivity calculated for an arbitrary system is shown in Fig. 1a for various values of Γ, the damping. It can be seen that there exists a region bounded by ω_0 and ω_l in which the reflectivity is high. The extent of this region is determined by the oscillator strength S, and the magnitude of the reflectivity is determined by Γ.

White light incident upon such crystals will therefore be spectrally separated according to the values of reflectivity, as shown in Fig. 1b. The region between ω_0 and ω_l may have 90% reflectivity, whereas outside this region, the reflectivity may be $\sim 5\%$. By using only the reflected radiation, some degree of filtering can be achieved. Usually, to obtain sufficiently high contrast, as number of reflections are required, which in turn reduces the level of wanted radiation.

Use of a variety of materials can produce a limited number of bandpass filters. This technique was widely used in the early days of infrared physics but finds little or no application today. However, the same process can be used to produce transmission filters.

B. NATURAL ABSORPTION FILTERS

In addition to the absorption of radiation by lattice vibrations, electronic transitions across the band gaps of semiconductors are also an important absorption mechanism. The other primary absorption process is that due to intraband electronic absorption. This latter process is relatively featureless, but the two mechanisms have well-defined absorption regions, which provide extremely useful bases for filtering. One advantage of these types of filters is that they are used in transmission. Useful lattice-vibrational absorption is found in a number of plastics as well as the ionic and covalent solids.

1. Semiconductor Interband Absorption

A semiconductor is so-called because of the position of the Fermi level with respect to filled and unfilled bands (Kittel, 1976). A semiconductor has the interesting property that there is a certain energy E_g (the energy

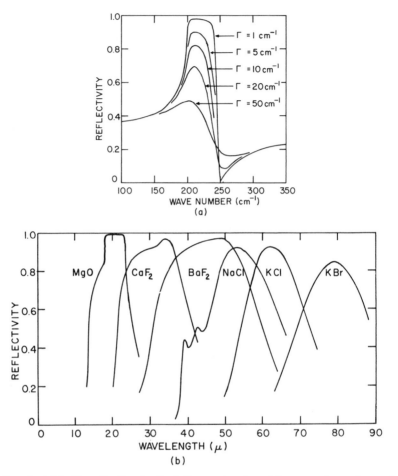

FIG. 1   (a) Lattice-vibrational reflectivity as a function of frequency for various values of the damping parameter Γ. Reststrahlen frequency is constant at 200 cm^{-1}. (b) Lattice reflectivity of various materials as a function of wavelength. For good filtering, more than one reflection would be needed. (Reproduced from Houghton and Smith, 1966.)

gap) above which photons can excite electron–hole pairs across the band gap and hence be absorbed and below which the photon is transmitted.

The transition from absorption to transmission takes place very rapidly, within a few milli-electron-volts, and the levels of absorption also change by many orders of magnitude. The absorption coefficient α is defined by

$$I(z) = I_0 e^{-\alpha z}(1 - R)^2/(I - R^2 e^{-2\alpha z}), \qquad (6)$$

where $I(z)$ is the intensity of radiation at a distance z inside the crystal, I_0 the incident intensity, α the absorption coefficient, and R the reflectivity.

In the case where the reflectivity is low (as in all the semiconductor edge filters discussed here in which the surfaces will be antireflective), Eq. (6) (for $R < 0.1$) reduces to

$$I(z) = I_0 e^{-\alpha z}, \tag{7}$$

where α can change from 10^4 cm^{-1} in the absorption region to <10 cm^{-1} in the transmission region.

Not all semiconductors can be used as low-pass filters as they were in the preceding discussion, which was concerned mainly with direct-gap transitions in which extrema in the band structure occur at the zone center (see Fig. 2a). The transition takes place vertically without the involvement of a quantum of vibrational energy (phonon). In so-called indicrect-gap semiconductors, the transition takes place between the maximum in the valence band, at the zone center, and the minimum in the conduction band, which is not at $k = 0$ (see Fig. 2b). This requires the involvement of phonons to ensure the conservation of momentum. Because of the extent of the phonon density of states, many phonons can be involved, which broadens out the absorption edge of the semiconductor, reducing its usefulness as a low-pass filter.

The four most commonly used semiconductors are Si, Ge, InAs, and InSb with edges at 1.1, 1.8, 1.3, and 7.3 μm, respectively, at room temperature. The spectral characteristics of these are shown in Fig. 3. The thickness of the crystals is kept below 150 μm to prevent excessive absorption in the passband, and the surfaces are all antireflective.

It is possible to modify the position of the edge of a semiconductor by

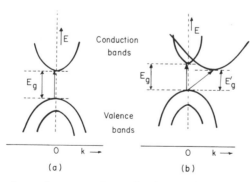

FIG. 2   Simple electronic energy scheme showing (a) direct and (b) indirect ($E_g^1 < E_g$) transitions for semiconductors.

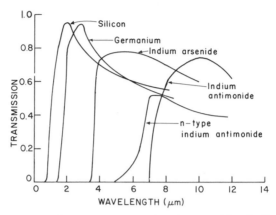

FIG. 3 Transmission of some common semiconductors with antireflection-coated surfaces. (Reproduced from Seeley and Smith, 1966.)

doping, i.e., the deliberate adding of impurities with excess carriers relative to the host lattice. For example, consider InSb, a III–V compound with a band gap of 7.3 μm in its intrinsic state. If atoms of a group IV compound, e.g., Te, were used to replace indium, there would be an excess of electrons, which is called an n-type extrinsic. Similarly, a surplus of holes (from a group II) is referred to as a p-type. The excess electrons in n-type InSb "fill up" the conduction band, so the first vacant site in the conduction band to which an electron in the valence band can be excited is now at a higher energy. Direct vertical transitions are still made, but now the effective band gap has been increased relative to that of intrinsic InSb. In this way, an absorption edge near 5.3 μm can be obtained.

These semiconductors are the only ones that have become standard edge filters. Although near their edge they have almost ideal filter characteristics, there are other considerations that have restricted the use of semiconductor absorption: the insufficient variation in band gaps; the considerable reststrahlen absorption in all binary compounds; and the free carriers' increasing absorption toward longer wavelengths. Although it is possible to alloy semiconductors and semimetals to produce compounds such as $Cd_xHg_{1-x}Te$ in which the band gap is a function of x, small band gaps are always accompanied by increased free-carrier absorption. Because, on a simple scattering picture, the free-carrier absorption coefficient is proportional to λ^2, this severely limits the useful transmission range.

Semiconductors such as those just mentioned are also used as substrates for interference filters, which will be discussed later. Semiconductor edge filters can be obtained that have edges between 1 μm and 7 μm. Natural absorption by lattice vibrations can also be used and will be discussed next.

2. *Filters Using Lattice-Vibrational Absorption*

In addition to interband electronic absorption in semiconductors, another important absorption process occurring in the infrared is that due to lattice or molecular vibrations. Many ionic and polymer materials exhibit absorption regions and transmission regions that can be used as high-pass or low-pass filters. A variety of materials are used, and in some cases it is advantageous to operate them at low temperatures.

Polymers such as black polyethylene and polytetrafluorethylene (PTFE) find significant use in the infrared. Black polyethylene cuts out the whole of the visible region and up to 4 μm without significant absorption loss above 60 μm, and PTFE has a natural absorption edge in the 50-μm area. The transmission of typical samples of these are shown in Fig. 4. Also shown in Fig. 4 is quartz, which has a useful absorption edge near 40 μm. Of course these materials do not reject all wavelengths below the edge, and in most cases it is necessary to use all of these together or in conjunction with other types of filters.

In general, most ionic materials have limited use as high-frequency pass filters. Before discussing these, the use of some of these at low temperatures as low-frequency pass filters will be considered. At low temperatures, the absorption bands are sharpened considerably; this is demonstrated most clearly for the cases of CsI at 4 °K (Armstrong and Low,

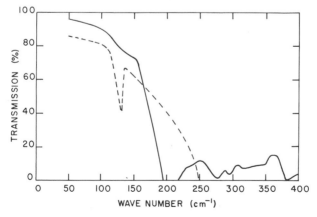

FIG. 4   Transmission of 950-m-thick PTFE (Teflon) (solid curve) and 1-mm-thick z-cut crystal quartz at room temperature. Note characteristic absorption line of quartz at 125 cm⁻¹.

1973; Hadni, 1967) shown in Fig. 5. Because most far-infrared detectors operate at 4 °K, it is not necessarily inconvenient to use cooled filters. Also shown in Fig. 5 is CaF_2, together with the effect of thickness and antireflection coatings (Armstrong and Low, 1973). These polyethylene antireflection coatings offer some improvement in performance. Again, however, short-wavelength blocking is needed. The performance of sapphire can also be improve by antireflection coatings. Reflection losses are important and any effort to reduce them should be made. Assuming negligible absorption, the relation between reflection and refractive index is given by

$$R = (n - 1)^2/(n + 1)^2 \qquad (8)$$

For $n = 2$, $R = 0.11$; $n = 3$, $R = 0.25$; $n = 4$, $R = 0.36$. This should be applied to both faces of the filter.

The simplest method for antireflecting a surface is a $\lambda/4$ thickness of a material whose refractive index is $n = (n_2 n_1)^{1/2}$, where n_2 and n_1 are the refractive indices of the media on each side of the antireflection layer. Usually, of course, $n_1 = 1$. It should also be appreciated that antireflection can only be optimized at wavelengths given by

$$\lambda = nd(4/p), \qquad p = 1, 3, 5, \ldots . \qquad (9)$$

Thus some fringing may occur with the minima having a reflectivity equal to nonantireflected surfaces. In the far-infrared, it is virtually impossible to use standard dielectric evaporation to produce the antireflection layer because of the difficulty of evaporating thick homogeneous

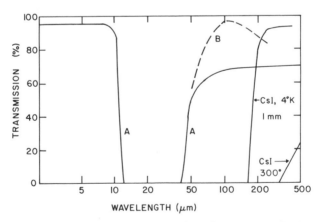

FIG. 5   Transmission of CaF_2 and CsI showing effects of antireflecting and cooling. Curves A are for 10-mm CaF_2 at 4 °K and curve B is for 10-mm CaF_2 plus 13.5-μm polyethylene. (Taken from Armstrong and Low, 1973 and Hadni, 1967.)

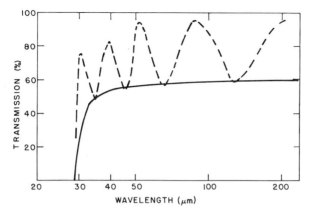

FIG. 6.   Effect of antireflecting sapphire. The solid curve is for 1-mm-thick synthetic sapphire at 4 °K; the dashed curve is for 1-mm-thick synthetic sapphire plus a 0.045-mm-thick antireflection coating of polyethylene. Thickness of coating gives rise to modulation of transmission depending on whether the optical thickness of the coating is λ/4. (Reproduced from Armstrong and Low, 1974.)

films with adequate adhesion. However, it has been shown (Armstrong and Low, 1974) that thermal bonding polyethylene on to the filter material can produce some reduction in reflection losses for a range of materials. The results are shown in Fig. 6.

Thus far we have discussed only low-frequency pass filters that make use of transmission regions at wavelengths longer than band gaps in semiconductors or longer than reststrahlen absorption in ionic crystals or polymer molecule absorption. It is also possible to use ionic materials as high-pass filters with transmission regions at wavelengths shorter than the reststrahlen. The transmission region is limited mainly by the onset of two-phonon absorption. Some of these materials, shown in Fig. 7, find important uses as window materials in the near- or mid-infrared, particularly NaCl (even though it suffers from hygroscopy) and KRS-5 (a mixture of thallous iodide and bromide.)

Although filters using natural absorption are simple to use, even at low temperatures, they are by no means the answer to filter problems for a number of reasons: insufficient variety of edge position, limited bandwidth, and poor blocking, at least for the ionic and polymer materials. Only black polyethylene and the semiconductor filters will reject the visible.

Some attempts have been made to extend the range by using a technique introduced by Yoshinaga; they are appropriately called *Yoshinaga filters* (Yamada *et al.*, 1962) and will be discussed next.

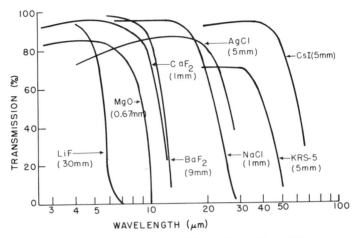

FIG. 7   Transmission of some materials useful as high-pass filters. The long-wavelength cutoff can be varied by changing the material thickness. (Reproduced from Houghton and Smith, 1966.)

C. YOSHINAGA FILTERS

Yoshinaga filters are produced by mixing powdered crystals together with a polymer powder, heating the two together, and pressing out a disk when cooled. Although these are, in essence, simple to make, there are a number of important factors in optimizing the transmission characteristics.

Examples of these filters have been given in considerable detail by Baldecchi and Melchiorri (1973). A few of the most useful ones are shown in Fig. 8. As in the previous section, the transmission shapes are not ideal. Here there is usually considerable structure in addition to poor rejection of the visible. A combination of soot, BeO, and fused SiO_2 does cut out most of the visible and near-infrared and is useful as a general-purpose blocking filter. Other similar combinations are also shown in Fig. 8. Again, these filters also suffer from inflexibility in design because band shapes and edges cannot be tailored to suit particular requirements.

The filtering mechanism is partially by scattering in addition to absorption by the powders. It has been concluded by Baldecchi and Melchiorri that powder size is important; the smaller the particle size, the sharper the cuton. The distribution of particles is also important.

At wavelengths small compared with the powder size and separation, the filtering is by diffraction, refraction (hence refractive index is important), and reflection. At wavelengths comparable to the dimensions, Mie scattering is dominant, and for long wavelengths, Rayleigh scattering is the only loss and this is reduced toward longer wavelengths.

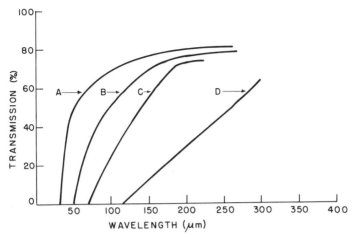

FIG. 8 Series of Yoshinaga filters. Composition: curve A, 1% soot, 25% BeO, 5% MnO$_2$, 5% CaF$_2$, thickness 200–230 μm; curve B, 3% soot, 25% BeO, 20% fused quartz, 10% NaF, 5% CaF$_2$, thickness 280–330 μm; curve C, 3% soot, 25% BeO, 20% fused quartz, 30% ZrO$_2$, thickness 200–280 μm; curve D, 3% soot, 13% BeO, 20% fused quartz, 40% Bi$_2$O$_3$, 25% BaCO$_3$, 25% PbCl$_2$, thickness 300 μm. (Curves reproduced from Baldecchi and Melchiorri, 1973.)

Scatter filters using diamond particles of various size embedded in polyethylene substrates have been described by Armstrong and Low (1974) and are shown in Fig. 9. The mechanisms for these scatter filters are the same as for Yoshinaga filters.

D. SELECTIVE CHOPPING

Many of the systems used in infrared instrumentation make use of phase-sensitive detection. Essentially, this consists of modulating the radiation observed by the detector at a frequency ω and the using electronic processing to amplify and detect only those signals at the frequency ω. This helps to reduce signals and noise from sources other than that under investigation. For example, although most infrared detectors respond to room lights, unless these lights have frequency ω contained in their output, they will be essentially rejected by the processing electronics.

This technique makes possible another use of natural absorption filters. In the normal situation, the radiation is modulated by a metal chopper of alternate metal blades and open spaces. This gives 100% modulation at all wavelengths. Consider, however, the situation where the metal blades are replaced by some absorber such as mica. The transmission spectrum of mica is shown in Fig. 10. With mica absorbing beyond 7 μm when used as

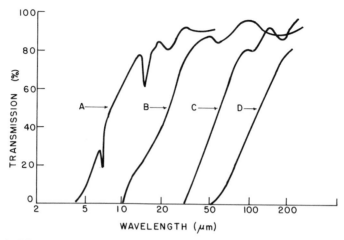

FIG. 9 Diamond-powder scatter filters. Curve A, industrial diamond, particles 1–5 μm in diameter, layer approximately two particles thick, deposited on polyethylene; curve B, 5–10-μm-diameter diamond on a polyethylene substrate; curve C, 15–25-μm-diameter diamond on a clear plastic substrate; curve D, 40–60-μm-diameter diamond on a clear plastic substrate. (Curves reproduced from Armstrong and Low (1974.)

a chopper blade, wavelengths shorter than 7 μm will, therefore, not be modulated or will be modulated to a lesser degree than with a metal blade and hence will not accepted by the phase-sensitive detector. Because this technique does not require significant modification to equipment, it is easy to use, although complete rejection is usually not possible, having rejection problems similar to those of reflection filters.

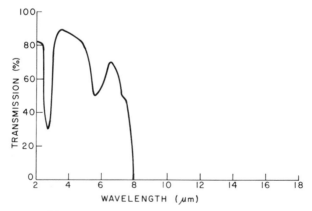

FIG. 10   Transmission of 0.06-mm-thick mica.

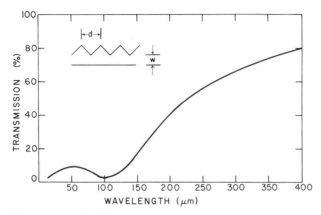

FIG. 11   Transmission of echelette grating on polyethylene, where d = 333 μm. (Reproduced from Houghton and Smith, 1966.)

E.   RULED POLYETHYLENE GRATINGS

Diffraction gratings are also useful as filters, and a technique based on them has been described by Möller and McKnight (1963). These filters consist of polyethylene sheets with gratings pressed into them. They are made by heating the polyethylene between two ruled formers. The shape of the groove cross section is right-angled, as shown in Fig. 11. The gratings on each side are orthogonal in order to remove the polarization effects associated with gratings. The performance of a typical filter is shown in Fig. 11. This is not a particularly useful filter because the cuton is too gradual and the rejection region may have leaks. It may be useful, however, when used in conjunction with other filters.

III.   Interference Filters

The previously discussed filtering techniques are clearly not ideal. What is required is a technique that can be used to make low-pass, high-pass, and bandpass filters at any predetermined position and steepness of edge, with high transmission in pass region and good rejection at all wavelengths outside the pass region. These criteria can be satisfied by interference filters, which have generally replaced most, if not all, of the previous methods. Although the actual process of producing the reflecting elements of interference filters depends on the wavelength, the design features are the same and will be discussed first.

It has been shown that any interference filter may be discussed in terms of only two effective interfaces (Smith, 1958). The two interfaces can be characterized by frequency-dependent reflection and transmission char-

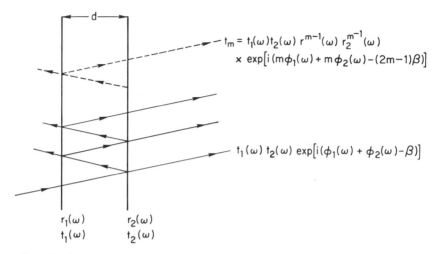

FIG. 12   Two effective interface analyses for multiple-beam interference, where t_m is the transmitted amplitude of the beam after m passages across the layer.

acteristics. These interfaces may themselves be a composition of many interfaces. Figure 12 shows the two effective interfaces, where the separation of the two interfaces is d, the refractive index of the medium between the interfaces is n, and the reflection and transmission coefficients of the two elements are $r_1(\omega)$, $r_2(\omega)$, $t_1(\omega)$, and $t_2(\omega)$, respectively.

Assuming that an electromagnetic field of unit amplitude is incident on the first effective surface at an angle θ to the surface normal, there will be a phase change between reflected and transmitted beams at each surface, $\phi_1(\omega)$ and $\phi_2(\omega)$. For dielectrics, with no absorption, $\phi(\omega) = \pi$ if there is an increase in refractive index across the surface and $\phi = 0$ if there is a decrease in refractive index. In the presence of absorption, $\phi \neq \pi$. As will be discussed later, the values of ϕ for filters other than dielectrics may be different from π or 0.

The transmitted amplitude at surface 1 is $t_1(\omega) \exp[i\phi_1(\omega)]$ and the transmitted amplitude at 2 after the first pass is

$$t_1(\omega)t_2(\omega) \exp\{i[\phi_1(\omega) + \phi_2(\omega) - \beta]\}, \tag{10}$$

where β represents the phase change due to the optical path across the structure with

$$\beta = (2\pi/\lambda)(\text{optical path}) = (2\pi/\lambda)(nd \cos \theta), \tag{11}$$

n the effective refractive index of the layer, and d the physical thickness.

Following the paths of successive reflected and transmitted paths, the transmitted beams have amplitudes

$$\tau_1 = t_1(\omega)t_2(\omega) \exp\{i[\phi_1(\omega) + \phi_2(\omega) - \beta]\},$$

$$\tau_2 = t_1(\omega)r_2(\omega)t_1(\omega)t_2(\omega) \exp\{i[\phi_1(\omega) + \phi_2(\omega)$$
$$+ \phi_1(\omega) + \phi_2(\omega) - \beta - \beta - \rho]\} \tag{12}$$
$$= t_1(\omega)r_2(\omega)r_1(\omega)t_2(\omega) \exp\{i[2\phi_1(\omega) + 2\phi_2(\omega) - 3\beta]\},$$

$$\tau_m = r_1^{m-1}(\omega)r_2^{m-1}(\omega)t_1(\omega)t_2(\omega) \exp\{i[m\phi_1(\omega) + m_2\gamma_2(\omega) - (2m - 1)\beta]\},$$

where m is the number of passages across the layer. To obtain the total amplitude of the transmitted beam, the previous terms are summed, giving

$$t(\omega) = \frac{t_1(\omega)t_2(\omega) \exp(-i\beta)}{1 - r_1(\omega)r_2(\omega) \exp\{i[\phi_1(\omega) + \phi_2(\omega) - 2\beta]\}}. \tag{13}$$

The total *intensity* of the transmitted beam is given by

$$T(\omega) = t^*(\omega)t(\omega)$$
$$= \frac{t_1^2(\omega)t_2^2(\omega)}{1 - r_1(\omega)r_2(\omega) - 2\cos[\phi_1(\omega) + \phi_2(\omega) - 2\beta] + r_1^2(\omega)r_2^2(\omega)}. \tag{14}$$

In order to simplify the following discussion, some assumptions that are usually found in real cases will be made, namely, $t_1(\omega) = t_2(\omega)$, $r_1(\omega) = r_2(\omega)$, $\phi_1(\omega) = 0$ and $\phi_2(\omega) = \pi$. Therefore,

$$T(\omega) = \frac{T_0^2(\omega)}{[1 - R_0(\omega)]^2} \left[1 + \frac{4R_0(\omega)}{[1 - R_0(\omega)]^2} \sin^2 \beta \right]^{-1}, \tag{15}$$

where $T_0(\omega)$ is the frequency-dependent intensity transmission coefficient for each surface and $R_0(\omega)$ the corresponding reflection coefficient.

Equation (15) contains all the information needed to discuss all types of interference filters. To continue the discussion, Eq. (15) will be considered in relation to the design of specific types of filters.

A. BANDPASS FILTERS

To discuss the use of Eq. (15) in designing various filters, some knowledge will be assumed, i.e., experience shows what conditions are needed to produced specific filter types. Consider the case where $R_0(\omega)$ is constant and reasonably high. Under these conditions, Eq. (15) gives rise to a series of peaks whose positions are given by $\sin^2 \beta = 0$, i.e.,

$$(2\pi/\lambda)(nd \cos \theta) = p\pi, \qquad p = 0, 1, 2, 3, \ldots, \tag{16}$$

where p is the order. Equation (16) is identical to that describing a Fabry–Perot etalon.

Equation (15) is plotted in Fig. 13 for various values of $R_0(\omega)$. The posi-

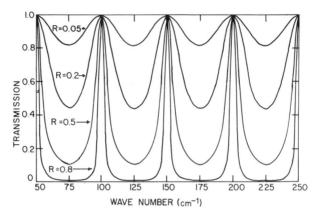

FIG. 13   Transmission of a Fabry–Perot with layer optical thickness 100 μm and zero absorption for various values of reflectivity.

tion of the peaks is given by $p\lambda = 2nd \cos \theta$, where usually, $\theta = 0$ and $p\lambda = 2nd$. The transmission at a maximum is given by

$$T_{\max} = T_0^2(\omega)/[1 - R_0(\omega)]^2. \tag{17}$$

In the absence of absorption, $T_{\max}$ is always unity, independent of $R_0(\omega)$. However, in the case of finite absorption, $T_{\max}$ is a function of $R_0(\omega)$. Returning to Fig. 13, it can be seen that, for high $R_0(\omega)$, the structure behaves as a filter, allowing only a series of small, well-defined ranges of wavelengths through, usually in filter design $p = 1$, i.e., first order is used.

Equation (15) can then be rewritten as

$$T(\omega) = T_{\max}(1 + F \sin^2 \beta)^{-1}, \tag{18}$$

where $F = 4T_0(\omega)/[1 - R_0(\omega)]^2$, and is referred to as the effective number of interfering beams. The function $(1 + F \sin^2 \beta)^{-1}$, which determines the shape of Eq. (18) is plotted in Fig. 14 as a function of β for various values of F. For large values of F, that is, high values of $R_0(\omega)$, the shape will change rapidly near $\beta = p\pi$. Therefore, the half-wave structure for $p = 1$ and $nd = \lambda/2$ will behave as an efficient bandpass filter if the reflectivity is high. Most bandpass filters are based on the half-wave structure.

From expression (16), which gives the positions of the peaks, $p\lambda = 2nd$, it is clear that the wavelength separation of the orders is given by

$$\Delta\lambda = 2nd/p(p + 1). \tag{19}$$

Therefore, the maximum separation occurs between first and second orders, and it is advantageous to work in first order to obtain the widest possible stopband. The wave number separation is independent of order [wave number $\bar{\nu} = 1/\lambda(\text{cm})$].

Having established a method for determining the peak position of a bandpass filter, i.e., $\lambda = 2nd$, the next problem is to determine the factors affecting the shape and maximum transmission of the bandpass filters.

The instrumental function of a filter represents the effect that the filter has on incident monochromatic radiation. The finesse of a filter is defined as the ratio of the separation of consecutive orders to the width of the peak $\delta\bar{\nu}$, i.e. (that is, the full width at half-height),

$$\mathcal{F} = \Delta\lambda/\delta\lambda = \Delta\bar{\nu}/\delta\bar{\nu}, \tag{20}$$

where $\Delta\bar{\nu}$ is the free spectral range (FSR).

Suppose that for order p the intensity falls to half its maximum at points corresponding to

$$\beta = p\pi \pm \Delta\beta, \tag{21}$$

where β allows variations in wavelength, refractive index, thickness, and angle of incidence to be considered as contributions and $\Delta\beta$ is the phase change due to any of these parameters. Near a peak, $\sin^2(p\pi + \Delta\beta)$ is small and, consequently,

$$\sin^2(p\pi + \Delta\beta) \simeq \sin^2 \Delta\beta \simeq \Delta\beta^2. \tag{22}$$

For the half-width $T = \frac{1}{2}T_{\max}$, this applies whether $T_{\max} = 1$ for zero absorption or $T_{\max} < 1$ for absorption. Therefore,

$$\tfrac{1}{2} = (1 + F \, \Delta\beta^2)^{-1} \tag{23}$$

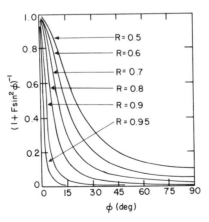

FIG. 14   Curve of $(1 + F \sin^2 \beta)^{-1}$ against $\theta \, [\equiv (2\pi/\lambda)nd]$ for values of $F \, [= 4R/(1 - R)^2]$ (and hence reflectivity). Note rapid increase in slope as R increases.

$$\Delta\beta = 1/\sqrt{F} = [1 - R_0(\omega)]/2[R_0(\omega)]^{1/2} \qquad (24)$$

if $\theta = 0$, $\beta = (2\pi/\lambda)nd$,

$$\Delta\beta/\delta\lambda = 2\pi nd/\lambda^2, \qquad (25)$$

and

$$\delta\lambda = (\lambda^2/2\pi nd)[(1 - R)/2\sqrt{R}], \qquad (26)$$

where $R = R_0(\omega)$ has been used to simplify the equations. The separation of consecutive orders corresponds to a phase change of 2π, therefore, the finesse $\mathscr{F}$ can be written in terms of β as

$$\mathscr{F} = 2\pi/2 \, \Delta\beta = \pi\sqrt{F}/2. \qquad (27)$$

This finesse has been calculated assuming that the two surfaces are perfectly flat and parallel over the width of the radiation beam. This is called the reflection finesse $\mathscr{F}_R$ and can be written as

$$\mathscr{F}_R = \pi\sqrt{R}/(1 - R). \qquad (28)$$

A similar exercise, starting from

$$\beta = (2\pi/\lambda)nd$$

and allowing nd, the optical thickness, to change results in the following expression for the flatness finesse:

$$\mathscr{F}_F = \tfrac{1}{2}s, \qquad (29)$$

where the surfaces are flat and parallel to λ/s. In the visible and near-infrared, the difficulty of obtaining surfaces that are "flatter" than, say, 0.25 μm, leading to a finesse at 5 μm of 20 results in $\mathscr{F}_F$ sometimes being called the limiting finesse. At longer wavelengths, especially beyond 100 μm, this no longer applies, and the reflectivity finesse is usually the limiting contribution.

The other main contribution to half-width is due to the use of noncollimated radiation and as such is not an inherent property of the filter but depends on the mode in which it is being used. In addition to introducing a broadening due to a finite range of angles of incidence, the use of collimated but nonnormal radiation will also move the peak wavelength. This question has been addressed by Pidgeon and Smith (1964) and by Baker and Yen (1967).

Rewriting Eq. (16) for first order, we have

$$\lambda = 2nd \cos\theta. \qquad (30)$$

If θ is increased from 0, using collimated radiation, they λ is decreased, i.e., the peak shifts to shorter wavelengths. This may often be used to "tune" a filter if it is produced at a slightly longer wavelength than required.

Returning to Eq. (16) and having radiation collimated with an angle of incidence normal to the surface, i.e., $\theta = 0$,

$$\lambda_0 = 2nd.$$

For collimated radiation at an angle $\theta \neq 0$,

$$\lambda_1 = 2nd \cos \theta.$$

The shift in wavelength $\lambda_0 - \lambda_1$ is given by $\lambda_0 - \lambda_1 = 2nd[1 - \cos \theta]$, leading to $(\lambda_0 - \lambda_1)/\lambda_0 = 1 - \cos \theta$. In terms of the wave number, which is a more popular unit,

$$(\bar{\nu}_0 - \bar{\nu}_1/\nu_0 = (\cos \theta)^{-1} - 1. \tag{31}$$

Of course, θ is the internal angle within the layer, and in practical terms, refraction must be taken into account and the external angle θ_e used, in which case Eq. (31) is modified to

$$(\bar{\nu}_0 - \bar{\nu}_1/\bar{\nu}_0 = (\cos \theta)^{-1} - 1 \simeq \tfrac{1}{2}[(\sin \theta_e)/n]^2, \tag{32}$$

where n is the refractive index of the layer.

In many practical cases, the incident radiation is in the form of a cone (i.e., focused) with a semiangle of θ_e. Following Pidgeon and Smith, the effect of such a geometry is to broaden and shift the peak according to

$$\delta\bar{\nu} = [(\Delta\bar{\nu}')^2 + (4\bar{\nu}_0/\pi^2 F^2)]^{1/2}, \tag{33}$$

and the half-width

$$\Delta\bar{\nu}' = \bar{\nu}'[(\cos \theta)^{-1} - 1], \qquad F = 2\bar{\nu}_0/\pi\delta\bar{\nu}_c, \qquad \bar{\nu}' = \bar{\nu}_0 + \tfrac{1}{2} \Delta\bar{\nu}'. \tag{34}$$

Equations (33) and (34) express the half-width and shift in peak position due to a cone of incident radiation of semiangle θ. As before, θ is the internal angle and should be corrected according to the refractive index. Note that F is related to the half-width $\delta\bar{\nu}_c$ for collimated radiation. The results for a dielectric multilayer Fabry–Perot-type filter are shown in Fig. 15.

Having discussed the shift and broadening due to finite angles of incidence, there is an additional effect due to noncollimated radiation, which is also demonstrated in Fig. 15. The peak transmission falls as the filter broadens. This situation is also true for the broadening due to reflectivity and flatness. Essentially, the area of the filter can be thought of as being nearly constant but the distribution of energy changes.

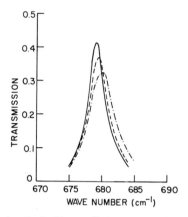

FIG. 15   Effect of varying the incident radiation cone angle on a narrow-band dielectric Fabry–Perot filter. Values for the cone angle are 3°(—), 8.5°(– – –), and 17.5°(– · – ·). Note that both half-width and transmission are degraded as the cone angle increases. (Reproduced from Pidgeon and Smith, 1964.)

1. Absorption Effects

Thus far in the discussion it has been assumed that the absorption in the layer is negligible; this is not the case in practice, however. Absorption gives rise to a loss in transmission of the overall filter.

In Eq. (17), T_{max} is given in terms of the transmission and reflection of each surface as

$$T_{max} = T_0^2(\omega)/[1 - R_0(\omega)]^2. \tag{17}$$

There is a slight problem about the origin of the absorption, i.e., whether it is in the layer or in the surface. This problem is removed if the absorption is assumed to be totally in the layer, and it is assumed that the first surface behaves if it has absorption, transmission, and reflection related by $A + T + R = 1$, i.e., $1 - R = A + T$. Substitution into Eq. (17) leads to

$$T_{max} = [T/(A + T)]^2 = (1 + A/T)^{-2}. \tag{35}$$

It is now important to consider the effect of absorption on T_{max} for various values of reflectivity. From Eq. (28), the reflection finesse is given by

$$\mathscr{F}_R = \Delta\bar{\nu}/\delta\bar{\nu} = \pi\sqrt{R}/(1 - R)$$

and is plotted in Fig. 16a. A useful value of finesse is ~ 30, i.e., $R \geq 0.91$. Suppose that the absorption is 0.01, i.e., for $R = 0.98$, $T = 0.01$ and for

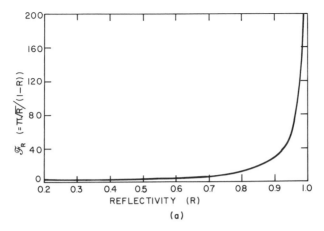

(a)

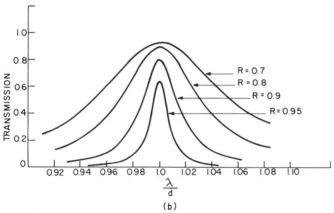

(b)

R	T_{max}	Half-width (%)
0.95	0.64	1.5
0.9	0.8	3.4
0.8	0.9	7.0
0.7	0.93	10.0

FIG. 16 (a) Variation of reflectivity finesse with reflectivity. This demonstrates the need for high reflectivity in Fabry–Perot filters to achieve narrow bandwidths and good rejection. (b) Effect of constant absorption on Fabry–Perot transmission for filters using different reflectivities. As the reflectivity increases, the half-width decreases; however, the effect of the absorption increases rapidly.

$R = 0.91$, $T = 0.08$. Substituting these sets of values into Eq. (35), it follows that

$$T_{max}(R = 0.98, A = 0.01) = 0.25,$$

$$T_{max}(R = 0.91, A = 0.01) = 0.79.$$

This shows that the effect of a small absorption is very important in decreasing T_{max}, increasingly so as R increases (see Fig. 16b).

To summarize the results for Fabry–Perot filters:

(a)   Peaks occur for a given optical separation of the interfaces d at wavelengths given by $p\lambda = 2d \cos \theta$, $p = 1, 2, 3, \ldots$.

(b)   Total half-width of the filter is given by

$$d\bar{\nu}_{total}^2 = d\bar{\nu}_R^2 + d\bar{\nu}_F^2 + d\bar{\nu}_\theta^2 + d\bar{\nu}_S^2,$$

where $d\bar{\nu}_R$ is the half-width determined by the reflectivity, $d\bar{\nu}_F$ that due to lack of parallelism and flatness of the interfaces, $d\bar{\nu}_\theta$ that due to the filter being used with noncollimated radiation, and $d\bar{\nu}_S$ that due to the resolution of the spectrometer being used to measure the filter. Clearly, the latter two half-widths are not due explicitly to the filter, but as we have seen, $d\bar{\nu}_\theta$ does depend to a certain extent on the effective refractive index of the filter.

(c)   If there is no absorption, the maximum transmission of the filter is unity. For the case where absorption is present either in the medium separating the interfaces or in the interfaces themselves, then the maximum transmission is given by

$$T_{max} = [1 + (A/T)]^{-2},$$

where A and T are, respectively, the effective absorptivity and transmittivity of the interface (including absorption between the interfaces). The effect of absorption is greater the higher the finesse and hence reflectivity.

2.   Double Half-Wave Structures

From the discussion in the previous section, it is clear that the Fabry–Perot filter, although capable of producing narrow bandwidths, does not have an ideal shape. Too much energy lies within the skirts of the filter and a squarer passband would be more desirable. Additionally, the high value of F (the effective number of interfering beams) causes the transmission to be reduced considerably in the passband when absorption is present, whereas in the rejection region, where F is low, absorption has little effect. The contrast is therefore quite low.

The performance in terms of shape and rejection can be improved by using a double half-wave structure, which, as its name implies, is two half-wave structures resonant together. Following Smith (1958), who showed that any interference filter can be analyzed in terms of two effective interfaces, we can write the transmission of the filter as

$$T(\omega) = T_1(\omega)T_2(\omega)/[1 - R(\omega)]^2[1 + F(\omega) \sin^2 \beta]^{-1}, \qquad (36)$$

where $R(\omega) = [R_1(\omega)R_2(\omega)]^{1/2}$, $T_1(\omega)$ and $R_1(\omega)$ are, respectively, the transmission and reflection of the single half-wave system, and $T_2(\omega)$ and $R_2(\omega)$ the coefficients for the single surface. It should be remembered that both these entities may be the effect of many layers of dielectric. For the sake of simplicity in the discussion, we shall assume $R_2(\omega)$ and $T_2(\omega)$ to be constant. Then,

$$T_{\max} = T_1(\omega)T_2/[1 - R(\omega)]^2. \qquad (37)$$

We should note that if $R_1 = R_2$ (and, therefore, $T_1 = T_2$, assuming no absorption), then $T_{\max}$ is unity. This is still similar to a single half-wave system. However, because the value of $R_1(\omega)$ i.e., of the single half-wave is a *minimum* at the resonant frequency and is rapidly varying, it follows that not only is R_1 also small but that there are three possible situations depending on the relationship of R_1 and R_2. The three situations are $R_1 = R_2$ at one point only, $R_1 = R_2$ at two frequencies, and $R_1 \neq R_2$ at any frequency. In the first case, we have one maximum peak; for case two, there are two maxima whenever $R_1 = R_2$, for the final case, where $R_1 \neq R_2$, there will be one peak whose transmission is not a maximum.

In order to discuss the salient features, the remaining discussion, following Smith, will specifically relate to a given multilayer filter design. In the usual notation, this is HHLHH, where H is a quarter-wave optical thickness of a high refractive index material and L a quarter-wave optical thickness of a low refractive index material.

The computed transmittance and reflectance curves for this filter are shown in Fig. 17. Curve $R_1(\omega)$ is the reflectance of the Fabry–Perot half-wave filter and hence has the frequency dependence as shown, i.e., at the resonant frequency ω_0, R_1 is a minimum. Curve $R_2(\omega)$ is the frequency-independent reflectance of the H-to-air interface. This particular filter has a number of frequencies where $R_1 = R_2$. At each position, $T(\omega)$ is unity. We are specifically interested in the region near ω_0. In Fig. 17, $R_1 = R_2$ for two values and from Eq. (36), this leads to a double peak. Ideally, of course, we would like R_1 to equal R_2 only at ω_0. Given that the reflectance curve R_2 is essentially the reciprocal of the transmission of the single half-wave, it is clear that the edges of the double half-wave are much steeper and hence more desirable.

The effect of absorption can be understood by considering the number

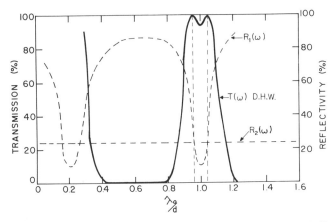

Fig. 17 Reflectivity and transmission of the components of a double half-wave filter. Note that at the resonance wavelength λ_0 the reflectivity of the components is low and hence the transmission is high because absorption effects are negligible.

of effective interfering beams F:

$$F = 4R/(1 - R)^2,$$

where $R = (R_1 R_2)^{1/2}$. Note: $R_1 R_2$ and F are frequency-dependent for the double half-wave (DHW) and we can take $R_1 = R_2$ to be less than 0.1. (This is in contrast to the single half-wave for which $R_1 = R_2$ is high, i.e., >0.9). Under these conditions, we find that $F \leq 1$, and so the effect of absorption is minimized. This, of course, holds only in the passband, so we should expect DHWs to be characterized by high transmission.

Away from the passband, R_1 (the reflectivity of the single half-wave) is high, say, 0.9. Then, assuming R_2 is still low, i.e., ~ 0.1, this leads to a value of F of 63, and hence the effect of absorption in the rejection region is to reduce the transmission, leading to an increase in the contrast ratio.

We see, therefore, that going from a single half-wave to a double half-wave leads to a much squarer passband, higher transmission, and better rejection (see Fig. 18). The width also increases, and hence in some cases where extremely small passbands are required, it is necessary to use single half-wave systems.

The approach using two effective interfaces to analyze any interference filter can be extended to more complex multi-half-wave structures, which are often used when wider passband filters are needed. Unless the effective reflectivities of each subelement of the system are exactly right for a single peak, the number of peaks will, in general, be equal to the number of effective half-waves.

All the filters discussed thus far have been bandpass filters based on the

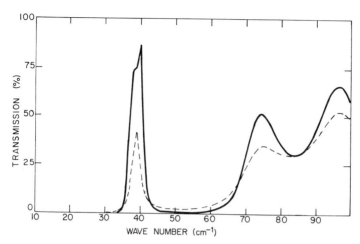

FIG. 18   Single half-wave (– – –) and double half-wave (—) composed of similar elements. Improvements of the double half-wave in transmission and rejection are clear. Half-width is degraded slightly.

basic equation

$$\sin^2 \beta = 0, \qquad p\pi = (2\pi/\lambda)2nd \cos \theta.$$

To consider other conditions, specifically $\sin^2 \beta = 1$, leads to the possibility of other filter designs; in this case, edge filters, either of low-frequency pass or high-frequency pass.

B.   LOW-FREQUENCY PASS-INTERFERENCE FILTERS

The previously discussed filters have all been bandpass filters based on the half-wave system where the optical separation of the reflecting interfaces is a half-wavelength of the wavelength of interest. Returning to Eq. (15), we note that half-wave filters are obtained for the phase condition $\sin^2 \beta = 0$. In order to discuss other types of filters, in this particular case low-pass filters, we find it instructive to reinterpret Eq. (15) for other phase conditions, specifically $\sin^2 \beta = 1$:

$$T = T_{\max}(1 + F \sin^2 \beta)^{-1}, \tag{15'}$$

where

$$\sin^2 \beta = 1, \qquad \beta(= (2\pi/\lambda)nd \cos \theta) = \tfrac{1}{2}(p + 1)\pi, \qquad p \text{ an integer.} \tag{38}$$

In Eq. (38), we have assumed, for ease of discussion, that phase contributions due to phase differences between transmitted and reflected beams at the interface are 0 or π. For values such that $\sin^2 \beta \simeq 1$, $\sin^2 \beta$ varies slowly with β and consequently the total reflectivity and transmission are relatively flat functions of β. For $p = 0$ in Eq. (38), $\beta = \frac{1}{2}\pi$, which, for $\theta = 0$, leads to

$$nd = \tfrac{1}{4}\lambda. \tag{39}$$

This system is naturally called a quarter-wave. The transmission is a minimum when $\sin^2 \beta = 1$, i.e., $\lambda = 4nd$. At wavelengths longer than $4nd$, the transmission rises and, except for varying degrees of modulation, never falls again.

Of course, the exact shape of the curve depends on the values of F, and in Fig. 14 we showed the function $(1 + F \sin^2 \beta)^{-1}$ as a function of β (and hence λ) for various values of F. From these curves, we can deduce that for low values of F at wavelengths greater than $4nd$ a relatively flat passband is produced. Ideally, this should be coupled with high reflectivity (high F) at wavelengths below $4nd$ to give good rejection in the stopband. We require, if possible, a system in which the value of F is frequency-dependent.

Thus far we have only discussed the basic theory in outline; the use of these principles will become more apparent as specific filters are discussed.

There are two main techniques for producing interference filters. The earliest technique is based on evaporated dielectric multilayers in which many layers of alternate high and low refractive index are deposited to produce the final filter. This technique can be used to make high-precision filters for the visible and near-infrared, i.e., $\lambda < 25$ μm. Above this wavelength, the problems associated with evaporating thick homogeneous layers of good adhesion that do not suffer from excessive absorption due to free carriers or lattice vibrations are restrictive. This technique will be discussed first.

IV. Evaporated Dielectric Multilayer Filters

The types of filters we shall be considering in this section will consist of a series of evaporated thin films of various materials with approximately $\lambda/4$ and $\lambda/2$ optical thicknesses. For an average refractive index of 3 and a maximum infrared wavelength of 30 μm, this implies that the maximum physical thickness of layer will be 5 μm and the minimum thickness corresponding to a quarter wave at 5-μm wavelength will be of the order of 5000 Å where we have arbitrarily defined the infrared region as extending from 5 to 30 μm. Because such films are not self-supporting, they have to be deposited on some substrate. The main requirement of such a substrate is that it should be transparent for the region of interest. The most common substrates for evaporated multilayers are semiconductors.

These have sharp transitions from an absorbing to a transmitting region and use of this can be made as auxiliary blocking. This will be demonstrated by considering a sequence of low-frequency pass filters.

A. Low-Frequency Pass Multilayer Filters

In Fig. 3 the transmission in the near-infrared, i.e., $\lambda < 20$ μm, of a series of semiconductors was shown. The reflection losses at both faces of a slice can be reduced using an antireflection coating. In its simplest form, this consists of a quarter-wavelength optical thickness layer whose refractive index is given by

$$n_{AR} = (n_1 n_2)^{1/2}, \tag{40}$$

where n_1 is the refractive index of the medium on the incident side (usually air) and n_2 the refractive index of the medium through which the radiation is passing.

This series of semiconductors produce a good series of low-pass filters. However, there are important gaps roughly at 2.6, 5, 12, and 20 μm. The 20-μm edge represents the limit as far as multilayer filters are concerned because most materials exhibit prohibitively large absorption to longer wavelengths. The rejection achieved using such filters, i.e., the ratio of transmission in the stopband to that in the passband can be as high as 10^6.

To produce filters with such high performance using multilayer techniques is a difficult problem, the solution of which requires filters with alternate quarter-wave thicknesses of high and low refractive index and many layers. These are the important parameters in determining the strength and breadth of the rejection region.

The breadth of the primary rejection region is given by Smith et al. (1972) as

$$(400/\pi) \text{ arc } \sin[(n_H - n_L)/(n_H + n_L)]\% \tag{41a}$$

and the maximum rejection is given by

$$(n_H/n_L)^x, \tag{41b}$$

where n_H and n_L are the values for the high and low refractive indices, respectively, and x the number of layers. We see, therefore, that to have good filters, i.e., large rejection, requires high index contrast and many layers. To achieve the best performance compatible with the problems of evaporating many relatively thick layers, it is advantageous to have as high an index ratio as possible, and therefore the minimum value of x, and to make use of semiconductor edge filters as auxiliary blocking. Detailed filter-profile calculations require a design system that optimizes the transmission in the passband while maintaining good rejection. Various approaches have been used, and a particularly useful method has been found to be based on Tchebyshev equal-ripple functions. This is detailed by Seeley and Smith (1966). The conclusions for good design of both low-pass and high-pass filters are as follows:

1. The refractive index of the substrate n_s should lie between n_H and n_L. The index seen at the front of the filter (requiring antireflection) will then be n_s for all frequencies in the pass region.

2. The central layers should be nearly equal in thickness to ensure that the rejection band maintains its maximum width.

3. The first layer at the substrate should be of high refractive index if $n_H/n_s > n_s/n_L$ or of low refractive index if $n_s/n_L > n_H/n_s$.

4. The fractional ripple is $[(n_H/n_s - n_s/n_L)/(n_H/n_s + n_s/n_L)]^2$.

5. The phase thicknesses defined at cutoff should be less than $\pi/2$ for low-pass and between $\pi/2$ and π for high-pass filters.

6. Because a suitable thickness of n_L is often good enough to antireflect n_s, the front layer in the filter itself must therefore be high.

The main considerations in the design of low-pass filters have been discussed, and it now remains to give examples of such filters.

The materials that have been used extensively in the near-infrared, for low-pass filters at 2.6, 5, and 12 μm are ZnS as the low-index ($n_L = 2.2$) and either PbTe or Ge as the high-index material ($n_H = 5.1$ or 4). All these materials can be deposited in the relatively thick films necessary. The performance of the filters are shown in Fig. 19. It should be clear that by using these materials filters at any predetermined wavelength between 2.5 and 20 μm can be produced, although at the long-wavelength end absorption severely limits their use, particularly that of ZnS.

Fourteen alternating layers of PbTe and ZnS are necessary to produce the filter at 12 μm. The layers are symmetrical but do not all have the same thickness. To remove the expected leak near 6 μm (i.e., the half-wave position for a quarter-wave at 12 μm), a thin (~ 100-μm) slice of antireflected InSb (with an edge at 7 μm) is used. The rejection bandwidth of the PbTe/ZnS system is large enough to link up with the InSb edge. Using 14 layers, a rejection of greater than 10^4 with a ripple of 0.04 is possible.

These materials, i.e., PbTe, Ge, and ZnS, are the semiconductors enabling a complete range of blocking filters up to 20 μm to be produced. The extension of these materials to longer wavelengths becomes increasingly difficult because of the free-carrier and lattice absorption that decrease the filter transmission beyond about 20 μm to an unacceptably low level. Materials that may, however, be useful for this range include the heavy alkali halides, i.e., those with lattice-absorption peaks beyond 100 μm. It has been found that cesium iodide is a useful low-index mate-

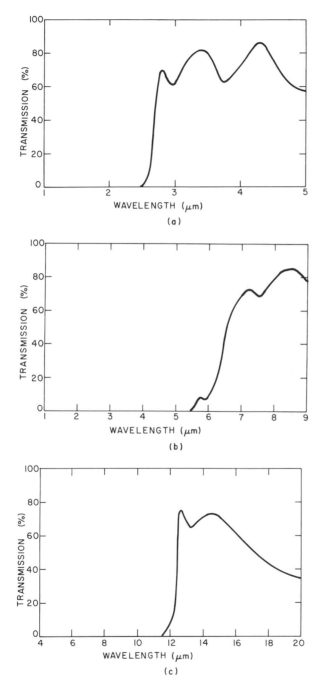

FIG. 19 Series of standard commercial dielectric multilayer low-pass filters: (a) 2.6-μm low-pass multilayer filter; (b) 5.6-μm low-pass multilayer filter; (c) 12-μm low-pass multilayer filter.

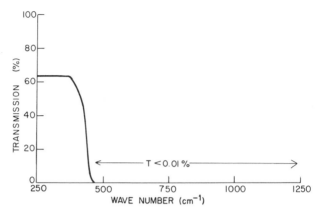

FIG. 20   Low-pass dielectric multilayer filter at 22 μm with CsI and PbTe layers. (Reproduced from Smith *et al.*, 1972.)

rial (n_L = 1.7) for filters beyond 20 μm. Using 30 alternate layers of PbTe and CsI, the filter in Fig. 20 is possible. Unfortunately, CsI is hygroscopic, and to overcome possible peeling of layers, the final filter is sealed in a thin film of polystyrene. Using this combination of materials, filters with edges out to 30 μm have been constructed.

Alternative materials can be used, and Fig. 21 shows a 24-μm edge filter available from Perkin Elmer. This filter is a far more complicated structure, utilizing 72 alternating layers of Te as the high-index material and TlCl as the low-index material. This filter is also protected by a plastic film.

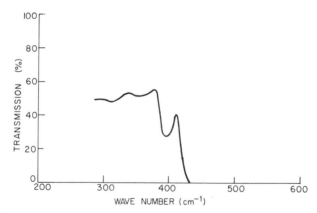

FIG. 21   Low-pass dielectric multilayer filter at 24 μm. Filter was made by Perkin Elmer Corporation and curve reproduced from technical literature. Note that 72 alternating layers of Te and TlCl were required to achieve this performance.

These filters represent the limit of present dielectric multilayer techniques as applied to low-pass filters. At longer wavelengths, it is possible to make use of interference filters based on metallic mesh. Before considering metal-mesh low-pass filters, narrow bandpass filters using dielectric multilayers will be discussed.

B. MULTILAYER BANDPASS FILTERS

To demonstrate the capabilities of multilayer narrow bandpass filters, it is instructive to consider the production of a filter pitched at 15 μm (668 cm^{-1}) with a bandwidth of 3 cm^{-1}, i.e., $\frac{1}{2}\%$. Such a bandwidth implies that the evaporated layers have to be controlled to better than $\frac{1}{6}\%$ and also be uniform to within $\frac{1}{10}\%$ over the filter aperture (Smith *et al.*, 1972). This probably represents the limit of present techniques. To manufacture filters with layers deposited to this high degree of control, considerable developments in thickness monitoring, substrate temperature control, and source–substrate conditions were required. To achieve this filter profile, a Fabry–Perot design (Single half-wave)

$$L\,|\,Ge\,|\,LHLHL\underline{HH}LHLH$$

can be used, in which L is a quarter-wave thickness of ZnS and H a quarter-wave thickness of PbTe. The index ratio for these materials is 2.5. This high index ratio enables this half-width to be achieved with only ten layers, whereas using Ge as the high index material would require double the number of layers. The profile is shown in Fig. 22. Such filters can be pitched to within ±1 cm^{-1}. In rare cases, this may not be sufficient pitching accuracy, but it can be improved by angle tuning, i.e., changing

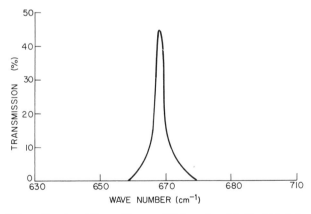

FIG. 22   Fabry–Perot multilayer filter at 667.9 cm^{-1} and half-width 3.3 cm^{-1}. (Reproduced From Smith *et al.*, 1972, with author's data.)

the angle of incidence of the incoming radiation. Tuning can also be achieved by varying the temperature of the filter. For this particular design, a 10 °C increase in temperature will increase the peak frequency by 0.7 cm^{-1}.

The dependence of the peak frequency on the angle of incidence is given by

$$\frac{dv}{v_0} \simeq \tfrac{1}{2}[(\sin \theta_e)/n^*]^2, \tag{32'}$$

where we have n^* as the effective refractive index for the particular filter. We should point out that not only is the angle of incidence important but also the form of the irradiance, i.e., whether it is collimated or Lambertian.

As indicated in previous sections, the Fabry–Perot filter is not the ideal design for a narrow bandpass filter because a large fraction of the transmitted energy lies outside the half-width of the filter. In addition, the peak transmission is very sensitive to absorption, increasingly so as the reflectivity, needed for narrow filters, increases. An alternative design is the double half-wave (DHW) filter.

Two examples are shown in Figs. 23 and 24. In both cases the substrates were germanium. The designs were as follows: DHW at 14 m was

$$\text{L}|\text{Ge}|\text{LHL } \underline{\text{HH}} \text{ LHL HL HL } \underline{\text{HH}} \text{ LG,}$$

where L = $\lambda/4$ at 14 μm of ZnS and H = $\lambda/4$ of PbTe. The transmission

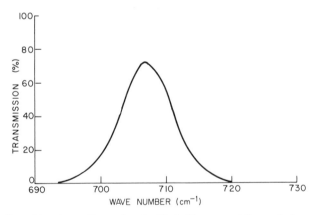

FIG. 23 Broad bandpass filter using double half-wave multilayer design with peak at 7068 cm^{-1} and half-width 9 cm^{-1}. (Reproduced from Smith *et al.*, 1972, with author's data.)

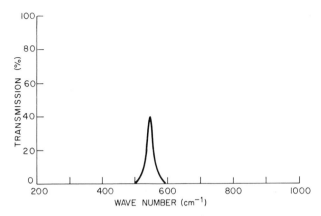

FIG. 24 Double half-wave multilayer dielectric filter. (Reproduced from Smith *et al.*, 1972, with author's data.)

at the peak is 72% with a half-width of 1.28%. The double half-wave at 18.6 μm was

$$Ge \,|\, \underline{LHH} \; LHLHL \; \underline{HH} \; LH,$$

where = λ/4 of ZnS and H = λ/4 of PbTe. The profile shows the combination of this filter together with its blocking filter.

C. WIDE BANDPASS FILTERS

In some cases it is necessary to have relatively wide bandpass filters, say, with bandwidths of the order of 15% or more. This can be achieved using multiple half-wave filters or, for very wide filters, a combination of low- and high-pass filters in series.

A triple half-wave filter can be made according to the following design:

$$Ge \,|\, HL \; \underline{HH} \; LHL \; \underline{HH} \; LHL \; HH$$

where L = λ/4 of ZnS and H = λ/4 of PbTe. The midpoint of the bandpass was 858 cm⁻¹ and the width was 80 cm⁻¹; the filter profile is shown in Fig. 25.

D. SUMMARY

Dielectric multilayer filter technology has reached a high level of sophistication and can routinely produce good-quality filters with low-frequency pass, high-frequency pass, and bandpass characteristics. These techniques are restricted to the infrared region of the spectrum below 25 μm by the absorption and adhesion properties of thick films.

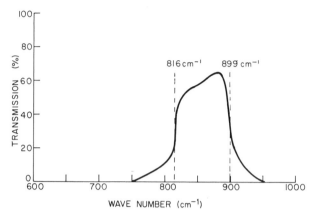

FIG. 25 Dielectric multilayer triple half-wave filter. (Reproduced from Smith *et al.*, 1972, with author's data.)

Although one or two filters have been made at longer wavelengths using these techniques, they cannot be produced on a routine basis. Some alternative process is required for long wavelengths. This is found in the use of metallic mesh as the reflecting surface of the interference filter and will be discussed in the next section.

V. Metallic Mesh Interference Filters

The first important utilization of metallic mesh in an infrared interference device was by Renk and Genzel (1962), who constructed a scanning Fabry–Perot interferometer. This development was extended using metallic mesh as the reflecting interface in an interference filter by coworkers of Genzel, especially Ulrich, whose articles in 1967 on bandpass and low-pass filters laid the groundwork for subsequent developments (Ulrich, 1967a,b).

There are two main types of metallic meshes. There is the type called *inductive,* which consists of periodic holes in thin metal foil; and the *capacitative* type, which has periodic metal squares. The latter obviously requires a substrate. Although there is no necessity for the holes or sections to be square, it is convenient for the arrangement to have $\pi/2$ symmetry to remove polarization effects. Various shapes have been used, but only the meshes with square holes or metal squares have found extensive use. The structure, dimensions, and optical properties of inductive and capacitative mesh are shown in Fig. 26. The theoretical details concerning the frequency dependence of the transmission and reflection of the mesh are

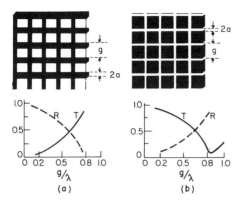

FIG. 26 Geometrical and optical characteristics of two types of metallic meshes used in far-infrared filters: (a) inductive; (b) capacitative. Note that the inductive mesh becomes completely reflective at long wavelengths and the capacitative mesh is completely transmittive at long wavelengths.

given elsewhere (Marcuvitz, 1951), and only the important points will be given here.

When a plane electromagnetic field is incident normally on the mesh, surface currents and charges are produced, which in turn act as secondary sources of electric and magnetic fields. In the nondiffraction region, i.e., where the wavelength of the radiation is much larger than the periodicity of the mesh, these sources give rise to zero-order reflected and transmitted beams of equal amplitude but with phases that may be different from the incident beam. On the transmitted side of the mesh, there will be two fields of the same frequency: the "part" of the original incident beam that "went through the holes" of the mesh and the zero-order transmitted part. These waves superimpose to give the final transmitted beam.

If the zero-order reflection coefficient has a complex amplitude $\Gamma(\omega)$, the total transmitted amplitude is given by

$$\tau(\omega) = 1 + \Gamma(\omega). \tag{42}$$

So that any mesh can be discussed, the frequency ω is usually normalized to the mesh periodicity g, i.e., $\omega = g/\lambda$, where λ is the wavelength. For a loss-free system,

$$|\tau(\omega)|^2 + |\Gamma(\omega)|^2 = 1. \tag{43}$$

The phase changes between the incident and transmitted beams and between the incident and reflected beams are

$$\sin^2 \psi_\tau(\omega) = 1 - |\tau(\omega)|^2 \tag{44}$$

and

$$\sin^2 \psi_\Gamma(\omega) = |\tau(\omega)|^2, \qquad (45)$$

respectively. These expressions, which are valid in the nondiffraction region and for thin meshes $a \gg t$, are for the inductive-type mesh. (The only requirement for the metal thickness is that it should be greater than the skin depth.) The expressions for capacitative mesh, which is the structure complementary to inductive mesh, are obtained using Babinet's principle, which give

$$\tau_i(\omega) + \tau_c(\omega) = 1, \qquad |\tau_i(\omega)|^2 + |\tau_c(\omega)|^2 = 1,$$

$$\tau_i(\omega) = -\Gamma_c(\omega), \qquad \tau_c(\omega) = -\Gamma_i(\omega), \qquad (46)$$

again for lossless systems.

The transmission of a thin capacitative mesh in the nondiffraction region, including ohmic losses, is given by

$$T_c = \{r^2 + Z_0^2[(\omega/\omega_0) - (\omega_0/\omega)]^2\}/\{(1 + r)^2 + Z_0^2[(\omega/\omega_0) - (\omega_0/\omega)]^2\}, \quad (47)$$

where r is the ohmic loss term due to surface currents flowing in the mesh and is related to the absorption by

$$A = |\Gamma|^2 2r \qquad \text{with} \qquad r = (c/\lambda\sigma)^{1/2}\zeta, \qquad (48)$$

where $|\Gamma|^2$ is the reflectivity, σ the dc bulk conductivity, and ζ a geometrical factor, which is effectively that part of the cross section of the mesh that may conduct current. For capacitative mesh,

$$\zeta = [1 - (2a/g)]^{-1}; \qquad (49)$$

for inductive mesh,

$$\zeta = g/2a.$$

To reduce absorption losses, the conductivity of the mesh should be as high as possible. Bearing the economics in mind, the most widely used metals have been copper, nickel, and aluminum. With ω_0 the normalized resonant frequency given by $\omega_0 = g/\lambda_0$, where λ_0 is the wavelength where the transmission of the inductive mesh is a maximum and that of the capacitative mesh a minimum, ideally, $\omega_0 = 1$, but, in practice, it turns out to be somewhat less.

The parameter Z_0 is known as the characteristic impedance of the mesh and has been given by Marcuvitz (1951) as

$$Z_{0c} = [\ln \csc(\pi a/2g)]^{-1} \qquad (50)$$

for capacitive mesh, which, together with Z_{0i} (for inductive mesh),

$$Z_{0c}Z_{0i} = 1, \qquad (51)$$

also gives the impedance for an inductive mesh of identical geometry. Curves of T_c for various values of Z_{0c} are shown in Fig. 27, from which it is clear that the higher the value of a/g, the steeper the profile near ω_0. Capacitative mesh, because it needs a supporting substrate for the metal squares, will have additional losses due to substrate absorption.

In principle, then, any desired value of reflectivity or transmittivity for a mesh can be obtained by the appropriate choice of g and a/g. In practice, of course, only a restricted range of g values is available. However, metallic meshes can be used as the reflecting surfaces in interference filters to produce devices in the far-infrared that are as good as those produced in the visible and near-infrared regions by evaporated multilayers.

Before discussing some examples, two further points are worthy of mention, both relating to capacitative mesh. The techniques used to produce inductive mesh are quite sophisticated, using precision electroforming on thin metal foil, and there are few suppliers. Those whose mesh has been used by the author include Buckbee-Mears, Dainippon, and E.M.I. For reasons not clearly understood at the moment, the two former manufacturers seem to have products with superior optical performance. The situation for capacitative mesh is somewhat different. There is, to the author's knowledge, only one supplier of capacitative mesh, Cambridge Consultants Ltd., and this is as a result of collaboration with researchers involved in filter developments.

It is of fundamental importance that the substrate be as highly transmitting as possible and also as rigid as possible, compatible with transparency. The rigidity criterion excludes the use of polyethylene. The two best choices would seem to be polypropylene and polyethyleneteraphthalate. The latter is more commonly known by tradenames such as Mylar, Melinex, and Hostaphan. Polypropylene is ideally the best material, but

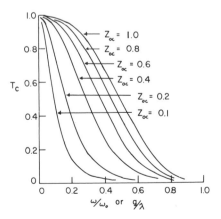

FIG. 27   Transmission of capacitative mesh as a function of Z_{0c} $\{=[\ln \csc(\pi a/2a)]^{-1}\}$.

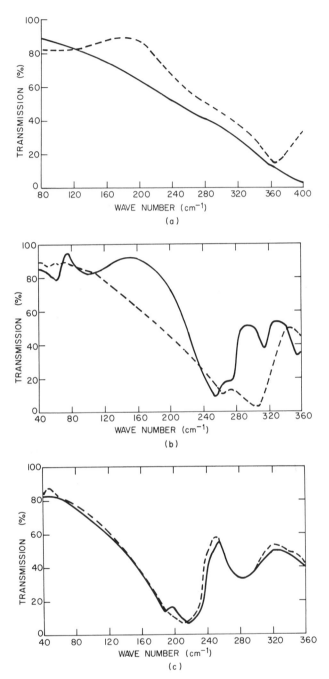

FIG. 28 Transmission of capacitative mesh using various substrates and metal thicknesses. (a) 18-μm capacitative mesh on 15-μm-thick polypropylene (– – –) and on 2.5-μm-thick Mylar (—); (b) 25-μm capacitative mesh on 2.5-μm-thick Mylar (– – –) and on 15-μm-thick polypropylene (—); (c) 37.5-μm capacitative mesh on 0.1 μm of Al (—) and on 0.25 μm of Al(– – –).

there is some difficulty in obtaining it in sufficiently thin film; presently the thinnest readily available is about 12 μm. This represents a problem in some cases because as discussed previously, a low-pass filter at, say, 60 μm would require a series of quarter-waves at 60 μm, i.e., an optical path of 15 μm. The optical path of 12-μm polypropylene is about 20 μm. Consequently, 2.5-μm Hostaphan from Kalle has been widely used for the substrate material. At long wavelengths, however, better performance may be achieved using polypropylene.

The condition for thickness of metal is that it must be greater than the skin depth. The inductive mesh commonly used is of the order of 5 μm. Because most techniques of fabricating capacitative mesh require evaporation of metal, which is difficult for films of 5-μm thickness, an additional process was initially used. Following the evaporation of a thin metal layer, increasing thickness was built up by electroplating. This proved inconvenient, and it was eventually found to be unnecessary. The thickness of metal used in commercial capacitative mesh is usually about 0.2 μm, which is thin enough to be evaporated easily and then photoetched to produce the required mesh geometry. The finest mesh available has a g value of 4 μm. Curves for capacitative meshes with various substrates and metal thicknesses are shown in Fig. 28 (Holah and Smith, 1972).

Although all the filters to be discussed use either capacitative mesh made as detailed here or electroformed inductive mesh, it is also possible to fabricate coarse meshes using woven wire cloth. The optical properties of such wire-cloth meshes are different from electroformed meshes of the same geometry, presumably because the condition $a \gg t$ is not followed for wire cloth, and in fact, the wire cloth will be quite thick as well as having a circular cross section (Ressler and Möller, 1967). There is no pronounced minimum for the wire cloth. The reflectivity falls to shorter wavelengths and then stays down at a low value, whereas the reflectivity of the electroformed mesh falls and then rises, although it never rises to a value found at long wavelengths. The wire-cloth mesh reflection characteristic makes it useful as a single-mirror surface to aid rejection of short waves. Unfortunately, the transmission is not complementary, so wire cloth does not find much application.

Having found an alternative reflecting surface to dielectric multilayers for the infrared, the utilization of the meshes can be discussed in relation to specific filters.

A. NARROW BANDPASS FILTERS

Historically, narrow bandpass filters were the first to be made using metallic mesh, and the design was always that of the single half-wave or Fabry–Perot filter. Various workers have been involved in producing good narrow-band interference filters; those worthy of mention include Ulrich (1967a), Lecullier (1974), Wijnbergen (1973), Rawcliffe and Randall (1967), Holah et al. (1978), and Holah and Morrison (1977).

The metal-mesh Fabry–Perot filter is the simplest possible interference device, being composed of only two reflecting surfaces. The problems arise when the tolerances required to produce good filters are considered. In order for a narrow-band filter to define adequately a range of frequencies together with sufficient rejection of unwanted radiation, a total finesse of at least 20 is required. From the discussions in previous sections, assuming that the ideal situation of equal flatness and reflection finesse is possible, this implies a flatness finesse of 28 and the surfaces have to be flat and parallel to within λ/m, where $m = 56$ ($\mathscr{F}_F = m/2$), i.e., at 50 μm, the surfaces have to be parallel to better than 1 μm over the whole area of the filter, a precision that is difficult using mechanical techniques.

Most Fabry–Perot filters consist of two inductive meshes separated by a metal annulus. This annulus serves to define the peak wavelength of the filter as well as the flatness finesse. Therefore, to produce a narrow-band filter at 50 μm requires a metal annulus whose thickness is less than 25 μm ($\lambda/2$) to allow for phase changes and whose thickness is uniform to better than 1 μm over the whole area. Of course, the flatness restrictions become progressively easier as the wavelength of the peak increases.

The mesh size is determined from the reflectivity finess. Given that, at long wavelengths at least, a flatness finesse of 30 is possible, the reflectivity finesse has to be greater than or equal to 30, i.e., $\pi\sqrt{R}/(1 - R) \geq 30$. Therefore, the value of reflection at the wavelength of the filter must be at least 0.9. Because it was shown that increased reflectivity increases the effect of absorption in Fabry–Perot filters, the optimum performance (halfwidth together with maximum transmission) occurs when $\mathscr{F}_R = \mathscr{F}_F$, where the value of R should not be higher than is necessary. Knowing the value of R at a given wavelength automatically selects the particular size of mesh.

In order to improve the pitching accuracy and flatness finesse, a technique was devised by the author to produce Fabry–Perot filters without the need for annular spacers (Holah and Morrison, 1977). The scheme is shown in Fig. 29. The instrument consists of a rigid 12-mm-thick duraluminum baseplate to which is attached a high-precision linear translation slide (1), machined from duraluminum and utilizing six hardened steel ballbearings per side, held and space by a perforated PTFE strip, the balls running between vertical double rails of stub steel. The outer rail track, attached to the moving top plate of the slide, is fixed, and the inner track on the fixed bottom slide is adjustable by a small cam operating against the steel holding blocks at each end of the slide; this provides adjustment for

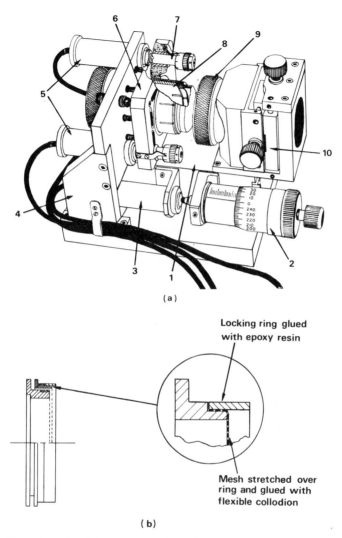

(a)

Locking ring glued
with epoxy resin

Mesh stretched over
ring and glued with
flexible collodion

(b)

FIG. 29   (a) Scanning Fabry–Perot used to make narrow-band spacerless filters. (b) Detail of mesh support system. See text for details. (Reproduced from Holah and Morrison, 1977.)

removing any transverse lash. The slide was actuated by a small, specially constructed, micrometer (2). It had a 0.25-mm pitch screw, and with 100 divisions on the thimble, it provided a movement of 2.5 μm/division. Finer adjustment was provided by a Burleigh PZ-40 piezoelectric transducer with a displacement of 1.5 μm/100 V (3). The bridge (4) attached to the baseplate over the slide carries two Burleigh PZ-40 transducers (5) for vertical and horizontal tilt of the kinematic plate (6), i.e., to ensure parallelism of filter elements. Crude adjustment for parallelism is provided for by having the piezoelectric transducers move against locating cups on 120-threads/in. screws (7) fitted to the kinematic plate. This plate also carries a stepped internal collet (8) for holding a ring, which has a mesh stretched across it. This will be one of the reflecting elements in the Fabry–Perot filter. The other filter ring is held in an external collet (9) attached to the linear translation slide through x and z translation stages (10). These stages provide for axial alignment of the two filter elements.

Initial parallelism is achieved using the reflected diffraction patterns from the two meshes. The source is a small He–Ne laser. The diffraction patterns, especially the bright central spots corresponding to the zero-order reflection, are overlapped using the coarse adjustments previously mentioned. The separation of the meshes at this stage is usually about 5 mm. They are then moved together until about 1 mm. The parallelism is rechecked using the laser. Generally the translation does not result in any observable changes in parallelism. The whole device is then placed inside the sample chamber of a Beckman-RICC FS-720 fourier transform spectrometer and all remaining adjustments made in situ. Spectra are recorded for evaluation the filter. The fine adjustments are made under vacuum using electrical leadthroughs for the three piezoelectric transducers, which provide for tilt and translation. The total precision to which a filter can be pitched is $\pm 1\ \mu$m and is limited by shrinkage in the epoxy, which is used to glue the rings together after all the adjustments are complete.

1. *Determination of Narrow Bandpass Filter Profiles*

Before continuing the discussion the of filter designs, it is constructive to examine the relationship among the experimentally determined profile, the filter parameters, and the experimental factors. Two factors that are important are the transmission and half-width of the filter profile. The half-width will be considered first.

The total half-width is a function of four contributions and is, in fact, the convolution of these contributions. The relation can be written as

$$d\bar{\nu}^2 = d\bar{\nu}_R^2 + d\bar{\nu}_F^2 + d\bar{\nu}_S^2 + d\bar{\nu}_\theta^2, \tag{52}$$

where $d\bar{\nu}$ is the total observed half-width, $d\bar{\nu}_R$ the half-width due to the reflectivity of the interfaces, $d\bar{\nu}_F$ the half-width due to the flatness or parallelism of the meshes, $d\bar{\nu}_S$ the resolution of the spectrometer used to measure the filter profile, $d\bar{\nu}_\theta$ a broadening due to the range of angles of incidence of the radiation incident on the filter, and $d\bar{\nu}_R$ is related to the reflection finesse $\mathscr{F}_R$ as discussed in previous sections. $\mathscr{F}_R = \Delta\bar{\nu}/d\bar{\nu}_R =$

$\pi\sqrt{R}/(1 - R)$, where $\Delta\bar{\nu}$ is the free spectral range and R is the reflection coefficient of the filter elements and $d\bar{\nu}_F$ has also been discussed previously and is related to the overall parallelism of the two meshes. The parallelism finesse is given by $\mathcal{F}_F = \Delta\bar{\nu}/d\bar{\nu}_F = \frac{1}{2}m$, where the meshes are parallel to λ/m. The contribution to the half-width due to the angles of incidence is given by $d\bar{\nu}_\theta = \frac{1}{2}\nu \sin^2 \theta$, where angles range from 0 to $\pm\theta$.

These parameters can best be considered for a real situation. Suppose that a narrow-band Fabry–Perot filter at 100 cm^{-1} is required with a half-width of 3 cm^{-1}. It has been shown by Jacquinot (1960) that the optimum situation is that in which $d\bar{\nu}_R = d\nu_F$. (Note that these are the only two half-widths inherent to the filter itself.) Therefore,

$$dv_{\text{inh}}^2 = dv_R^2 + dv_F^2, \tag{53}$$

where $d\bar{\nu}_{\text{inh}}$ is the inherent half-width, i.e., 3 cm^{-1}. From this, $d\bar{\nu}_R = d\bar{\nu}_F = 2.12$ cm^{-1}. If the filter is used in first order, the free spectral range is 100 cm^{-1}, and, therefore, $100/2.12 = \pi\sqrt{R}/(1 - R) = \frac{1}{2}m$. Thus the reflectivity should be 0.94, and the meshes have to be parallel to $\lambda/100$ at 100 cm^{-1}, i.e., to within 1 μm.

In the FS-720 interferometer, the standard cone semiangle of the radiation incident on any sample is 15°, which gives a half-width $d\bar{\nu}_\theta = 3.35$ cm^{-1}. (Note that this is already the largest contribution to the observed profile. If the filter were a delta-function, it would be measured, even with infinite resolving power, as having a half-width of 3.35 cm^{-1} with such a cone semiangle.)

With the FS-720, a resolution $d\bar{\nu}_S$ of 1.5 cm^{-1} is easily and conveniently achieved. The observed half-width is therefore given by

$$d\bar{\nu}^2 = (2.12)^2 + (2.12)^2 + (1.5)^2 + (3.35)^2 = (4.74)^2,$$

i.e., the half-width is 4.74 cm^{-1}, of which the dominant contribution is from the range of angles of incidence. All the narrow-band filters shown in this chapter were measured using the FS-720 spectrometer and the profiles are experimentally observed and have not been corrected to allow for $d\bar{\nu}_\theta$ and $d\bar{\nu}_S$.

An additional effect of the cone angle is the shift in the peak wavelength or wave number. The shift is given by

$$\bar{\nu}_{\text{obs}} = \bar{\nu}_0 + \frac{1}{2}d\bar{\nu}_\theta, \tag{54}$$

where $\bar{\nu}_{\text{obs}}$ is the wave number of the observed peak and $\bar{\nu}_0$ the true wave number of the filter. For the situation as just detailed, i.e., $d\bar{\nu}_\theta = 3.35$ cm^{-1}, the shift to higher wave numbers is 1.675 cm^{-1}. This is quite significant because it is of the same order as the pitching accuracy. Ideally, of course, the filters should be measured with collimated radiation.

The transmission at the peak of a filter profile is given by

$$T_{max} = T_A T_P T_S T_\theta,\tag{55}$$

where T_A is the transmission factor that appears in the design equations as $T_A = T^2/(1 - R)^2$, which is, of course, unity if $T = 1 - B$, i.e., there is no absorption. If absorption is present, with $1 - R$ now equal to $A + T$, the T_A reduces to $[1 + (A/T)]^{-2}$, where R, T, and A are the reflection, transmission, and absorption of the single meshes, respectively. The factor T_P depends on the parallelism of the meshes. It should be obvious that any process that increases the width of the filter profile must also decrease the transmission; from energy conservation, the area of the profile must be constant. The factor T_S results from resolution broadening, and T_θ is due to the angles of incidence.

The effect of absorption is dramatically increased as the reflectivity is increased. If the absorption is assumed constant at 1% and the reflectivity varied from 90 to 98%, the value of T_A changes from 0.81 to 0.25. For the example in which $R = 0.94$ and $T_A = 0.69$, this is the maximum transmission observed even if all the other factors had zero contribution.

The factor T_P can best be evaluated as follows. Suppose that the half-width of the filter due only to the reflectivity is x cm^{-1} and that there is no absorption, i.e., $T_A = 1.0$. Because of the lack of parallelism, there will be a contribution to the half-width of y cm^{-1}, where now the total half-width z cm^{-1} is given by

$$z^2 = x^2 + y^2.$$

From the condition for equal areas and approximating the filter profile to a triangular function,

$$x \cdot 10 = z T_P.$$

For the same example, i.e., $x = y = 2.12$ cm^{-1} and $z = 3$ cm^{-1}, $T_P = 0.71$. Then T_θ can be obtained by similar reasoning to give 0.53. Also, T_S is obtained as 0.82. Therefore, the transmission at the peak of the filters is

$$T_{max} = 0.69 \times 0.71 \times 0.53 \times 0.82 = 0.21,$$

and the inherent transmission of the filter is 0.49. These calculations are very simplified, but they do give a clear indication of the importance of experimental conditions on the observed profiles.

2. Metallic Mesh Fabry–Perot Narrow Bandpass Filters

The filter profiles shown in this section will demonstrate clearly the state of the art. The two techniques for producing Fabry–Perot filters

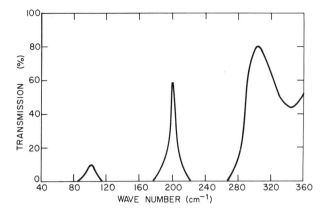

FIG. 30   Mesh Fabry–Perot filter showing second-order performance at 200 cm⁻¹. (Reproduced from Holah and Smith, 1972.) *Journal of Physics D: Applied Physics* **5**, 469. Copyright of The Institute of Physics.)

have been used successfully, as will be shown. The wavelength region extends from 50 to almost 2000 μm. In the main, these filters will be used in first order. However, in some cases, higher order filters will be used. Unless stated otherwise, the filters shown will have been made by the author.

The filter graphed in Fig. 30 was made for use as part of a radiometer, which was carried on a NIMBUS weather satellite. The pitching accuracy required for this was ± 1 μm. The filter was extensively vibration-tested and temperature-cycled between $+ 30$ and $- 5$ °C and withstood the rigorous testing. The meshes used for this filter are 1500-line/in. inductive meshes; the periodicity is 16 μm. It was found that this filter worked more easily in second order, and consequently, the spacing was about 42 μm. Problems sometimes occurred when metal spacers of total thickness less than 25 μm were used.

Because the reflectivity of the mesh increases to longer wavelengths, the effect of absorption also becomes more significant toward long wavelengths. This limits the useful working range of any mesh. For filters at wavelengths between 50 and 100 μm, mesh with a periodicity of 25 or 20 μm is required.

An example of a filter at 67 μm, made using the spacerless filter jig, is shown in Fig. 31a. It was made using 1270 lines/in. (i.e., 20 μm). The performance shown here occurred after the filter had been fixed together using epoxy. Some shrinkage occurred, but the measured profile had a half-width of 5.4 cm⁻¹, a peak at 130 cm⁻¹, and a transmission of 0.6. Considering the experimental contributions to the profile as discussed

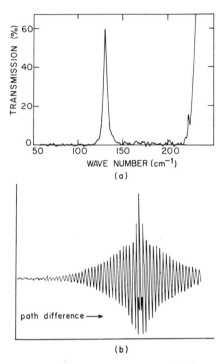

FIG. 31   (a) Mesh Fabry–Perot filter near 130 cm⁻¹. (b) Corresponding interferogram obtained with fourer transform spectrometer.

previously, we were lead to an inherent half-width of 4 cm⁻¹ and a transmission of about 0.8.

Further profiles for filters made without spacers are shown in Fig. 32, together with the interferograms for various stages of the construction. Because the filters have such a well-defined profile, it is possible to use the interferogram to deduce the peak of the filter. The adjustment and measure technique used with the spacerless filter jig would be much lengthier if it were always necessary to use a computer for Fourier transformation.

Having calibrated the interferogram in terms of mirror displacement speed and chart recorder speed, it is possible to deduce the peak wavelength to within 10% at wavelengths down to 100 μm. Final adjustment is achieved using a transform. The shape of the filter can also be deduced from the interferogram. The narrower the filter profile, the more extensive the interferogram, i.e., for a first-order filter, the interferogram approaches that of cos² fringes. Figure 32 was obtained using 25-μm mesh.

A graph of a filter made by a French worker, Lecullier (1974) is shown

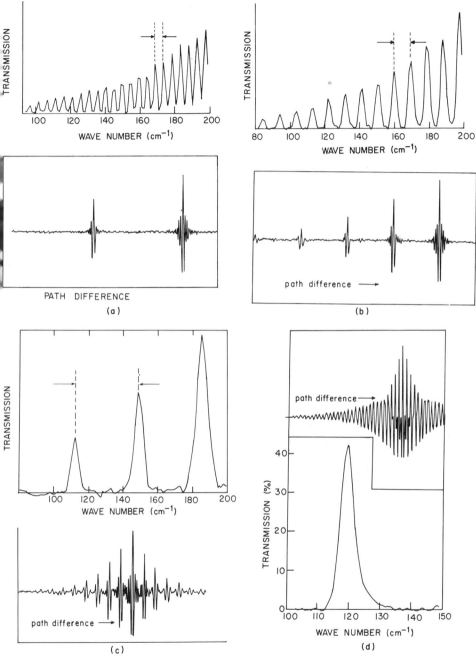

FIG. 32   Series of transmission spectra and interferograms at various stages of construction of a Fabry–Perot filter using the device shown in Fig. 29: (a) FSR = 5 cm^{-1}; (b) FSR = 9 cm^{-1}; (c) FSR = 37 cm^{-1}; (d) FSR = 120 cm^{-1}. Note how the interferogram tends toward a "damped" cosinelike curve as the number of wavelengths passed by the Fabry–Perot is reduced.

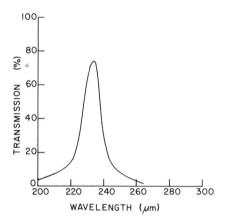

FIG. 33 Mesh Fabry–Perot filter near 235 μm. Only the first order is shown. Note the extensive skirting associated with the relatively broad filter. The reflectivity of the meshes is too low to give good rejection. (From Lecullier, 1974.)

in Fig. 33. The filter was made from two meshes with a periodicity of 50 μm and a spacing of 94 μm. The a/g value for this mesh was 0.22, and the reflectivity of the mesh was ~0.9. The absorption was calculated to be 0.006, leading to a value for T_A of 0.895. The measured transmission was 0.74, which is consistent with the measured broadening of 1.8 cm^{-1}. Figure 34 shows three orders of Fabry–Perot made from 500 lines/in. ($g = 50$ μm) with a spacing of 150 μm. The first-order peak is at 31 cm^{-1}. The difference in height of the three orders is due to the changing reflectivity together with a constant absorption of 0.01. The first two orders of a calculated design are also shown in the figure. This is not a completely realistic calculation because effects due to angles of incidence, flatness, and

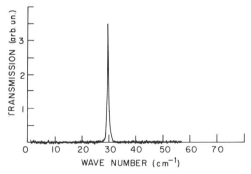

FIG. 34 Mesh Fabry–Perot at 31 cm^{-1} made by author. The measured half-width is less than 3%. (Reproduced from Holah, Morrison and Goebel, 1979a.)

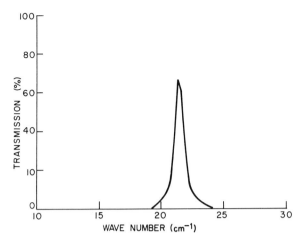

FIG. 35   Mesh Fabry–Perot at 21.9 cm⁻¹, with half-width measured is 3.4%.

resolution have not been taken into account. The high reflectivity of the individual meshes below 50 cm⁻¹ accounts for the good rejection as well as the low value of peak transmission at 31 cm⁻¹. At 70 cm⁻¹, the reflectivity is low, hence the rejection is not good. At the wavelength of the peak, ~320 μm, the flatness finesse is much higher than any other contribution. The remaining contributions, $d\bar{\nu}_R$, $d\bar{\nu}_\theta$, and $d\bar{\nu}_S$, are approximately equal at 1 cm⁻¹, leading to the observed half-width of 1.7 cm⁻¹. The inherent transmission is approximately 0.5. This demonstrates clearly the difficulties of measuring narrow bandwidth filters.

Further Fabry–Perot filter profiles are shown in Figs. 35–37. The curves clearly show the possibilities of mesh Fabry–Perot filters. Resultant finesses of more than 30 and transmissions of the order of 0.5 are available throughout the wavelength range from 50 to 2000 μm.

Before leaving Fabry–Perot filters, their performance at low temperatures will be considered. Of the two techniques used in the manufacture of filters, the spacerless one is preferable in terms of pitching accuracy. Unfortunately, the filters made using this process do not seem to perform well at temperatures below 77 °K (Holah and Morrison, 1977).

There is an increased requirement for infrared filters to function at low temperatures because many observations are of blackbody equivalents at temperatures down to 3 °K or less. The performance of a filter made without spacers as a function of temperature is shown in Fig. 38. The peak wavelength varies as some undetermined function of temperature. What is shown clearly is the variation of overall transmission with frequency. This is due partly to the change in the reflectivity of the mesh with fre-

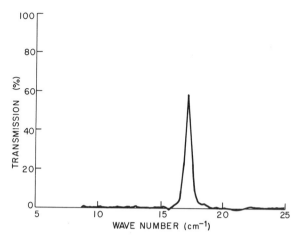

FIG. 36 Mesh Fabry–Perot at 588 μm. The half-width is slightly larger than in Fig. 35 because of the lower reflectivity of meshes.

quency; the meshes used were 1270 lines/in. ($g = 20$ μm). It is surprising that the maximum transmission can vary so much over a relatively small spectral range. It is instructive, therefore, to examine this situation. The highest peak occurs near 140 cm^{-1} with a transmission of about 0.68, and the lowest peak is near 120 cm^{-1} with a transmission of about 0.54. At these two frequencies, the reflectivities of the 1270-lines/in. mesh are 0.92 and 0.94, respectively. If we assume a constant absorption of 0.15 and ignore any other effects, T_{max} for $\bar{\nu} = 120$ cm^{-1} is calculated to be 0.56 and that at 140 cm^{-1} is 0.66. This is clearly in the right ballpark.

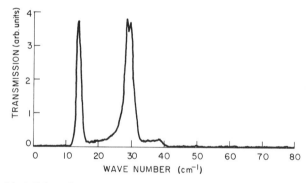

FIG. 37 Mesh Fabry–Perot at 14 cm^{-1} and 300 °K, showing two orders. The half-width corresponds to a reflectivity that theoretically should not give rise to the apparent leakage between the two orders.

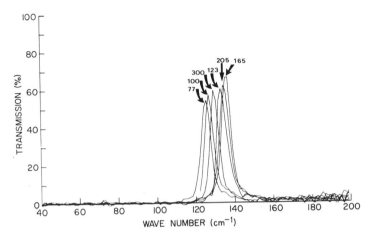

FIG. 38 Mesh narrow bandpass Fabry–Perot filter made without spacers and demonstrating temperature dependence. Values shown are in degrees Kelvin.

Although this filter performed well at these temperatures, degradation did set in at lower temperatures, and the large shifts with temperature were unacceptable. A series of filters using metal spacers was made with Invar frame rings and Torr-Seal as the adhesive. These performed satisactorily at 1.4 °K, even after repeated thermal cycling (Holah *et al.*, 1979a). Their spectral performances are shown in Figs. 39–41. From this data it is reasonable to assume that narrow bandpass filters using the Fabry–Perot design can be made to perform adequately at low temperatures with little change in their spectral profiles relative to room temperature.

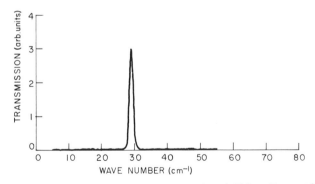

FIG. 39 Low-temperature (1.4 °K) performance of mesh Fabry–Perot at 29 cm⁻¹ after two temperature cycles. The change in performance compared with room temperature (cf. Fig. 34) is relatively small. (Reproduced from Holah, Morrison, and Goebel, 1979.)

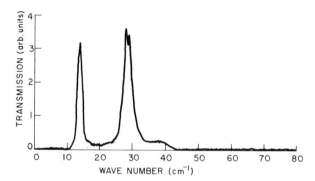

FIG. 40   Low-temperature (1.4 °K) performance of filter shown in Fig. 37. Little change in performance from Fig. 37 although apparent interorder leak is still present. (Reproduced from Holah, Morrison, and Goebel, 1979a.)

B. DOUBLE HALF-WAVE FILTERS

It has already been discussed how an improved spectral profile can be achieved if the Fabry–Perot design, or single half-wave, is replaced by the double half-wave. For this design, because the reflectivity of the two effective surfaces—one the combined single half-wave and the other a surface spaced a half-wave distance from the combined single half-

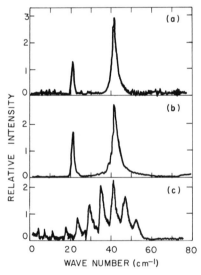

FIG. 41   Temperature dependence of a Fabry–Perot filter that did not work at low temperatures and suffered permanent degradation: (a) 300 °K; (b) 100 °K; (c) 300 °K after helium cycle. (Reproduced from Holah, Morrison, and Goebel, 1979a.)

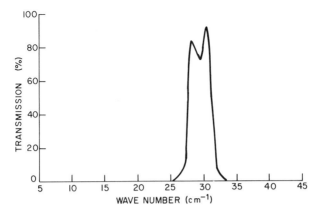

FIG. 42 Double half-wave mesh filter used to isolate the HCN laser line at 29.7 cm^{-1}. Note the sharp cuton and cutoff and high transmission. (Reproduced from Holah and Smith, 1972.) *Journal of Physics D: Applied Physics* **5**, 496. Copyright of The Institute of Physics.)

wave—is a minimum, the number of effective interfering beams is low as is the effect of absorption on the overall transmission. Additionally, in the rejection regions, the reflectivities are high, leading to enhanced constrast ratios.

Although possessing desirable features, the double half-wave structure is correspondingly more difficult to make, and until recently, only one or two isolated attempts had been made to produce DHW filters using metallic mesh (Holah *et al.*, 1979b).

Figure 42 shows a DHW at 29.7 cm^{-1}, which clearly demonstrates the salient features of this design. The three meshes used in this symmetrical filter were $g_1 = g_3 = 100$ μm and $g_2 = 50$ μm, and the spacings were 150 μm each. The double-peak behavior, typical for a DHW in which the reflectivities of the effective surfaces are equal at more than one frequency, is clearly visible. The shape of the filter and its high transmission demonstrate the improved performance.

Figures 43–45 show further DHW filters between 10 and 100 cm^{-1}. It is clear that whenever the proper conditions exists for making DHW filters, then they provide better bandshapes. There is one disadvantage with DHW, which is that the half-width is broader than that achievable with a Fabry–Perot. For this reason, whenever high resolution <3% is required, the filter usually has to be made with a Fabry–Perot (FP). It is, of course, possible to build a DHW to aid the blocking of a Fabry–Perot, i.e., by using an FP and a DHW, pitched at the same frequency in series, to achieve the narrowness of an FP with the rejection of a DHW. There will, however, be a small decrease in overall transmission.

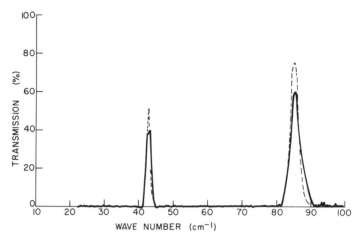

Fig. 43   Theoretical (– – –) and experimental (—) curves for a mesh double half-wave filter using three inductive meshes. Design: $g_1 = g_3 = 59$ μm, $g_2 = 33$ μm, and $S_1 = S_2 = 100$ μm. (Reproduced from Holah, Davis, and Morrison, 1979b.)

One problem that both DHW and FP filters have is the presence of higher order peaks. Usually a bandpass filter is used in first order, $\lambda_0 = 2d$, and some means of rejecting the shorter wavelengths must be incorporated to define the spectral bandwidth of radiation incident on the detector. This necessitates low-pass filters.

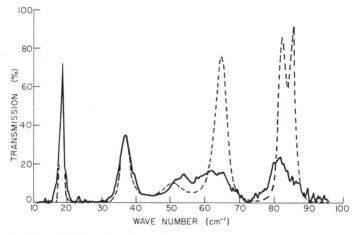

Fig. 44   Theoretical (– – –) and experimental (—) curves for a mesh double half-wave filter. Design: $g_1 = g_3 = 167$ μm, $g_2 = 50$ μm, and $S_1 = S_2 = 225$ μm. Note that, although the positions of the higher orders are reasonably well predicted, the transmission values do not correlate well because the theory does not hold for $\lambda < g$. (Reproduced from Holah, Davis, and Morrison, 1979b.)

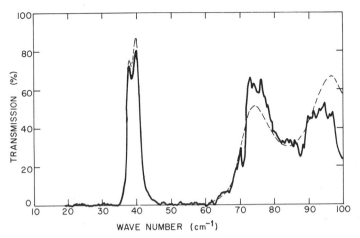

FIG. 45 Theoretical (– – –) and experimental (—) curves for a mesh double half-wave filter demonstrating the two-peak appearance characteristic of having $R_1 \neq R_2$. Design: $g_1 = g_3 = 100 \ \mu m$, $g_2 = 50 \ \mu m$, and $S_1 = S_2 = 100 \ \mu m$. (Reproduced from Holah, Davis, and Morrison, 1979b.)

C. LOW-PASS METALLIC MESH INTERFERENCE FILTERS

The filters discussed thus far have all made use of inductive mesh, that is, thin metal foil with a regular array of holes. They could also have been constructed using the complementary structure, that is, capacitative mesh. The capacitative mesh, because it consists of metal squares, needs a substrate to support them. This is usually in the form of very thin Mylar. The additional dielectric losses and problems associated with the mechanical stability make it unnecessarily complicated to use capacitative mesh in half-wave bandpass filters. The frequency dependence of the optical properties of capacitative mesh does, however, lend itself very readily to the use of these structures in low-frequency pass filters. In these filters, mechanical tolerances, such as defining the parallelism of the meshes, are usually not as severe as in bandpass filters, and in fact, some small decrease in parallelism may help to improve the rejection width.

All the low-pass filters discussed have been made from at least four capacitative meshes spaced with metal annuli of essentially quarter-wave thickness, e.g., for a low-pass filter at 100 μm, the total effective inter-mesh spacing will be of the order of 25 μm. This includes the optical path difference of the substrate (about 4 μm), as well as the optical path due to phase and metal spacers.

In designing low-pass filters, what is required is rapid transition from stopband to passband, high transmission in the passband, and as large and

as good a rejection region as possible. Because the long-wavelength reflectivity of capacitative mesh is usually very small, high transmission at
long wavelengths is not difficult to achieve. The criterion that determines
the particular mesh size used, i.e., the g value, is the reflectivity required
to give a rapid onset of transmission, i.e., at the edge, the reflection must
be high. The sharpest edge will be obtained if all meshes are identical. It
must be remembered, however, that each quarter-wave at λ will behave
as a half-wave at $\lambda/2$, and some strategy must be used to remove or reduce the associated bandpass at $\lambda/2$. The usual design procedure, therefore, is to use, say, two meshes to give the required transition from stop to
pass and then to add on other meshes whose g values are such that they
have transmission minima near $\lambda/2$. The associated very high reflectivity
will ensure that the overall transmission is low at $\lambda/2$. Further meshes
with progressively smaller g values can be added to expand the rejection
region.

The ability to design out the half-wave leak is not readily available in
multilayer filters, and the standard procedure is to use a series of blocking
filters, thereby reducing the overall transmission in the passband.

Using the previously discussed techniques, filters have been made with
edges from 50 to 2000 μm (Holah and Auton, 1974). In some cases, the
slight decrease in the rate of change from pass- to stopband associated
with using different meshes may not be acceptable, in which case the possible half-wave leak must be blocked with additional low-pass filters.

The performance of the multielement low-pass filters is demonstrated in
Figs. 46–52. The shortest wavelength edge filter successfully produced by
the author is shown in Fig. 46. The construction was $g_1 = 25$ μm, $g_4 =$

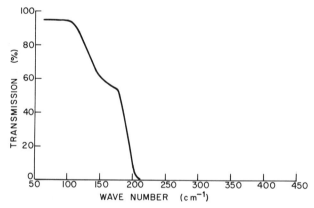

FIG. 46 Mesh low-pass filter near 200 cm^{-1}. This design was used for capacitative
meshes. (Reproduced from Holah and Auton, 1974.)

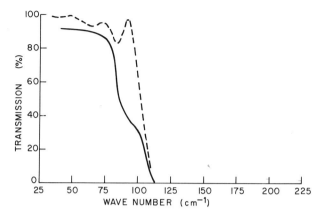

FIG. 47 Experimental (—) and calculated (– – –) curves for four capacitative mesh low-pass filters: near 100 cm^{-1}. This filter was used successfully on the NIMBUS V meteorological satellite. (Reproduced from Holah and Smith, 1972.) *Journal of Physics D: Applied Physics* **5**, 496. Copyright of The Institute of Physics.)

25 μm, and g_2 and $g_3 = 33$ μm, with three identical spacers of 8 μm. The design was not perfect, and the constructional difficulties associated with such small spacers reduced the expected performance, particularly with regard to the rejection region. Fortunately, it is possible to back the filter with either quartz or PTFE, which results in good overall performance. In general, the edge position can be determined to within ± 5%.

A four-element low-pass filter near 100 μm is shown in Fig. 47. This was one of the earliest designs and was produced specifically for a satellite radiometer experiment. It was important that the 50% of maximum transmission occur at 100 ± 3 cm^{-1}. The design used here was $g_1 = 45$ μm, $g_2 = g_3 = 52$ μm, and $g_4 = 34$ μm, with spacers of 24 μm achieving the specification. With this combination of meshes, it would have been possible to have moved the edge within certain limits and still maintained good rejection at the $\lambda/2$ point. As with most of the meshes used in this work, the a/g ratio of the meshes was 0.2, corresponding to an impedance of 0.94. Note that it is usual to use a thin piece of black polyethylene to ensure complete rejection of wavelengths below 4 μm.

Figure 48 shows a low-pass filter with an edge near 75 μm. The design for this is $g_1 = 25$ μm, $g_2 = g_3 = 42$ μm, and $g_4 = 33$ μm, with metallic spacers of 15 μm. It should be mentioned that the effective g value, i.e., the wavelength at which the transmission is a minimum, is *not* the same value as the physical g value, g_{eff} being greater than g_{real} by approximately 8 μm until $g_{real} \sim 90$ μm and is roughly 10% greater thereafter. When designing filters, the effective value of g is used. It is probably timely to

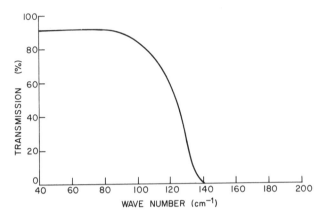

FIG 48   Low-pass filter near 140 cm^{-1}. (Reproduced from Holah and Auton, 1974.)

point out that the parameters g_{eff}, Z_0, and absorption used to describe the meshes' frequency-dependent reflection and transmission are all determined experimentally by fitting observed single-mesh data.

The design for the 67-cm^{-1} edge filter shown in Fig. 49 is $g_1 = 50$ μm, $g_2 = g_3 = 100$ μm, and $g_4 = 42$ μm, with spacers of 26 μm. The steepness of the edge is determined primarily by the central meshes, with the half-wave rejection being determined mainly by the 50-μm mesh. Ideally, the g value for this mesh should have been closer to 80 μm. However, the cost of producing a large variety of g values is prohibitive, and meshes are made that can be used in various combinations to provide as wide a range of low-pass filters as possible without compromising per-

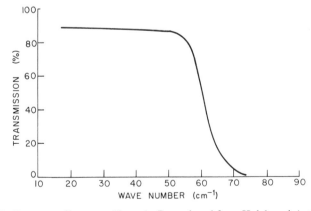

FIG. 49   Low-pass filter near 67 cm^{-1}. (Reproduced from Holah and Auton, 1974.)

formance. If cost is not important, it is always possible to produce a "perfect" filter, i.e., one with rapid cuton and good rejection.

A longer wavelength filter is shown in Fig. 50. This has an edge near 33 cm^{-1} with a design consisting of $g_1 = 42$ μm, $g_2 = 100$ μm, $g_3 = g_4 = 200$ μm, and $g_5 = 50$ μm, with spacers of 72 μm. The combination of the 200-μm and 100-μm meshes ensured that the half-wave point would not be troublesome. The use of additional meshes helped to extend the blocking range.

The steepness of the edge can be increased by increasing the reflectivity of the meshes. If, however, the reflectivity is very high near the edge, it will also be reasonably high in the passband, which will give rise to some oscillation of overall transmission in the passband.

Although the transmission of four- and five-element mesh filters is greater than 90% in the passband, the use of additional blocking reduces this. Usually with a filter completely blocked down to and including the visible, a transmission of 60% is achieved in the passband. Again, if cost is not restrictive, all the blocking can be incorporated into the low-pass filter using additional meshes that do not significantly reduce the passband transmission.

A six-element low-pass filter with an edge near 10 cm^{-1} is shown in Fig. 51. The design for this is $g_1 = 200$ μm, $g_2 = 400$ μm, $g_3 = g_4 = 600$ μm, $g_5 = 300$ μm, and $g_6 = 100$ μm, with spacers equal at 380 μm. The design for this filter is not ideal because to have removed the leak at 20 cm^{-1} would have required a mesh with a g value of 500 μm. The rejection at wave numbers greater than 25 cm^{-1} is quite good. As the filter is cooled to 77 °K, the performance improves, the leak at 20 cm^{-1} is reduced to 1%,

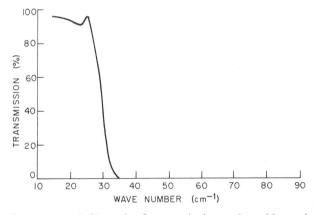

FIG. 50 Low-pass mesh filter using five capacitive meshes with an edge at 33 cm^{-1}. (Reproduced from Holah and Auton, 1974.)

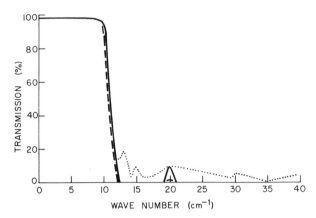

FIG. 51 Temperature dependence of six-element low-pass mesh filters at 10 cm^{-1}: 300 °K(—), 77 °K(– – –), 1.4 °K(· ·). Degradation occurred at low temperatures. The filter returned to its original performance after repeated cycling. (Reproduced from Holah and Auton, 1974.)

and there is no measurable leak elsewhere. This measurement was performed using a lamellar grating Fourier spectrometer. Unfortunately cooling to 1.8 °K resulted in degradation of the filter profile due to the loss of interference between the beams. This was caused by the meshes losing parallelism because of the wrinkling in the plastic substrates.

This section on low-pass filters will be completed with the ultra-long-wavelength edge filters shown in Fig. 52. These are filters with edges near 4.3 cm^{-1} and 6.3 cm^{-1} and were made for use in a tokamak plasma diagnostic interferometer. In this case, the position and shape of the edge were of primary importance and the long-wavelength performance above 12 cm^{-1} was unimportant. Therefore, a design using just two different meshes was used. In order for both filters to be made from the same mesh, meshes with g values of 900 and 600 μm were used, giving adequate reflection at 4.3 and 6.3 cm^{-1} to ensure rapid transition from pass- to stopband and at the same time to keep the transmission in the rejection region down to an acceptable level, especially in the case of the 6.3-cm^{-1} filter.

The measurement of filters at such long wavelengths represents a severe test for any spectroscopic system. The filter at 6.3 cm^{-1} was measured on a conventional Beckman-R.I.I.C. FS-720 Fourier spectrometer, using a standard quartz window Golay cell. The measurement was possible because all the shorter wavelength radiation beyond 30 cm^{-1} was removed by a series of low-pass interference filters. A single transmission point was obtained using a 5-mm microwave system. The Fourier measurement was a single run; no averaging of spectra was therefore performed.

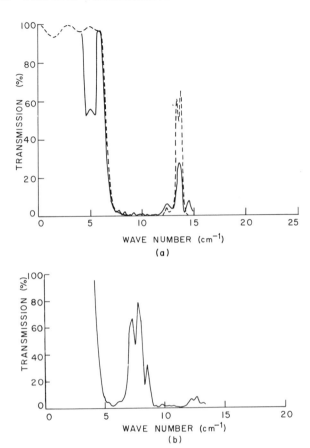

FIG. 52   Theoretical (– – –) and experimental (—) curves for two very long wavelength low-pass filters with edges out to 2300 μm: (a) edge at 6.3 cm⁻¹; (b) edge at 4.3 cm⁻¹. (Unpublished data.)

Further measurements on an N.P.L. Fourier system using cooled germanium bolometers confirmed the curve.

The 4.3-cm⁻¹ filter was even more difficult, and only the leak could be observed using the Beckman system. However, if it had been possible to use either the 6.3-cm⁻¹ or, ideally, the 10-cm⁻¹ blocking filter, then there is every reason to believe that the passband could also have been determined using the Beckman instrument. The curve in Fig. 52 was obtained from the more sophisticated N.P.L. system. Because it was possible to measure the leak, a theoretical fit to the leak enabled the complete theoretical curve, allowing for losses, to be obtained that agreed very closely with the final experimental curve.

VI. Summary

This chapter has discussed in some detail the most important techniques for optical filtering of radiation between 2 and 2000 μm, with particular emphasis on the region beyond 50 μm. It is clear that the interference filter approach using either evaporated dielectric multilayers or metallic mesh can produce controllable high-performance filters for most of the region between 2 and 2000 μm. At the present time, the least well served region is between 30 and 50 μm because of the difficulty in producing the relatively thick dielectric layers or the relatively thin metal spacers needed for mesh filters. It is, however, probable that within the next year the mesh-filter technology will entend down to 30 μm.

The low-temperature performance of all the interference filters has not been sufficiently studied with the exception of metallic mesh bandpass filters, and here the results would indicate that they perform well. Unfortunately, low-pass filters of both metallic and dielectric multilayers have had limited success and development is still under way.

Temperature effects notwithstanding, it is possible, given sufficient funding, to produce low-pass and narrow bandpass metallic mesh filters at any wavelength and to almost any specification. Any nonideal filter profile is usually due to the large cost in obtaining the exact mesh size and having to compromise on performance.

ACKNOWLEDGMENTS

The author would like to gratefully acknowledge the invaluable help given over many years by Drs. J. P. Auton and Roger Miller of Cambridge Consultants Ltd., who have produce the capacitative meshes and made all the low-pass mesh filters shown in this chapter.

The precision instrumentation expertise of Neil Morrison of the Physics Department, Heriot Watt University, who designed and made the jigs for both spaced and spacerless mesh filters, has been fundamental to the successful conclusion of narrow-band filters. Finally, I should like to thank the assistance of my graduate student Brian Davis, who improved the computation procedures, making the design of mesh filters much more efficient, and also measured many of the filters shown.

REFERENCES

Armstrong, K. R., and Low, F. J. (1973). *Appl. Opt.* **12,** 2007–2009.
Armstrong, K. R., and Low, F. J. (1974). *Appl. Opt.* **13,** 425–430.
Baker, M. L., and Yen, V. L. (1967). *Appl. Opt.* **6,** 1343–1351.
Baldecchi, M. G., and Melchiorri, B. (1973). *Infrared Phys.* **13,** 189–211.
Chantry, G. W. (1971). "Submillimetre Spectroscopy." Academic Press, New York.
Hadni, A. (1967). "Essentials of Modern Physics Applied to the Study of Study of the Infrared." Pergamon, Oxford.
Holah, G. D., and Auton, J. P. (1974). *Infrared Phys.* **14,** 217–229.
Holah, G. D., and Morrison, N. D. (1977). *J. Opt. Soc. Am.* **67,** 971–974.

Holah, G. D., and Smith, S. D. (1972). *J. Phys. D* **5**, 496–509.

Holah, G. D., and Smith, S. D. (1977). *J. Phys. E* **10**, 101–111.

Holah, G. D., Davis, B., and Morrison, N. D. (1978). *Infrared Phys.* **18**, 621–625.

Holah, G. D., Morrison, N. D., and Goebel, J. G. (1979a). *Appl. Opt.* **18**, 3526–3532.

Holah, G. D., Davis, B., and Morrison, N. D. (1979b). *Infrared Phys.* **19**, 639–647.

Houghton, J. T., and Smith, S. D. (1966). "Infrared Physics." Oxford Univ. Press, London and New York.

Jacquinot, P. (1960). *Rep. Prog. Phys.* **23**, 267–312.

Kittel, C. (1976). "Introduction to Solid State Physics." Wiley, New York.

Lecullier, J.-C. (1974). *Nouv. Rev. Opt.* **5**, 313–317.

McLeod, H. A. (1969). "Thin Film Optical Filters." Am. Elseier, New York.

Marcuvitz, N. (1951). "Waveguide Handbook." McGraw-Hill, New York.

Möller, K. D., and McKnight, R. V. (1963). *J. Opt. Soc. Am.* **53**, 760.

Pidgeon, C. R., and Smith, S. D. (1964). *J. Opt. Soc. Am.* **54**, 1459–1466.

Rawcliffe, R. D., and Randall, C. M. (1967). *Appl. Opt.* **6**, 1353–1358.

Renk, K. F., and Genzel, L. (1962). *Appl. Opt.* **5**, 643–648.

Ressler, G. M., and Möller, K. D. (1967). *Appl. Opt.* **6**, 893–896.

Seeley, J. S., and Smith, S. D. (1966). *Appl. Opt.* **5**, 81–85.

Smith, S. D. (1958). *J. Opt. Soc. Am.* **48**, 43–50.

Smith, S. D., Holah, G. D., Seeley, J. S., Evans, C., and Hunneman, R. (1972). *In* "Infrared Detection Techniques for Space Research" (V. Manno and J. Ring, eds.), pp. 199–218. Reidel Publ., Dordrecht, Netherlands.

Ulrich, R. (1967a). *Infrared Phys.* **7**, 37–55.

Ulrich, R. (1967b). *Infrared Phys.* **7**, 65–74.

Ulrich, R. (1968). *Appl. Opt.* **7**, 1987–1996.

Wijnbergen, J. J. (1973). *Astron. Astrophys.* **29**, 159–161.

Yamada, Y., Mitsuishi, A., and Yoshinaga, H. (1962). *J. Opt. Soc. Am.* **52**, 17–19.

INDEX

411